ELEMENTARY TOPOLOGY

SECOND EDITION

MICHAEL C. GEMIGNANI

University of Houston—Clear Lake

DOVER PUBLICATIONS, INC.

NEW YORK

Published in Canada by General Publishing Company, Ltd., 30 Lesmill Road, Don Mills, Toronto, Ontario.
Published in the United Kingdom by Constable and Company, Ltd., 3 The Lanchesters, 162–164 Fulham Palace Road, London W6 9ER.

This Dover edition, first published in 1990, is an unabridged and corrected republication of the second edition (1972) of the work originally published in 1967 by Addison-Wesley Publishing Company, Inc., Reading, Mass.

Manufactured in the United States of America
Dover Publications, Inc., 31 East 2nd Street, Mineola, N.Y. 11501

Library of Congress Cataloging-in-Publication Data

Gemignani, Michael C.
Elementary topology / Michael C. Gemignani.
p. cm.
Reprint. Originally published: Reading, Mass. : Addison-Wesley, 1972.
Includes bibliographical references and index.
ISBN 0-486-66522-4
1. Topology. I. Title.
QA611.G45 1990
514—dc20 90-42722
CIP

To my wife, Carol

PREFACE

This book is intended as a first text in topology. If the reader has completed at least three semesters of a calculus and analytic geometry sequence, he should have sufficient background to understand this book. The reader might gain a deeper appreciation of the contents, however, if he has also had at least one semester of real analysis, or its equivalent.

If anyone asks, "What is topology?", the most correct answer is, "Topology nowadays is a fundamental branch of mathematics and like most fundamental branches of mathematics does not admit of a simple concise definition." Topologists are indeed investigating widely different problems and are using a multitude of techniques. Topology is today one of the most rapidly expanding areas of mathematical thought.

Historically, however, topology has its roots in geometry and analysis, that is, the study of real and complex functions. Geometrically, topology was the study of properties preserved by a certain group of transformations, the homeomorphisms. Geometry itself can be considered as the study of properties preserved by certain types of functions; e.g., Euclidean metric geometry is the study of properties preserved by rigid (that is, distance-preserving) transformations (known sometimes as *congruences*). (Of course, as with topology, it is somewhat unfair to try to define geometry as the study of one particular thing.)

Certain of the notions of topology are also abstractions of concepts which are classical in the study of real or complex functions. Open sets, continuity, metric spaces, etc., were a basic part of analysis before being generalized in topology.

Topology, then, has its roots in at least two areas of mathematics; but now topology has reached the point where a mathematician engaged in topological research is not only justified in calling himself a topologist, but he must specify whether he is a point set topologist, differential topologist, algebraic topologist, or some other topological specialist.

Since topology is now a study in its own right, we would be justified in merely introducing topological concepts without giving some idea of where or how these concepts arose. Since this is an introductory text, however, we have tried to motivate the concepts introduced so that the reader can see where these concepts came from originally. Where a concept is primarily geometric, we have tried to treat it geometrically; where analysis is the inspiration for a concept, an analytic approach is used. It is hoped that the reader of this book will not only learn the fundamentals of topology, but will appreciate how abstract topological notions developed from classical mathematics.

Topology, the child of geometry and analysis, now serves as a powerful tool not only in these areas, but in almost all areas of mathematical study. But besides being an instrument for use elsewhere, topology has a beauty and a content of its own. Topology is valuable in its own right in so far as any well-developed mathematical study is valuable, or, in fact, any aesthetically pleasing creation of the human mind is valuable.

In this second edition we have attempted to correct all errors, typographical and otherwise, found in the first edition. Numerous exercises have been added as well as a section dealing with paracompactness and complete regularity. The Appendix on infinite products has been extended to include the general Tychonoff Theorem; a proof of the Tychonoff theorem which does not depend on the theory of convergence has also been added in Chapter 7.

Northampton, Mass. M. G.
September 1971

CONTENTS

1 PRELIMINARIES

1.1 SETS AND FUNCTIONS

It is assumed that any reader of this book has already had some experience with sets; hence most of what is said in this section will be for the sake of review rather than for the purpose of presenting new material.

We will not deal with sets axiomatically. A *set* will be taken to be any well-defined collection of objects; the objects in a set are called *elements*, or *points*, of the set. If x is an element of the set S, we write $x \in S$. We denote the phrase *is not an element of* by $\notin$.

Sets may be denoted either by explicitly listing their elements inside of braces (for example, $\{1, 2, 3\}$ is the set having 1, 2, and 3 as elements) or by giving the rule by which a typical object of the set is determined (for example, $\{x \mid x \text{ is a red schoolhouse}\}$ is the set of all red schoolhouses or, alternatively, the set of all x such that x is a red schoolhouse).

A set S is said to be a *subset* of a set T if each element of S is an element of T. We usually denote *S is a subset of T* by $S \subset T$. Two sets S and T are *equal* if they contain exactly the same elements; that is, $S = T$ if $S \subset T$ and $T \subset S$. The phrase *is not a subset of* is denoted by $\not\subset$.

The *empty set*, that is, the set which contains no elements whatsoever, is denoted by ϕ.

If S and T are any two sets, then the *complement of S in T* is the set of all elements of T which are not elements of S; we denote the complement of S in T by $T - S$. Similarly, the *complement of T in S*, denoted by $S - T$, is the set of all elements of S which are not elements of T.

The two most basic set operations are *union* and *intersection*. If $\{S_i\}$, $i \in I$, is any family of sets indexed by some set I, then the *union* of this family of sets is $\{x \mid x \in S_i \text{ for at least one } i \in I\}$. (We will rigorously define the notion of an index set later in this section; for now the reader can consider I to be merely a set of labels distinguishing the various members of the family of sets.) The union of $\{S_i\}$, $i \in I$, may be denoted by $\bigcup_I S_i$, or $\bigcup \{S_i \mid i \in I\}$. The *intersection* of this family of sets is $\{x \mid x \in S_i$ for every $i \in I$, that is, x is an element of every member of the family of sets$\}$. The intersection of $\{S_i\}$, $i \in I$, may be denoted by $\bigcap_I S_i$, or

$\bigcap\{S_i \mid i \in I\}$. Where only a few sets are involved, say $\{S_1, S_2, S_3\}$, the intersection and union of these sets may be denoted by $S_1 \cap S_2 \cap S_3$ and $S_1 \cup S_2 \cup S_3$, respectively.

It is assumed that the reader is moderately familiar with these set operations, at least so far as any finite family of sets is concerned. We now prove the *DeMorgan formulas* for an arbitrary family of sets.

Proposition 1. Suppose $\{S_i\}$, $i \in I$, is a family of subsets of some set T. Then

a) $\bigcup_I (T - S_i) = T - \bigcap_I S_i$;
b) $\bigcap_I (T - S_i) = T - \bigcup_I S_i$.

Proof. a) To prove that any two sets are equal, we must show that they contain the same elements. Suppose $x \in \bigcup_I (T - S_i)$; then $x \in T - S_i$ for at least one $i \in I$. Therefore $x \notin S_i$ for at least one $i \in I$. Then x is not in S_i for every $i \in I$; hence $x \notin \bigcap_I S_i$. Consequently, $x \in T - \bigcap_I S_i$. We have thus proved that every element of $\bigcup_I (T - S_i)$ is an element of $T - \bigcap_I S_i$; that is,

$$\bigcup_I (T - S_i) \subset T - \bigcap_I S_i.$$

Suppose $x \in T - \bigcap_I S_i$. Then there is some $i \in I$ for which $x \notin S_i$ (or else x would be an element of $\bigcap_I S_i$). Therefore $x \in T - S_i$ for some $i \in I$. Then $x \in \bigcup_I (T - S_i)$. Consequently, $T - \bigcap_I S_i \subset \bigcup_I (T - S_i)$; hence

$$T - \bigcap_I S_i = \bigcup_I (T - S_i).$$

The proof of (b) is left as an exercise.

If S and T are any two sets, then the *Cartesian product* of S and T is defined to be the set of all ordered pairs (s, t) such that $s \in S$ and $t \in T$. The Cartesian product of S and T is denoted by $S \times T$.

If S and T are any sets, then a subset R of $S \times T$ is said to be a *relation between S and T*. A subset of $S \times S$ is said to be a *relation on S*. If R is a relation between S and T, that is, if $R \subset S \times T$, then if $(s, t) \in R$, we may also write sRt, or say that s and t are *R-related*. Some special types of relations will be discussed in the next section. Although strictly speaking a relation is a set, at times a phrase or symbol defining the relation will be used in place of the actual set. For example, although *is equal to* defines a relation on the collection of subsets of some set, we usually write simply $S = T$ if S and T are equal subsets, rather than explicitly refer to any relation.

A *function* f from a set S into a set T is a relation between S and T such that each element of S is f-related to one and only one element of T.

If $(s, t) \in f$, then we may write $t = f(s)$. Functions are usually defined by giving a rule which enables us to find $f(s)$ whenever s is given. Again, rarely is explicit mention made of the fact that a function is a set. Functions are also called *maps* or *mappings*.

If f is a function from S into T, then S is called the *domain* of f, T the *range* of f, and $\{t \in T \mid t = f(s) \text{ for some } s \in S\}$ the *image* of f. The image of f may be denoted by $f(S)$.

If f is a function from S into T and $W \subset S$, then the *restriction of f to W*, denoted by $f \mid W$, is a function from W into T defined by $f \mid W(w) = f(w)$ for each $w \in W$.

If f is a function from S into T, we may write $f: S \to T$. If $f(S) = T$, then f is said to be *onto*. If $f(s) = f(s')$ implies $s = s'$ for any $s, s' \in S$, then f is said to be *one-one;* that is, f is one-one if each element of T is the image of at most one element of S.

Suppose $f: S \to T$. If $t \in T$, then

$$f^{-1}(t) = \{s \in S \mid f(s) = t\}.$$

If $U \subset T$, then

$$f^{-1}(U) = \{s \in S \mid f(s) \in U\}.$$

By f^{-1} we mean $\{(t, s) \mid (s, t) \in f\}$. Note that f^{-1} is a relation between T and S, called the *inverse relation of f*, and that it is a function from T to S if and only if f is one-one and onto.

Suppose $f: S \to T$ and $g: T \to W$. Then $g \circ f$ is defined by

$$\{(s, w) \mid s \in S,\ w \in W, \text{ such that there is some } t \in T \text{ with } t = f(s) \text{ and } w = g(t); \text{ that is, } w = g(f(s)) \text{ for some } s \in S\}.$$

$g \circ f$ is a function from S into W and is called the *composition of g with f*.

There are two special types of functions, *sequences* and *indices*, which the reader should already have encountered at least informally. A *sequence* u in a set S is any function from the set N of positive integers into S. If u is a sequence in S, then $u(n)$ is usually denoted by u_n; the sequence itself may be denoted by u, $\{u_n\}$, $n \in N$, or $\{u_1, u_2, u_3, \ldots\}$.

Sometimes the elements of one set are used to label the elements of another set, this often being a convenient way to express a collection of objects, or sets. For example, the elements of $\{1, 2, 3\}$ are used to label the elements of $\{t_1, t_2, t_3\}$. A one-one and onto function f from some set I onto a set S for the purpose of labeling the elements of S is called a *system of indices* for S, and I is called the *set of indices*, or the *index set*. The set S is said to be *indexed* by I, and we may represent this relationship by writing S as $\{s_i\}$, $i \in I$.

EXERCISES

1. Express each of the following in words.

   a) $\{A, B, C\}$ b) $\{A, B, \{C\}\}$

   c) $\{n \mid n$ is an integer and n is a multiple of $2\}$

   d) $\{\phi\}$ e) $h \in \{h, k, m\}$ f) $h \notin \{\{h\}\}$

   g) $\{h, k\} \subset S \cup T$ h) $S \not\subset T \cup W \cup \{a, b, c\}$

2. Prove each of the following.

   a) If $S \subset T$, then $T - (T - S) = S$.

   b) If S is any set, then $\phi \subset S$.

   c) $S \subset T$ if and only if $S \cap T = S$.

   d) If S and T are subsets of W, then $S \subset T$ if and only if $W - T \subset W - S$.

   e) If $\{U_i\}$, $i \in I$, is any family of sets, then
   $$T \cup (\bigcup_I U_i) = \bigcup_I (T \cup U_i)$$
   for any set T.

   f) If $f: S \to T$ and A and B are subsets of T, then
   $$f^{-1}(A \cup B) = f^{-1}(A) \cup f^{-1}(B).$$

   g) If $f: S \to T$ and $A \subset T$, then $f(f^{-1}(A)) \subset A$.

   h) If $f: S \to T$ and $\{U_i\}$, $i \in I$, is any family of subsets of T, then
   $$f^{-1}(\bigcap_I U_i) = \bigcap_I f^{-1}(U_i).$$

   i) If $f: S \to T$ and $g: T \to W$, then $(g \circ f)^{-1}(A) = f^{-1}(g^{-1}(A))$ for any $A \subset W$.

   j) If $f: S \to T$, then f^{-1} is a function from T into S if and only if f is one-one and onto.

3. Prove (b) of Proposition 1.

4. Suppose S is a set of n elements. How many subsets does S have? How many functions are there from S into S? How many one-one functions? [*Hint*: We may assume $S = \{1, \ldots, n\}$. To each subset W of S we may assign a string of length n consisting of the letters Y and N. The ith letter of the string will be Y if $i \in W$, and N if $i \notin W$. Use this to compute the number of subsets of S. In order to find the number of functions from S into S, consider the $n \times n$ matrix $((i, j))$, $i, j = 1, \ldots, n$. There will be a function from S into S for every distinct choice we can make of n elements from the matrix, never taking more than one element from each column.]

1.2 PARTIAL AND TOTAL ORDERINGS; EQUIVALENCE RELATIONS

The reader is already familiar with the ordering of the real numbers by *less than or equal to*. If R is the set of real numbers and less than or equal to is denoted by $\leq$, then $\leq$ defines a relation on R having the properties that

P1) $x \leq x$, for any $x \in R$;

P2) $x \leq y$ and $y \leq x$ implies $x = y$, for any $x, y \in R$; and

P3) $x \leq y$ and $y \leq z$ implies $x \leq z$ for any x, y, and z in R.

Any relation on any set S which shares properties P1 through P3 is called a *partial ordering* on S. If S has a partial ordering defined on it, then S is said to be a *partially ordered set.* We may denote a set S with partial ordering $\leq$ by $S, \leq$.

Example 1. Let $P(S)$ be the family of subsets of a set S. Then $\subset$ defines a partial ordering on $P(S)$. We verify that $\subset$ satisfies P1 through P3.

P1) If W is any subset of S, then $W \subset W$.

P2) If W and T are any two subsets of S such that $W \subset T$ and $T \subset W$, then $W = T$.

P3) If W, T, and Z are any sets such that $W \subset T$ and $T \subset Z$, then each element of W is an element of T. But since $T \subset Z$, each element of T is also an element of Z; therefore each element of W is an element of Z, that is, $W \subset Z$.

Note that in Example 1, it is not true that given any two subsets W and T of S, either $W \subset T$ or $T \subset W$. It is true, however, that given any two real numbers s and t, either $s \leq t$ or $t \leq s$. In $P(S)$, $\subset$, any two elements are not necessarily comparable, whereas in R, $\leq$, any two elements are comparable. If S is any set with a partial ordering $\leq$, then the partial ordering of S is said to be a *total ordering* if given any elements s and t of S, either $s \leq t$ or $t \leq s$. The partial ordering $\leq$ on the set of real numbers is a total ordering, but $\subset$ does not define a total ordering on $P(S)$ in Example 1.

Since the less than or equal to relation on the set of real numbers is the prototype of a partial ordering, we will generally denote a partial ordering by $\leq$, unless there is a special symbol called for.

Suppose S is a set partially ordered by $\leq$, and $W \subset S$. Then W can also be considered to be partially ordered by $\leq$ through the device of letting $w \leq w'$ for any two elements of W if and only if $w \leq w'$ considering w and w' as elements of S. We say the ordering $\leq$ on S *induces* an ordering on W.

Let S, $\leq$ be any partially ordered set, and suppose $W \subset S$. An element u of S is said to be an *upper bound* for W if $w \leq u$ for each $w \in W$. An element v of S is said to be a *lower bound* for W if $v \leq w$ for each $w \in W$.

It is not necessarily true that every nonempty subset of a partially ordered set S, $\leq$ has an upper or a lower bound.

Example 2. Let $S = \{x \mid 0 < x < 1\}$ be partially ordered by $\leq$. Then if $W = S$, W has no upper bound, nor any lower bound. Suppose that u is an upper bound for W; then $0 < u < 1$. Therefore

$$0 < u < (u + 1)/2 < 1.$$

Hence $(u + 1)/2$ is an element of W which is greater than u; thus u could not be an upper bound for W. Similarly, W has no lower bound.

The partially ordered set $R, \leq$ of real numbers has neither an upper nor a lower bound since, given any real number, we can find both a larger real number and a smaller real number.

Suppose that W is a subset of a partially ordered set $S, \leq$. Then an element U of S is said to be a *least upper bound* for W if U is an upper bound for W and $U \leq u$ if u is any upper bound of W. An element L of S is said to be the *greatest lower bound* of W if L is a lower bound for W and if v is any lower bound for W, then $v \leq L$. The least upper bound and greatest lower bound for W may be denoted by lub W and glb W, respectively.

It is not always true that any nonempty subset of $S, \leq$ which has an upper bound has a least upper bound.

Example 3. Let Q be the set of rational numbers partially ordered by $\leq$. Let W be the set of rational numbers less than $\sqrt{2}$. Then 3 is an upper bound of W; but since $\sqrt{2}$ is an irrational number, it can be shown that W has no least upper bound in Q. Note that W does have a least upper bound in the full set of real numbers, namely, $\sqrt{2}$.

Every nonempty subset of the set of real numbers which has an upper bound (lower bound) has a least upper bound (greatest lower bound).

Example 4. Let $P(S)$, $\subset$ be the partially ordered set described in Example 1. Suppose $\{U_i\}$, $i \in I$, is any collection of subsets of S. Then this collection has a least upper bound $\bigcup_I U_i$ and a greatest lower bound $\bigcap_I U_i$. Note that glb $P(S) = \phi$ and lub $P(S) = S$.

Let $S, \leq$ be any partially ordered set, and let $W \subset S$. An element M of W is said to be *maximal in* W if $M \nleq w$ for each $w \in W - \{M\}$. An element m of W is said to be *minimal in* W if $w \nleq m$ for each $w \in W - \{m\}$. An element M of S is said to be *maximal* (*minimal*) if M is maximal (minimal) in S.

Example 5. Let R be the set of real numbers and $W = \{1, 2, 4\}$. Then 4 is maximal in W and 1 is minimal in W. R contains no maximal or minimal element.

Suppose $P(R)$ is the collection of subsets of R partially ordered by $\subset$. Let $W = \{\{1\}, \{2\}, \{4\}\}$. Then each element of W is both maximal and minimal in W. R is a maximal element and ϕ is a minimal element of $P(R)$.

If a subset W of a partially ordered set S, $\leq$ is *totally* ordered by the ordering on W induced by $\leq$, then W is said to be a *chain* in S. That is, $W \subset S$ is a chain in S if given any two elements w and w' of W, either $w \leq w'$, or $w' \leq w$.

Example 6. Let $S = \{1, 2, 3, 4, 5\}$ and $P(S)$ be the family of subsets of S partially ordered by $\subset$. Then

$$\{\{1\}, \{1, 2\}, \{1, 2, 3\}, \{1, 2, 3, 4\}, S\}$$

is an example of a chain in $P(S)$.

One of the fundamental axioms in the theory of sets (and hence in mathematics) is the *axiom of choice.* As its name implies, the axiom of choice is a true axiom, assumed and not proved, although there are different ways in which it can be formulated. The axiom of choice properly so-called is stated as follows.

> **The axiom of choice.** Suppose $\{S_i\}$, $i \in I$, is a family of nonempty sets. Then there is a function f from I into $\bigcup_I S_i$ such that $f(i) \in S_i$ for each $i \in I$.

The axiom of choice essentially says that given any collection of nonempty sets, it is possible to form a set by choosing one element from each set in the collection. It all sounds simple enough, but it is hardly simple; it has stemmed from and led to some of the deepest thinking in the foundations of mathematics. The purpose of this book is not to delve into this problem, however.

The axiom of choice has several apparently different but actually equivalent formulations. The particular formulation we will be interested in later in this book is known as *Zorn's lemma.*

> **Zorn's lemma.** Suppose S, $\leq$ is a partially ordered set with the property that every chain in S has an upper bound. Then S contains a maximal element.

A partial ordering is an example of a special kind of relation that can be defined on a set. Another particularly important type of relation is an *equivalence relation.* The prototype for an equivalence relation is $=$, just as $\leq$ is the prototype for a partial ordering. Since ambiguity is likely to result if $=$ is used to denote an arbitrary equivalence relation, E will be used instead. A relation E on a set S is said to be an *equivalence* relation on S if E satisfies the following properties:

- ***E*1)** sEs for any $s \in S$.
- ***E*2)** If s and s' are any elements of S such that sEs', then $s'Es$.
- ***E*3)** If s, s', and s'' are any elements of S such that sEs' and $s'Es''$, then sEs''.

Compare $E1$ through $E3$ with the properties of $=$. Note that the only difference between a partial ordering on S and an equivalence relation on S is that property $P2$ has been replaced by property $E2$.

Example 7. Let T be the set of all plane triangles. Then *is similar to* defines an equivalence relation on T. An equivalence relation on T is also defined by *is congruent to;* still another equivalence relation on T is defined by *has the same area as.*

The most important property of an equivalence relation is given in the following proposition.

Proposition 2. Let S be any set. A *partition* $\mathcal{P}$ of S is any collection of nonempty subsets of S such that each element of S is contained in one and only one member of $\mathcal{P}$. Suppose E is an equivalence relation on S. For each $s \in S$, set $\bar{s} = \{t \in S \mid sEt\}$. Then the collection of $\bar{s}$ for all $s \in S$ is a partition of S, called the *partition induced by E.* Moreover, given any partition $\mathcal{P}$ of S, there is an equivalence relation E on S such that $\mathcal{P}$ is the partition induced by E.

Proof. Suppose that E is an equivalence relation on a set S. We must show that $\{\bar{s}\}$, $s \in S$, is a partition of S. Since sEs for each $s \in S$ by $E1$, then $s \in \bar{s}$; hence each element of S is contained in at least one member of $\{\bar{s}\}$, $s \in S$. We now must show that each element of S is contained in only one member. Suppose that $s \in \bar{s}$ and $s \in \bar{t}$. Choose any $s' \in \bar{s}$. Then sEs'; also tEs since $s \in \bar{t}$. By $E3$, tEs and sEs' implies tEs'; hence $s' \in \bar{t}$. Therefore $\bar{s} \subset \bar{t}$. A similar argument, however, shows that $\bar{t} \subset \bar{s}$, and hence $\bar{s} = \bar{t}$. Thus $\bar{s}$ is the only member of $\{\bar{s}\}$, $s \in S$, which contains s for each $s \in S$. Therefore $\{\bar{s}\}$, $s \in S$, is a partition of S.

Suppose that $\mathcal{P}$ is a partition of S. Define a relation E on S by letting sEs' if and only if s and s' are contained in the same member of $\mathcal{P}$ for any s and s' in S. It is left as an exercise to prove that E is an equivalence relation on S. By definition of E, $\mathcal{P}$ is clearly the partition induced by E.

If E is an equivalence relation on a set S, then if sEs', s and s' are said to be *E-equivalent,* or simply, *equivalent.* The set of elements of S which are equivalent to an element s of S is said to be the *E-equivalence class of s,* or simply the *equivalence class of s.* It is the collection of E-equivalence classes which forms the partition of S induced by E.

Example 8. Let f be a function from a set S into a set T. Define sEs' if $f(s) = f(s')$ for any s and s' in S. Then E is an equivalence relation on S. Denote the set of E-equivalence classes by S/E; if $s \in S$, denote the equivalence class of s by $\bar{s}$. We may associate with f a function $\bar{f}: S/E \to T$, defined by $\bar{f}(\bar{s}) = f(s)$ for any $\bar{s} \in S/E$. Since $\bar{s}' = \bar{s}$ if and only if $f(s) = f(s')$, $\bar{f}$ is well defined. Note that whereas f may not have been one-one, $\bar{f}$ is one-one.

EXERCISES

1. Prove that E in the second part of the proof of Proposition 2 is an equivalence relation on S.
2. The following refer to Example 8.
   a) Verify that E is an equivalence relation on S.
   b) Prove that $\bar{f}$ is well defined, that is, single valued, and one-one.
3. Suppose that N is the set of positive integers. Let $n \mid m$ denote n divides m, that is, $m = nk$ for some positive integer k. Prove that $\mid$ defines a partial ordering on N. Does N contain a maximal element (with respect to this partial ordering)? a minimal element?
4. Let $N, \mid$ be the partially ordered set described in Exercise 3.
   a) Prove that any two-element subset of N has a greatest lower bound and a least upper bound.
   b) Which of the following subsets of N are chains in N? Find a maximal and a minimal element, an upper and a lower bound, and a least upper bound for each subset:
      i) $\{1, 2, 4, 6, 8\}$, ii) $\{1, 2, 3, 4, 5\}$,
      iii) $\{3, 6, 9, 12, 15, 18\}$, iv) $\{4, 8, 16, 32, 64, 128\}$.
5. A subset W of the set Z of integers is said to be closed under addition if given any elements w and w' of W, $w + w' \in W$. Prove that there is a maximal subset of Z which is closed under addition and does not contain 9. Do this using Zorn's lemma.

1.3 CARDINALITY

Two sets S and T are said to have the *same number of elements*, or to have the *same cardinality*, if there is a one-one function f from S onto T. That is, S and T have the same cardinality if the elements of S can be put into one-one correspondence with the elements of T.

A set S is said to be *finite* if S has the same cardinality as ϕ, or if there is a positive integer n such that S has the same cardinality as $\{1, 2, \ldots, n\}$. Otherwise, S is said to be *infinite*. Furthermore, a set S is said to be *countable* if S has the same cardinality as a subset of N, the set of positive integers. Otherwise, S is said to be *uncountable*. Thus any finite set is certainly countable.

Proposition 3

a) Any subset of a finite set S is finite.
b) Any subset of any countable set S is countable.

Proof

a) Since S is finite, either $S = \phi$ or there is a positive integer n such that S has the same cardinality as $\{1, 2, \ldots, n\}$. If $S = \phi$, then the only subset of S is ϕ, which is finite. Suppose that $S \neq \phi$. Then there is a one-one function f from S onto $\{1, 2, \ldots, n\}$ for an appropriate n. Suppose $W \subset S$. If $W = \phi$, then W is finite. If $W \neq \phi$, let $i_1, i_2, \ldots, i_m$ be the elements of $\{1, 2, \ldots, n\}$ in the image of W. Then defining $g: W \to \{1, 2, \ldots, m\}$ by $g(w) = j$, where $f(w) = i_j$, for each $w \in W$, we see that W is finite.

The proof of (b) is left as an exercise.

Proposition 4. Let $\{A_n\}$, $n \in N$, be a countable collection of countable sets. Then $\bigcup_N A_n$ is also countable (N represents the set of positive integers).

Proof. We may enumerate the elements of each of the A_n in an array as shown.

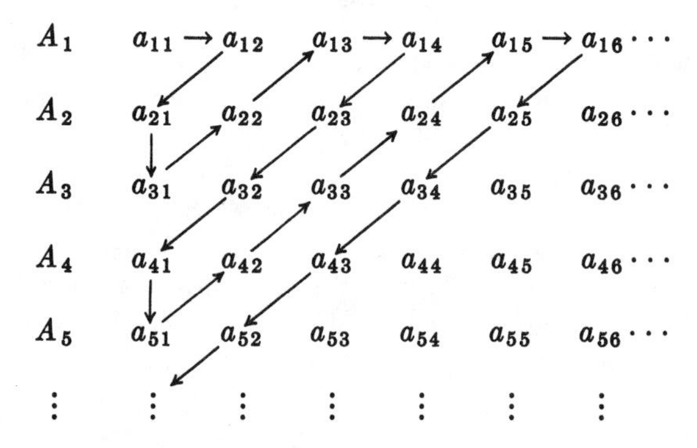

The element a_{nm} is the mth element of A_n. If we run out of elements in any set, i.e., if any of these sets are finite, we just put down x's in the spot where an element should go.

We now must find a one-one function f from $\bigcup_N A_n$ onto some subset of the set N of positive integers. Set $f(a_{11}) = 1$, $f(a_{12}) = 2$, and $f(a_{21}) = 3$. In general, follow the path indicated in the diagram and correspond the kth element reached with k. Eventually every element of $\bigcup_N A_n$ will be reached; hence $\bigcup_N A_n$ can be put in one-one correspondence with a subset of N, and is therefore countable.

Corollary 1. If A and B are countable sets, then $A \times B$ is countable.

Proof. Let $A = \{a_1, a_2, a_3, \ldots\}$ and $B = \{b_1, b_2, b_3, \ldots\}$. Set

$$A_n = \{(a_n, b) \mid b \in B\} \qquad \text{for each} \quad n \in N.$$

Then each A_n has the same cardinality as B, and hence is countable. Therefore, by Proposition 4, $\bigcup_N A_n$ is a countable set. But then $A \times B = \bigcup_N A_n$, as the union of a countable number of countable sets, is countable.

Corollary 2. The set Q of rational numbers is countable.

Proof. If q is any positive rational number, then we may consider q as the quotient of two positive integers m/n, where the fraction m/n is in lowest terms. Associate q with the ordered pair (m, n). We then see that the positive rational numbers may be associated with a subset of $N \times N$, where N is the set of positive integers. But $N \times N$ is countable by Corollary 1. Therefore, by Proposition 3, the set of positive rational numbers is countable. The set of negative rational numbers, however, has the same cardinality as the set of positive rational numbers (corresponding q with $-q$, where q is any positive rational number), and hence the set of negative rational numbers is countable. But Q is the union of $\{0\}$, the set of positive rational numbers, and the set of negative rational numbers, all three of which are countable sets; therefore, by Proposition 4, Q is countable.

Corollary 3. The set Z of integers is countable.

Proof. Z is a subset of Q, the set of rationals, and hence is countable by Proposition 3.

Although we now have a goodly number of sets we know to be countable, we have not yet shown that any set is uncountable. The following example shows that the set of real numbers is uncountable.

Example 9. The set S of unending decimals between 0 and 1 which contain only 0 or 1 as digits is uncountable. For proof, suppose that S is countable. Then we can find a one-one correspondence between S and the set N of positive integers; hence we can make a table like the following, in which the first column gives a positive integer and the second column the element of S associated with it by a suitable function f.

n	$f(n)$
1	$0.011010\cdots$
2	$0.1110111\cdots$
3	$0.101101\cdots$
4	$0.0000000111\cdots$
$\vdots$	$\vdots$

We now form an element $.x_1x_2x_3x_4\cdots$ of S as follows: If the first digit of $f(1)$ is 0, let $x_1 = 1$, and if the first digit of $f(1)$ is 1, let $x_1 = 0$. Similarly, if the second digit of $f(2)$ is 0, let $x_2 = 1$, and if the second digit of $f(2)$ is 1, let $x_2 = 0$. In general, if the nth digit of $f(n)$ is 0, let x_n, the nth digit of our new element of S, be 1, and if the nth digit of $f(n)$ is 1, let $x_n = 0$. Then $.x_1x_2x_3\cdots$ could not be $f(n)$ for any positive integer n, since $.x_1x_2x_3\cdots$ differs from each $f(n)$ at least in the nth digit because of the way it has been constructed. Hence f could not be onto, and S is therefore uncountable.

But S is a subset of R, the set of real numbers. If R were countable, then, by Proposition 3, S would also be countable; therefore R is uncountable.

We have only gone as far in our discussion of cardinality as it was felt necessary to go in order that the reader understand the contents of this book. The discussion has been somewhat informal and much has been left unsaid. For a more complete discussion of cardinality and cardinal numbers, the following texts are recommended.

1. J. L. KELLEY, *General Topology*, Van Nostrand, New York, 1955. The appendix to Kelley gives a concise axiomatic treatment of set theory and ordinal and cardinal numbers. It may be a bit too concise for the reader.
2. G. BIRKHOFF and S. MACLANE, *A Survey of Modern Algebra*, Macmillan, New York, 1953. Chapter XII gives a nice introduction to cardinal numbers and their arithmetic.
3. E. KAMKE, *Theory of Sets*, Dover, New York, 1950. This book is one of the classics in set theory and is a must in the library of any serious mathematician.

EXERCISES

1. Let $\mathcal{A}$ be the class of all sets. (Technically, the collection of all sets is not a set, so in such cases we use some word like *class*.) Prove that *has the same cardinality as* defines an equivalence relation on $\mathcal{A}$. An equivalence class is called a *cardinal number*.
2. Prove (b) of Proposition 3.
3. Prove that no finite set S has the same cardinality as one of its proper subsets W. (A subset W of S is said to be proper if $W \neq S$.) Does this remain true if S is infinite? Prove that any two infinite subsets of N, the set of positive integers, have the same cardinality.
4. The cardinality of a set S is said to be *strictly greater* than the cardinality of a set T if there is a subset W of S which has the same cardinality as T, but no subset of T which has the same cardinality as S.
   a) Prove that the cardinality of any uncountable set is strictly greater than the cardinality of any countable set.
   b) Let S be any set and let $P(S)$ denote the collection of subsets of S. Prove that the cardinality of $P(S)$ is strictly greater than the cardinality of S.
   c) Show that given any set whatsoever, there is a set of strictly greater cardinality.
5. Prove: The set R of real numbers has the same cardinality as a subset of the set $P(N)$ of all subsets of the positive integers. Prove that $P(N)$ has the same cardinality as a subset of R. It can then be shown that R and $P(N)$ have the same cardinality.

6. Prove or disprove: If S is an uncountable set and C is a countable subset of S, then $S - C$ is uncountable.

1.4 GROUPS

It is hoped that the reader will already have had some group theory, but this will not be essential for a comprehension of this text, since we will define in this section those notions which the reader will have to know in order to understand Chapter 11. It is, however, required that the reader have some familiarity with the structure and operations of the set of real numbers. It would be best if the reader of this book has already taken at least one semester of real analysis, or is taking such a course concomitantly with this one. If the reader has been fortunate enough to have had a solid course in calculus complete with some theoretical discussion of the foundations of the calculus, this should suffice.

Let S be any set. An *operation on* S is a function # from $S \times S$ into S. If $\#(s, s') = s''$, we usually write $s \# s' = s''$. Thus addition and multiplication are examples of operations on the set of real numbers.

An operation # on a set S is said to be *associative* if for any elements s_1, s_2, and s_3 of S,

$$(s_1 \# s_2) \# s_3 = s_1 \# (s_2 \# s_3).$$

We thus recognize addition and multiplication of real numbers to be associative operations.

A set S with an operation # may be denoted by $S, \#$. An element k of $S, \#$ is said to be an *identity with respect to* # if

$$s \# k = k \# s = s$$

for any $s \in S$. Thus 0 is an identity with respect to addition of real numbers, and 1 is an identity with respect to multiplication of real numbers.

Suppose $S, \#$ has an identity k with respect to #. An element t of S is said to be an *inverse* (with respect to #) for an element s of S if

$$s \# t = t \# s = k.$$

If # is associative and t is an inverse for s, then it can be shown that t is the only inverse for s. If # is associative and s has an inverse, the (unique) inverse of s is generally denoted by s^{-1}.

If r is any real number, then the inverse of r with respect to $+$ is $-r$. If $r = 0$, then r has no inverse with respect to multiplication, but if $r \neq 0$, then the inverse of r with respect to multiplication is $r^{-1} = 1/r$.

A set S with an operation # is said to be a group if (1) # is associative, (2) S has an identity k with respect to #, and (3) each element of S has an inverse with respect to #.

Example 9. Let R be the set of real numbers. Then $R, +$ is a group with identity 0. R is not a group with respect to multiplication, since 0 has no inverse with respect to multiplication; but $R - \{0\}$ is a group with multiplication as the operation and 1 as the identity.

Example 10. There are groups which contain only a finite number of elements. The table below gives the "multiplication table" for a group of only four elements:

#	s_1	s_2	s_3	s_4
s_1	s_1	s_2	s_3	s_4
s_2	s_2	s_1	s_4	s_3
s_3	s_3	s_4	s_1	s_2
s_4	s_4	s_3	s_2	s_1

It would actually take a great deal of computation to verify directly that this is indeed the operation table for a group; therefore, if the reader does not immediately recognize this group, he will more or less have to accept its being a group on faith. Note that the identity of this group is s_1 and that each element of the group is its own inverse.

Suppose that $S, \#$ and $T, \$$ are groups. There may be many functions from S into T, but perhaps only a few of these are related in any way to the group structures of S and T. When studying groups, however, we wish to consider functions which somehow respect the operations of the groups; such functions are called homomorphisms. More formally, a function $f: S \to T$ is called a *homomorphism* if

$$f(s_1 \# s_2) = f(s_1) \$ f(s_2)$$

for any elements s_1 and s_2 of S. If f is a one-one and onto function as well as being a homomorphism, then f is said to be an *isomorphism*, and the groups $S, \#$ and $T, \$$ are said to be *isomorphic*. Isomorphic groups have essentially the same group properties.

If $S, \#$ and $T, \$$ are any groups, then we can define an operation & on $S \times T$ as follows: If (s, t) and (s', t') are any elements of $S \times T$, define

$$(s, t) \ \& \ (s', t') = (s \# s', t \$ t').$$

We call the group thus formed the *direct sum* of $S, \#$ and $T, \$$ (see Exercise 5). We denote the direct sum of $S, \#$ and $T, \$$ by $S \oplus T$.

Again, we have only set forth as much about groups as will be required to understand the text. For a more complete treatment of the theory of groups, the following books are suggested.

1. G. BIRKHOFF and S. MACLANE, *A Survey of Modern Algebra,* Macmillan, New York, 1953. Chapter VI is a good introduction to groups. Chapters I and II can also be used as a reference on the structure of the real numbers.
2. W. LEDERMANN, *Introduction to the Theory of Finite Groups,* Oliver and Boyd, London, 1961. This is another excellent book that should be in anyone's mathematics library.

EXERCISES

1. Let S be any set. Prove that the set of one-one functions from S onto S is a group with composition as the group operation. Suppose that f and g are any one-one functions from S onto S. Is it necessarily true that $f \circ g = g \circ f$?
2. Suppose f to be a homomorphism from the group S, # into the group T, \$. Prove that $f(S)$, \$ is a group contained in the group T, \$. (If S, # is any group and W is a subset of S such that W, # is also a group, then W, # is said to be a *subgroup* of S, #.) Let k' be the identity of T, \$ with respect to \$. Prove that $f^{-1}(k')$ is a subgroup of S, #.
3. Let $Z, +$ be the additive group of integers. Prove that any subgroup of $Z, +$ consists of all the multiples of some fixed integer n; that is, if W is a subgroup of $Z, +$, then there is an integer n such that $W = \{nz \mid z \in Z\}$.
4. Suppose S, # is a group and $\{T_i\}$, $i \in I$, is a family of subgroups of S. Prove that $\bigcap_I T_i$ is a subgroup of S.
5. Prove that if S, # and T, \$ are groups, then $S \oplus T$, the direct sum of S and T, is also a group. Prove that S, # and T, \$ are each isomorphic to some subgroup $S \oplus T$. Find a homomorphism f from $S \oplus T$ onto S, #.
6. Suppose a set S with operation # has an identity k with respect to #. Prove that k is the only identity in S with respect to #. Prove that if # is associative and $s \in S$ has an inverse t, then t is the only inverse of s in S.

2
METRIC SPACES

2.1 THE NOTION OF A METRIC SPACE

Most of the important notions in point set topology are generalizations of concepts which were first studied in the context of metric spaces. By way of motivation, and also because metric spaces are still extremely important in their own right, we would do well to consider the fundamental properties of metric spaces.

A metric space is a set in which we have a measure of the closeness or proximity of two elements of the set, that is, we have a distance defined on the set. A metric is nothing more than the ordinary notion of distance. More precisely, we make the following definition.

Definition 1. Let X be any set. A function D from $X \times X$ into R, the set of real numbers, is said to be a *metric* on X if

i) $D(x, y) \geq 0$, for all $x, y \in X$;

ii) $D(x, y) = D(y, x)$, for all $x, y \in X$;

iii) $D(x, y) = 0$ if and only if $x = y$; and

iv) $D(x, y) + D(y, z) \geq D(x, z)$, for all $x, y, z \in X$.

A set X with metric D is said to be a *metric space*, and may be denoted by X, D.

Note that the metric D has properties that we intuitively associate with distance; in fact, as has already been remarked, a metric is merely a formalized expression of distance.

Example 1. Let R be the set of real numbers. One possible metric for R is the absolute value metric; that is, define $D(x, y) = |x - y|$.

Example 2. The metric in Example 1 should already have been familiar to the reader (although he may not have called it a metric). Another metric usually encountered in more elementary courses is the "Pythagorean" metric on the coordinate plane R^2. If (x_1, y_1) and (x_2, y_2) are any two points of R^2, then we can define a metric D on R^2 by setting

$$D((x_1, y_1), (x_2, y_2)) = \sqrt{(x_1 - x_2)^2 + (y_1 - y_2)^2}.$$

Example 3. The metric defined on R^2 in Example 2 is not the only metric which can be defined for R^2. Again letting (x_1, y_1) and (x_2, y_2) be any two points of R^2, we can define metrics D_1, D_2, and D_3 on R^2 as follows:

$$D_1((x_1, y_1), (x_2, y_2)) = |x_1 - x_2| + |y_1 - y_2|;$$

$$D_2((x_1, y_1), (x_2, y_2)) = \begin{cases} 0, & \text{if } (x_1, y_1) = (x_2, y_2); \\ 1, & \text{if } (x_1, y_1) \neq (x_2, y_2); \end{cases}$$

$$D_3((x_1, y_1), (x_2, y_2)) = \max (|x_1 - x_2|, |y_1 - y_2|).$$

Example 4. If X is any metric space with metric D, and Y is any subset of X, then Y can also be considered to be a metric space using the same metric as X. More precisely, Y with metric $D \mid Y$(i.e. D defined for pairs of elements of Y) is a metric space. Y, $D \mid Y$ is said to be a *subspace* of the metric space X, D.

Example 5. This example is given to illustrate that a metric can be defined on a set which is neither R^n for some n nor a subset of R^n. Let X be the set of all functions from the closed interval $[0, 1]$ into itself. If f and g are any such functions, define

$$D(f, g) = \text{least upper bound } \{|f(x) - g(x)| \mid x \in [0, 1]\}.$$

Since any subset of the real numbers which has an upper bound has a least upper bound and

$$0 \leq |f(x) - g(x)| \leq 1 \qquad \text{for all } f, g \in X \text{ and } x \in [0, 1],$$

then $D(f, g)$ is defined for all $f, g \in X$. We will now show that D is a metric for X by showing that D satisfies each of the properties required for a metric in Definition 1:

i) $D(f, g) \geq 0$ for all $f, g \in X$. Since each element of

$$\{|f(x) - g(x)| \mid x \in [0, 1]\}$$

is greater than or equal to 0, the least upper bound of this set, $D(f, g)$, is greater than or equal to 0.

ii) $D(f, g) = D(g, f)$ for all $f, g \in X$. This follows at once from the fact that $|f(x) - g(x)| = |g(x) - f(x)|$.

iii) $D(f, g) = 0$ if and only if $f = g$. If $f = g$, then $f(x) = g(x)$ for all $x \in [0, 1]$, and hence $\{|f(x) - g(x)| \mid x \in [0, 1]\} = \{0\}$. Therefore it follows that $D(f, g) = 0$. On the other hand, if $D(f, g) = 0$, then

$$\text{lub } \{|f(x) - g(x)| \mid x \in [0, 1]\} = 0.$$

Since $|f(x) - g(x)|$ is always greater than or equal to 0, it follows that

$$\{|f(x) - g(x)| \mid x \in [0, 1]\} = \{0\};$$

hence $|f(x) - g(x)| = 0$ for all $x \in [0, 1]$. Therefore $f(x) = g(x)$ for all $x \in [0, 1]$; that is, $f = g$.

iv) $D(f, g) + D(g, h) \geq D(f, h)$ for all $f, g, h \in X$. This inequality follows from

$$|f(x) - h(x)| \leq |f(x) - g(x)| + |g(x) - h(x)| \qquad \text{for all} \quad x \in [0, 1].$$

The details are left as an exercise.

EXERCISES

1. Prove that D_1, D_2, and D_3 as defined in Example 3 are really metrics for R^2.
2. Let (x_1, y_1) and (x_2, y_2) be any points of R^2. Which of the following do not define metrics for R^2? Explain your answer in each case.
   a) $D((x_1, y_1), (x_2, y_2)) = \min(|x_1 - x_2|, |y_1 - y_2|)$.
   b) $D((x_1, y_1), (x_2, y_2)) = (x_1 - x_2)^2 + (y_1 - y_2)^2$.
   c) $D((x_1, y_1), (x_2, y_2)) = D_1((x_1, y_1), (x_2, y_2)) - D_3((x_1, y_1), (x_2, y_2))$, where D_1 and D_3 are as defined in Example 3.
   d) $D((x_1, y_1), (x_2, y_2)) = |x_1| + |x_2| + |y_1| + |y_2|$.
3. Suppose X, D is a metric space. We may define a metric D' for $X \times X$ as follows: If (x, y) and (x', y') are any elements of $X \times X$, set

$$D'((x, y), (x', y')) = D(x, x') + D(y, y').$$

   Prove that D' is really a metric for $X \times X$. Define a metric for X^n, that is, $X \times \cdots \times$ (n times) $\times X$.
4. Suppose X, D is a metric space. If x and y are any elements of X, which of the following define metrics on X?
   a) $D_1(x, y) = kD(x, y)$, where k is any positive real number.
   b) $D_2(x, y) = kD(x, y)$, where k is any real number.
   c) $D_3(x, y) = D^n(x, y)$, where n is any positive integer.
   d) $D_4(x, y) = D^r(x, y)$, where $0 < r < 1$.
5. Supply the details for the proof of Example 5(iv).
6. For any two distinct points P_1 and P_2 of R^2, set

$$M(P_1, P_2) = \{P \in R^2 \mid d(P_1, P) = d(P_2, P)\},$$

   where d is a metric on R^2. Describe geometrically $M(P_1, P_2)$ for each of the metrics on R^2 introduced in Example 3 as well as for the Pythagorean metric.

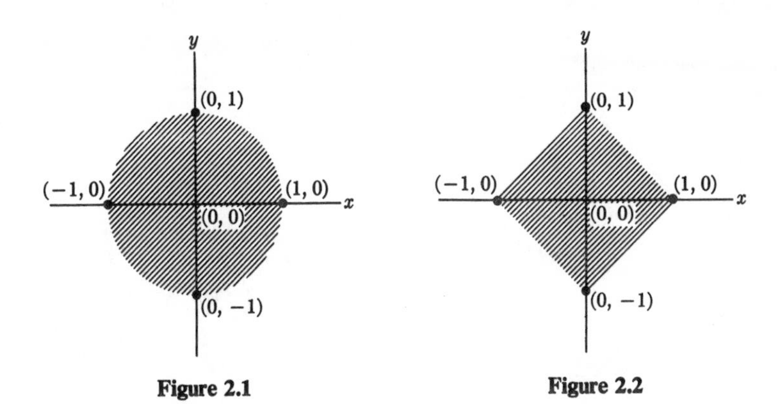

Figure 2.1 **Figure 2.2**

2.2 NEIGHBORHOODS

Let X, D be a metric space. If x is any point of X, then we may want to consider all the points of X within a certain distance of x, that is, the set of points of X which are within some degree of nearness to x.

Definition 2. If X, D is a metric space, $x \in X$, and p is any positive real number, then the *D-p-neighborhood* of x is defined to be the set of all points y of X such that $D(x, y) < p$; that is, the D-p-neighborhood of x is defined to be

$$\{y \in X \mid D(x, y) < p\}.$$

Where there is no danger of ambiguity, the D-p-neighborhood of x will be called the p-neighborhood of x and will be denoted by $N(x, p)$.

Example 6. Let R^2 be the coordinate plane. Figures 2.1 through 2.4 illustrate the 1-neighborhoods of (0, 0) with respect to the metrics D,

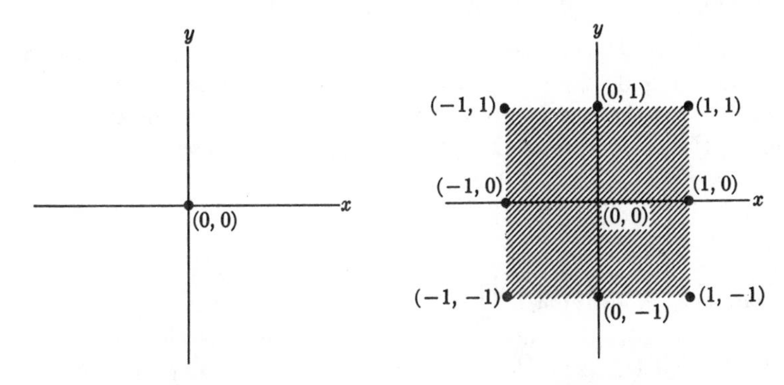

Figure 2.3 **Figure 2.4**

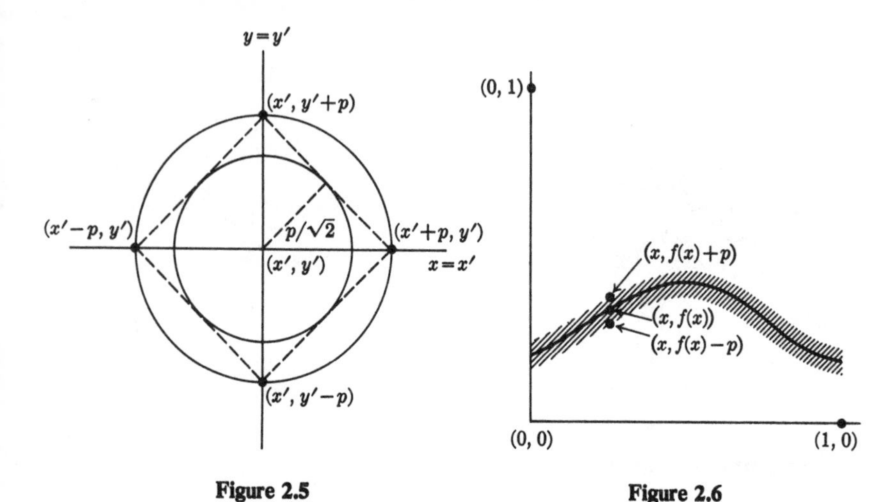

Figure 2.5 **Figure 2.6**

D_1, D_2, and D_3 of Examples 2 and 3. The reader should be sure to verify these figures.

Note that the D_1-1-neighborhood of $(0, 0)$ is a subset of the D-1-neighborhood of $(0, 0)$. Since

$$D((x_1, y_1), (x_2, y_2)) \leq D_1((x_1, y_1), (x_2, y_2))$$

for any two points (x_1, y_1) and (x_2, y_2) of R^2 (Exercise 2), the D_1-p-neighborhood of any point (x', y') of R^2 is a subset of the D-p-neighborhood for any positive real number p. However, a simple calculation shows that the D-$p/\sqrt{2}$-neighborhood of (x', y') is a subset of the D_1-p neighborhood of (x', y') (Fig. 2.5). Note, however, that if $p \leq 1$, there is no positive number q such that either the D-q-neighborhood of (x', y') or the D_1-q-neighborhood of (x', y') is a subset of the D_2-p-neighborhood of (x', y') which consists of (x', y') alone.

Example 7. Let X, D be the metric space described in Example 5. Suppose $f \in X$ and $p > 0$. We may draw a *p-collar* about the graph of f as shown in Fig. 2.6. Then the D-p-neighborhood of f will consist of all functions from [0, 1] into [0, 1] whose graphs lie within the p-collar of f.

EXERCISES

1. Confirm Figs. 2.1 through 2.4.
2. Prove that $D((x_1, y_1), (x_2, y_2)) \leq D_1((x_1, y_1), (x_2, y_2))$ for any two points (x_1, y_1) and (x_2, y_2) in R^2 as claimed in Example 6. Also in Example 6, carry out the computation which shows that the D-$p/\sqrt{2}$-neighborhood of (x, y) is a subset of the D_1-p-neighborhood of (x, y) for any $(x, y) \in R^2$.

3. Let (x, y) be any point of R^2, and p be any positive number. What is the relation of the D_3-p-neighborhood of (x, y) to the D-p-neighborhood of (x, y)? to the D_1-p-neighborhood of (x, y)? Prove that every D- and D_1-neighborhood of (x, y) contains a D_3-neighborhood of (x, y) and that each D_3-neighborhood of (x, y) contains both a D- and a D_1-neighborhood.

4. Let X, D be the metric space as described in Example 5. Sketch the graphs of the elements of X defined by each of the following, and sketch the p-collars (Example 6) for each of these elements.

   a) $f(x) = 1$, for all $x \in [0, 1]$.
   b) $f(x) = x$, for all $x \in [0, 1]$.
   c) $f(x) = x^3$, for all $x \in [0, 1]$.
   d) $f(x) = \begin{cases} 1, & \text{if } 0 \leq x < \frac{1}{2}; \\ x, & \text{if } \frac{1}{2} \leq x \leq 1. \end{cases}$

5. Let R^2 have metric D_1. Define the *circle* σ with center P_0 and radius $r > 0$ to be $\{P \in R^2 \mid D_1(P, P_0) = r\}$. Graph the circle with radius r and center P_0. List some properties that the circles defined here share with the usual geometric circles of R^2. List some properties which usual geometric circles have which the circles defined in this exercise do not have.

2.3 OPEN SETS

Closely associated with the concept of a p-neighborhood is that of an open set.

Definition 3. Let X be a set with metric D. A subset U of X is said to be *D-open* (or merely *open* where there is no danger of ambiguity) if given any point x of U, there is a positive number p such that the D-p-neighborhood of x is a subset of U.

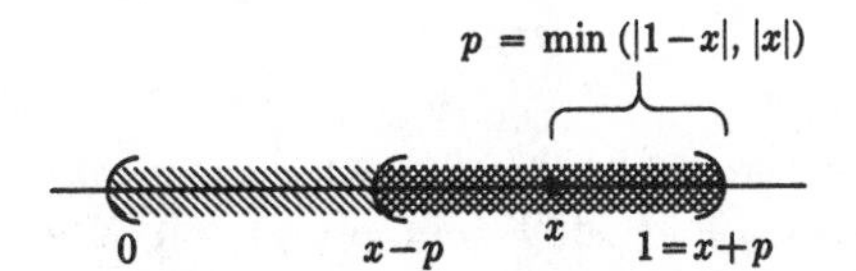

Figure 2.7

Example 8. Let R be the set of real numbers with the absolute value metric. The open interval $(0, 1)$ is an open set. For proof, let $x \in (0, 1)$ and $p = \min(|1 - x|, |x|)$. Then the p-neighborhood of x is a subset of $(0, 1)$ (Fig. 2.7). Note that the value of p depends upon the value of x.

The following propositions give the fundamental properties of open sets in metric spaces.

Proposition 1. If X, D is a metric space and p is any positive number, then the D-p-neighborhood $N(x, p)$ of any $x \in X$ is an open set.

Proof. Let w be any point of $N(x, p)$ (see Fig. 2.8). To prove that $N(x, p)$ is open, we must produce $q > 0$ such that $N(w, q) \subset N(x, p)$. Set $q =$

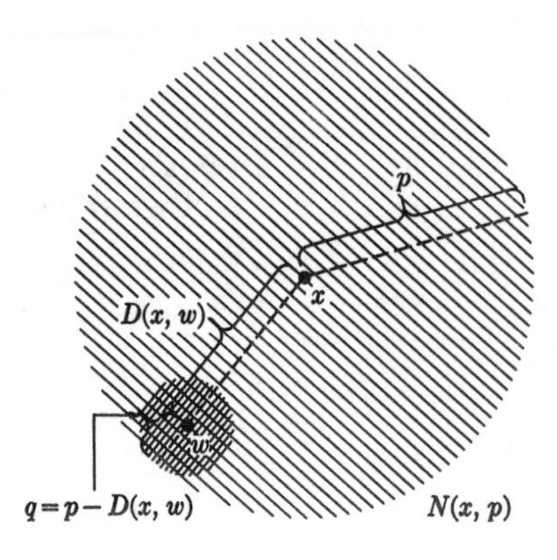

Figure 2.8

$p - D(x, w)$. Then if $z \in N(w, q)$, we have $D(w, z) < p - D(x, w)$. Therefore

$$D(x, z) \leq D(x, w) + D(w, z) < D(x, w) + p - D(x, w) = p.$$

We thus have that $z \in N(x, p)$; hence $N(w, q) \subset N(x, p)$. Therefore $N(x, p)$ is an open set.

Proposition 2. Let X, D be a metric space. Then

a) X and ϕ are both (D-) open sets,
b) the intersection of any two open sets is again an open set, and
c) the union of any family of open sets is again an open set.

Proof

a) If $x \in X$ and $p > 0$, then $N(x, p) \subset X$. Therefore X is an open set. Since ϕ contains no points whatsoever, it is true that for each $x \in \phi$ (there is no such x) and any $p > 0$, $N(x, p) \subset \phi$; hence ϕ is also an open set.
b) Suppose U and V are open subsets of X and $x \in U \cap V$. Since U is open, there is a positive number p_1 such that $N(x, p_1) \subset U$. Since V is open, there is a positive number p_2 such that $N(x, p_2) \subset V$. Set $p = \min(p_1, p_2)$. Then $N(x, p) \subset U \cap V$; therefore $U \cap V$ is open.
c) Let $\{U_i\}$, $i \in I$, be any family of open subsets of X and $x \in \bigcup_I U_i$. Then $x \in U_i$ for some i. Since U_i is open, there is a positive number p such that $N(x, p) \subset U_i$. But then $N(x, p) \subset \bigcup_I U_i$; hence $\bigcup_I U_i$ is open.

Proposition 3. Let X be a set with metric D. A subset U of X is open if and only if U is the union of a family of p-neighborhoods.

Proof. Assume U to be the union of a family of p-neighborhoods. Since each p-neighborhood is open by Proposition 1, U is the union of a family of open sets. Therefore U is open by Proposition 2(c).

Suppose that U is an open subset of X. Then for each $x \in U$, we can find at least one $N(x, p)$ such that $N(x, p) \subset U$. Since $N(x, p) \subset U$ for each $x \in U$, $\bigcup_U N(x, p) \subset U$. On the other hand, each $x \in U$ is an element of at least $N(x, p)$; hence $U \subset \bigcup_U N(x, p)$. Therefore $U = \bigcup_U N(x, p)$.

If X is any set, then X and ϕ have been shown to be D-open sets for any metric D which can be defined on X. If D_1 and D_2 are any two metrics for X, it is not necessarily true that each D_1-open set is D_2-open, or that each D_2-open set is D_1-open.

Example 9. Let D and D_2 be the metrics defined on R^2 as in Examples 2 and 3. For any $(x, y) \in R^2$, the D_2-1-neighborhood of (x, y) is precisely $\{(x, y)\}$, since (x, y) is the only point which is less than D_2-distance 1 from itself. Therefore $\{(x, y)\}$ is D_2-open, since it is a D_2-neighborhood. But $\{(x, y)\}$ is not D-open, since for any $p > 0$, the D-p-neighborhood of (x, y) contains infinitely many points besides (x, y). (See Exercise 4 also.)

EXERCISES

1. Let X, D be a metric space. Suppose that x and y are two distinct points of X. Prove that there are open sets U and V in X such that $x \in U$, $y \in V$ and $U \cap V = \phi$. [*Hint:* Let $U = N(x, \frac{1}{2}D(x, y))$.]
2. Determine which of the following subsets of the plane R^2 with the Pythagorean metric are open.
   a) $\{(x, y) \mid x < 0\}$
   b) $\{(x, y) \mid x + y > 5\}$
   c) $\{(x, y) \mid x^2 + y^2 < 1, \text{ or } (x, y) = (1, 0)\}$
   d) $\{(x, y) \mid x > 2 \text{ and } y \leq 3\}$
3. Let p be any positive number. Prove that any D-p-neighborhood of R^2 is D_1-, D_2-, and D_3-open.
4. Prove that any subset of R^2 which is D-open is D_1-open and, conversely, that any subset of the plane which is D_1-open is D-open.
5. Make appropriate sketches for the proofs of (b) and (c) in Proposition 2.
6. Let D_1 and D_2 be possible metrics for a set X. D_1 and D_2 are said to be *equivalent* if every D_1-open set is D_2-open and every D_2-open set is D_1-open. Prove that D_1 and D_2 are equivalent if and only if, given any $x \in X$ and any $p > 0$, there are positive numbers p_1 and p_2 such that the D_1-p_1-neighborhood of x is a subset of the D_2-p-neighborhood of x, and the D_2-p_2-neighborhood of x is a subset of the D_1-p-neighborhood of x.
7. Let L be any straight line in R^2. Prove that $R^2 - L$ is open with respect to all metrics introduced on R^2 thus far in the text. Try to find a metric on R^2 for which $R^2 - L$ is not necessarily open.

2.4 CLOSED SETS

Definition 4. Let X, D be a metric space. A subset F of X is said to be *closed* if F is the complement in X of an open set; that is, $F = X - U$, where U is open.

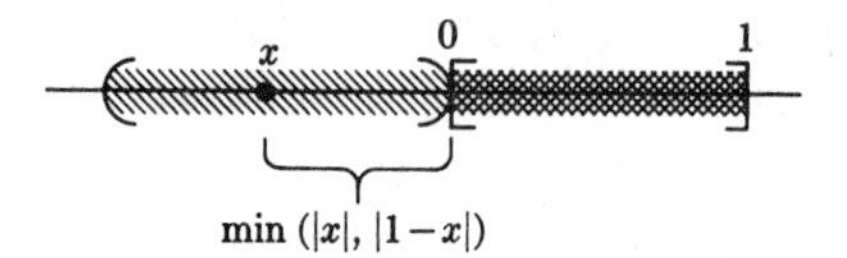

Figure 2.9

Example 10. The closed interval $[0, 1]$ is a closed subset of the real line R with the absolute value metric. For suppose $x \in R - [0, 1]$. Set $p = \min(|1 - x|, |x|)$. Then $N(x, p) \subset R - [0, 1]$. Therefore $R - [0, 1]$ is open, and hence $[0, 1]$ is closed (Fig. 2.9).

Example 11. If X is a set with metric D, if $x \in X$, and if p is any positive number, then the *closed p-neighborhood* of x, denoted by $\text{ClN}(x, p)$, is defined to be the set of all $y \in X$ such that $D(x, y) \leq p$, that is,

$$\text{ClN}(x, p) = \{y \in X \mid D(x, y) \leq p\}.$$

It is left as an exercise to show that $\text{ClN}(x, p)$ is a closed subset of X. Note in Example 10 that $[0, 1] = \text{ClN}(\frac{1}{2}, \frac{1}{2})$; therefore the fact that $[0, 1]$ is closed follows from the more general considerations of this example.

The following proposition gives the basic properties of closed sets in a metric space.

Proposition 4. Let X, D be a metric space. Then

a) X and ϕ are closed subsets of X,
b) the union of any two closed sets is closed, and
c) the intersection of any family of closed sets is again a closed set.

Proof

a) $X = X - \phi$. Since ϕ is an open set, X is the complement of an open set and hence is closed. Now, $\phi = X - X$, hence ϕ is also the complement of an open set, and is therefore closed. (Note that X and ϕ are *both* open and closed. It is quite possible for a set to be both open and closed; the complement of such a set would also have the property of being both open and closed.)

b) Let F and F' be any closed subsets of X. Then $F = X - U$ and $F' = X - U'$, where U and U' are open subsets of X. Then

$$F \cup F' = (X - U) \cup (X - U') = X - (U \cap U').$$

But $U \cap U'$ is open by Proposition 2(c); therefore $X - (U \cap U') = F \cup F'$ is closed.

c) Let $\{F_i\}$, $i \in I$, be any family of closed subsets of X. Then $F_i = X - U_i$, where U_i is an open subset of X for each $i \in I$. It follows that

$$\bigcap_I F_i = \bigcap_I (X - U_i) = X - \bigcup_I U_i.$$

Since $\bigcup_I U_i$ is open, $\bigcap_I F_i$ is closed.

Proposition 5. If X is a set with metric D and $x \in X$, then $\{x\}$ is a closed subset of X.

Proof. Since $\{x\} = X - (X - \{x\})$, if we show that $X - \{x\}$ is open, we will have shown that $\{x\}$ is closed. Suppose $y \in X - \{x\}$. Set $p = D(x, y)$. Then $N(y, p) \subset X - \{x\}$; hence $X - \{x\}$ is open.

EXERCISES

1. Show that the union of an arbitrary family of closed subsets of a metric space need not be closed. Find an example to show that the intersection of any family of open sets need not be open. [*Hint:* Use Proposition 5.]
2. Let X, D be a metric space and Y, $D \mid Y$ be a metric subspace of X (see Example 4). Prove each of the following.
   a) A subset W of Y is open in Y (that is, is $D \mid Y$-open) if and only if $W = Y \cap U$, where U is an open subset of X.
   b) A subset C of Y is closed in Y if and only if $C = Y \cap F$, where F is a closed subset of X.
   c) If Y is an open subset of X, then a subset of Y is open in Y if and only if it is open (in X).
   d) If Y is a closed subset of X, then a subset of Y is closed in Y if and only if it is closed (in X).
   e) A subset of Y may be open or closed in Y without being open or closed in X.
3. Prove that a subset F of a metric space X, D is closed if and only if $X - F$ is open.
4. Decide which of the following subsets of R^2 with the Pythagorean metric are closed.
   a) $\{(x, y) \mid x = 0, y \leq 5\}$
   b) $\{(x, y) \mid x = 2 \text{ or } x = 3, y \text{ is an integer}\}$
   c) $\{(x, y) \mid x^2 + y^2 < 1, \text{ or } (x, y) = (1, 0)\}$
   d) $\{(x, y) \mid y = x^2\}$
5. Suppose that $\{F_i\}$, $i \in I$, is a family of closed subsets of a metric space X, D with the property that given any $x \in X$, there is $p > 0$ such that $N(x, p)$ intersects finitely many of the F_i. Prove that $\bigcup_I F_i$ is closed. Try to find and prove an analogous statement for a family of open subsets of X.

6. Prove that a straight line in R^2 is closed with respect to all of the metrics on R^2 introduced thus far. Prove a circle in R^2 (circle in the usual geometric sense) is closed with respect to all of these metrics. List some other standard geometric objects which are always closed.

2.5 CONVERGENCE OF SEQUENCES

The reader should already have been introduced in previous courses to the notions of convergence of sequences, limits, and continuity, at least as far as the real numbers with the absolute value metric is concerned. We now extend these ideas to general metric spaces.

Definition 5. Let X, D be a metric space and $S = \{s_n\}$, $n \in N$, be a sequence in X. (The capital N will be used almost exclusively in this text to denote the set of positive integers.) Then S is said to *converge* to a point y of X if given any positive number p, there is a positive integer M such that if $n > M$, then $s_n \in N(y, p)$ (Fig. 2.10). If S converges to y, then we may write $s_n \to y$; y is said to be the *limit* of S.

Definition 5 could be restated as follows: $s_n \to y$ if all but a finite number of the s_n are in $N(y, p)$ for any positive number p. Or again, $s_n \to y$ if all but finitely many of the s_n are closer to y than any given distance.

Example 12. Consider the sequence defined by $s_n = (1, 1/n)$ in the coordinate plane (Fig. 2.11). If any of the metrics D, D_1, or D_3 are used, then this sequence converges to $(1, 0)$. For in these cases,

$$D(s_n, (1, 0)) = D_1(s_n, (1, 0)) = D_3(s_n, (1, 0)) = 1/n$$

for any $n \in N$. Given any positive number p, if we let M be any integer greater than $1/p$, if $n > M$, then $D(s_n, (1, 0)) < p$. On the other hand, this sequence does not converge to $(1, 0)$ with respect to the metric D_2.

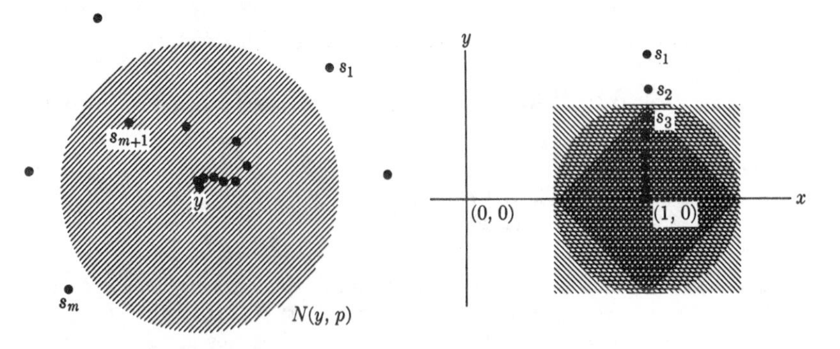

Figure 2.10 **Figure 2.11**

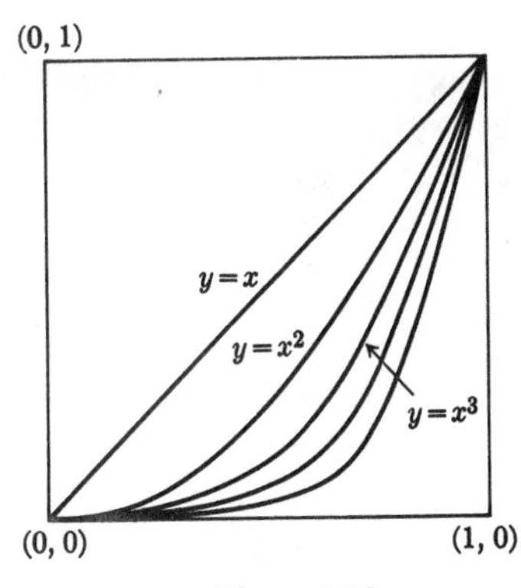

Figure 2.12

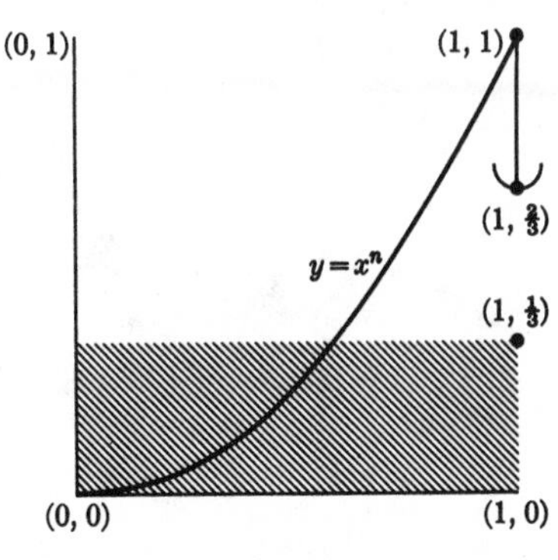

Figure 2.13

For if $p < 1$, then the D_2-p-neighborhood of $(1, 0)$ contains only $(1, 0)$, and hence excludes all of the points of the sequence.

Example 13. Let X, D be the metric space of functions described in Example 5. Let S be the sequence in X defined by $s_n(x) = x^n$. If we were to plot the graphs of s_n for successively greater n (Fig. 2.12), it would appear that the sequence S converges to the function $f \in X$ defined by

$$f(x) = \begin{cases} 0 & \text{if } x \neq 1, \\ 1 & \text{if } x = 1. \end{cases}$$

Such is not the case, however. For if we draw a p-collar about the graph of f for $p = \frac{1}{3}$ (Fig. 2.13), we see that no s_n has its graph wholly within the collar; therefore S cannot converge to f. Note, however, that S converges "pointwise" to f; that is, for each $x \in [0, 1]$, $s_n(x) \to f(x)$, where $\{s_n(x)\}$, $n \in N$, is considered as a sequence in the plane with the Pythagorean metric.

Proposition 6. If $S = \{s_n\}$, $n \in N$, is a sequence in a metric space X, D such that $s_n \to y$ and $s_n \to y'$, then $y = y'$. That is, a sequence in a metric space can converge to at most one limit.

Proof. We will suppose $y \neq y'$, and prove a contradiction. Set

$$p = \tfrac{1}{2}D(y, y')$$

(Fig. 2.14). Then

$$N(y, p) \cap N(y', p) = \phi.$$

For if $w \in N(y, p) \cap N(y', p)$, then we have

$$D(y, y') \leq D(y, w) + D(w, y') < p + p = D(y, y'),$$

a contradiction. But since $s_n \to y$ and $s_n \to y'$, both $N(y, p)$ and $N(y', p)$

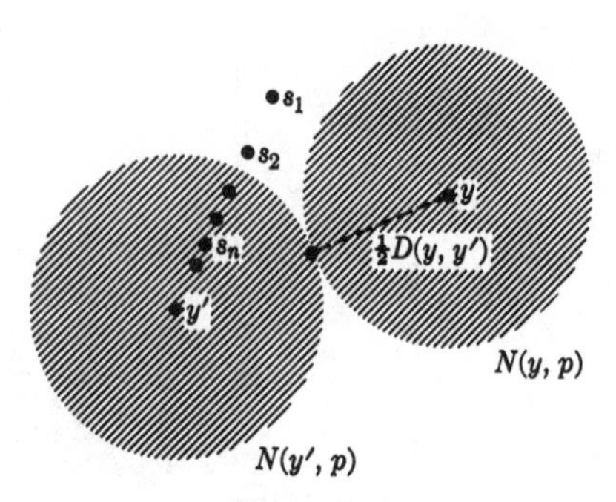

Figure 2.14

each contain all but finitely many of the s_n. Therefore some s_n must be in both $N(y, p)$ and $N(y', p)$, contradicting the fact that $N(y, p)$ and $N(y', p)$ have no points in common. Therefore $y = y'$.

Open sets were defined in terms of p-neighborhoods. Since convergence is also defined using p-neighborhoods, we might suspect that convergence can also be characterized solely in terms of open sets (rather than p-neighborhoods). The following proposition gives such a characterization.

Proposition 7. A sequence $S = \{s_n\}$, $n \in N$, in a metric space X, D converges to y if and only if any open set which contains y contains all but finitely many of the s_n.

Proof. Suppose S converges to y and U is any open set which contains y. Since U is open, there is $p > 0$ such that $N(y, p) \subset U$. But since $s_n \to y$, all but finitely many of the s_n are elements of $N(y, p)$; therefore all but finitely many of the s_n are elements of U.

Conversely, suppose that given any open set U which contains y, all but finitely many of the s_n are elements of U. Let p be any positive number. Then $N(y, p)$ is an open set which contains y; hence all but finitely many of the s_n are elements of $N(y, p)$. Therefore $s_n \to y$.

EXERCISES

1. Discuss the convergence of each of the following sequences in the spaces indicated.

   a) $s_n = 1 + 1/n$, in the space of real numbers with the absolute value metric
   b) $s_n = (2, 2)$, in the plane R^2 with the metric D_2 (Example 3)
   c) $s_n = (2, n)$, in the plane R^2 with metric D_3

d) s_n defined by

$$s_n(x) = \begin{cases} 0 & \text{if } x \leq 1/n, \\ 1 & \text{if } x > 1/n, \end{cases}$$

in the space X, D of Example 5

e) s_n defined by $s_n(x) = (1/n)x$, in the space X, D of Example 5

2. Let R^2 have metric D_3. Prove that the sequence defined by $s_n = (x_n, y_n)$ converges to (x, y) if and only if $x_n \to x$ and $y_n \to y$ in the space R of real numbers with the absolute value metric.

3. Let X, D be a metric space and Y, $D \mid Y$ be a subspace of X. Prove that if some sequence in Y converges to a point y of Y, then that same sequence considered as a sequence in X also converges to y. Show by example that a sequence in Y might not have a limit, but that the same sequence considered as a sequence in X might have a limit.

4. Suppose that X, D is a metric space with the property that for any $x \in X$, $\{x\}$ is an open set. An example of such a space is R^2 with metric D_2. Prove that a sequence $\{s_n\}$, $n \in N$, in X converges if and only if there is a positive integer M such that $s_{M+1} = s_{M+2} = \ldots$; that is, from some point on, the sequence is constant.

5. Suppose that convergence of sequences is characterized solely by Proposition 7. Discuss the convergence of sequences in the plane R^2 if R^2 has only the open sets in each of the following.

a) R^2 and ϕ

b) R^2, ϕ, and $\{(0, 0)\}$

c) R^2, ϕ, and all sets of the form $\{(x, y) \mid x^2 + y^2 < p\}$, where p is a positive number.

6. Let $\mathcal{L}$ be the set of lines of R^2. A sequence $\{L_n\}$, $n \in N$, in $\mathcal{L}$ will be said to *converge* to a line L if there exist sequences s and t such that s_n and t_n are distinct points of L_n for each $n \in N$ and $s_n \to x \in L$ and $t_n \to y \in L$, $x \neq y$. Suppose L_n has the equation $a_n x + b_n y + c_n = 0$. Prove or disprove: $L_n \to L$ if and only if $a_n \to a$, $b_n \to b$, and $c_n \to c$ in the usual space of real numbers, where L has the equation $ax + by + c = 0$.

2.6 CONTINUITY

Another concept which the reader has encountered before is continuity. Intuitively, a continuous function is a function which preserves "nearness"; it can be defined more formally as follows.

Definition 6. Let X, D and Y, D' be metric spaces (Fig. 2.15). A function f from X to Y is said to be *continuous* if, given any point $f(a) \in Y$ and any positive number p, there is a positive number q such that if $x \in N(a, q)$, then $f(x) \in N(f(a), p)$.

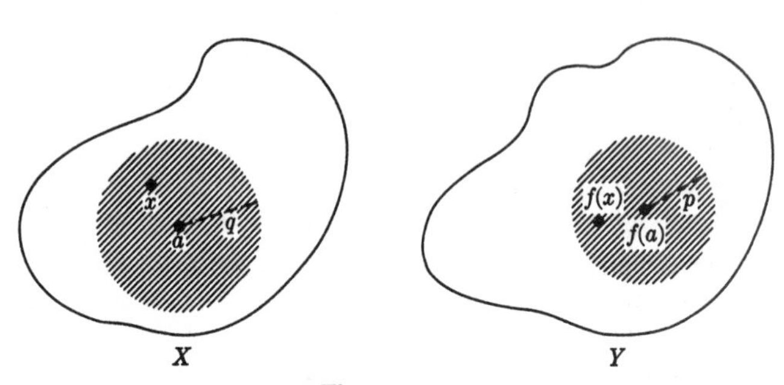

Figure 2.15

Example 14. Let R be the space of real numbers with the absolute value metric. Then the function defined by $f(x) = 2x + 3$ from R to R is continuous. A simple calculation shows that $f(N(a, p/2)) \subset N(f(a), p)$, for any $a \in R$.

It is usually quite awkward to prove the continuity of a function directly from Definition 6. We therefore need propositions which will help us determine whether or not a function is continuous, but which are, in general, easier to apply than Definition 6.

> **Proposition 8.** Let f be a function from the metric space X, D into the metric space Y, D'. Then f is continuous if and only if, given any open set U of Y,
>
> $$f^{-1}(U) = \{x \in X \mid f(x) \in U\}$$
>
> is an open subset of X.

Proof. First suppose that f is continuous and that U is an open subset of Y. Let $x \in f^{-1}(U)$; then $f(x) \in U$. Since U is open, there is a positive number p such that $N(f(x), p) \subset U$. Since f is continuous, there is a positive number q such that

$$f(N(x, q)) \subset N(f(x), p) \subset U.$$

Therefore $N(x, q) \subset f^{-1}(U)$. Since, for $x \in f^{-1}(U)$, we have found $q > 0$ such that $N(x, q) \subset f^{-1}(U)$, then $f^{-1}(U)$ is open.

Suppose, on the other hand, that $f^{-1}(U)$ is an open subset of X whenever U is an open subset of Y. We have previously shown that if $f(a) \in Y$ and if p is any positive number, then $N(f(a), p)$ is an open subset of Y; therefore $f^{-1}(N(f(a), p))$ is an open subset of X. There is therefore a positive number q such that

$$N(a, q) \subset f^{-1}(N(f(a), p)).$$

We have then that for this q, $f(N(a, q)) \subset N(f(a), p)$; hence f is continuous.

The following proposition is quite similar to Proposition 8, but is often easier to apply.

Proposition 9. Let f be a function from the metric space X, D into the metric space Y, D'. Then f is continuous if and only if given any D'-p-neighborhood U in Y, $f^{-1}(U)$ is an open subset of X.

Proof. Suppose f is continuous. If U is any D'-p-neighborhood in Y, then U is an open subset of Y. Therefore $f^{-1}(U)$ is an open subset of X by Proposition 8.

Suppose that $f^{-1}(U)$ is an open subset of X whenever U is a D'-p-neighborhood in Y. Let V be any open subset of Y. By Proposition 8 we will have shown that f is continuous if we show that $f^{-1}(V)$ is an open subset of X. Now V is the union of D'-p-neighborhoods (Proposition 3), say $V = \bigcup_I U_i$, where each U_i is a D'-p-neighborhood. Then

$$f^{-1}(V) = f^{-1}\left(\bigcup_I U_i\right) = \bigcup_I f^{-1}(U_i).$$

But each $f^{-1}(U_i)$ is by hypothesis an open subset of X; hence $f^{-1}(V)$ is the union of a family of open subsets of X and consequently is open. Therefore f is continuous.

Example 15. Using Proposition 9, we will show that the function f from R^2 with the Pythagorean metric to R, the set of real numbers with the absolute value metric, defined by $f(x, y) = x$, is continuous. If $a \in R$ and p is any positive real number, then $N(a, p)$ is the open interval $(a - p, a + p)$. Then $f^{-1}(N(a, p))$ is easily seen to be

$$\{(x, y) \mid a - p < x < a + p\}$$

(Fig. 2.16), an open subset of R^2. Therefore, by Proposition 9, f is continuous.

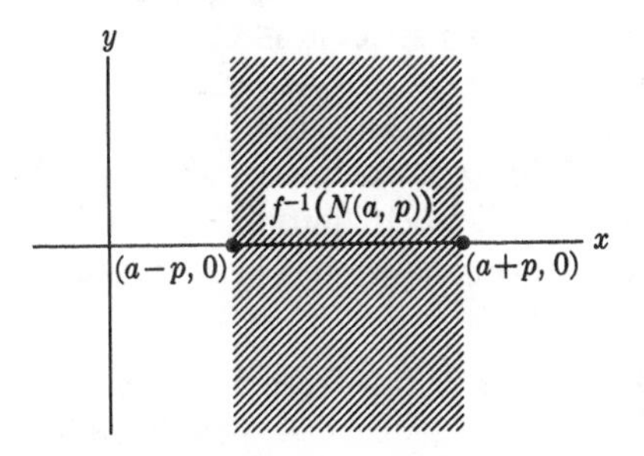

Figure 2.16

Example 16. The identity function i [defined by $i(x, y) = (x, y)$] from R^2 with metric D onto R^2 with metric D_1 is continuous, since any D-open subset of R^2 is D_1-open, and conversely (Section 2.3, Exercise 4). That is, if U is any open subset of R^2, D_1, then $i^{-1}(U) = i(U) = U$ is also an open subset of R^2, D; hence i is continuous. Since $i^{-1} = i$, and since each D-open set is also D_1-open, we see that i^{-1} is continuous as a function from R^2, D_1 onto R^2, D. It is quite possible, however, for a one-one function from a metric space, X, D onto a metric space Y, D' to be continuous without f^{-1} being continuous, as will be demonstrated in Example 17.

The following proposition relates continuity with the convergence of sequences.

Proposition 10. Let f be a function from the metric space X, D into the metric space Y, D'. Then f is continuous if and only if given any sequence $S = \{s_n\}$, $n \in N$, in X such that $s_n \to y$,

$$f(S) = \{f(s_n)\}, \qquad n \in N, \qquad \text{converges to } f(y) \text{ in } Y.$$

Proof. Suppose that f is continuous, but that there is a sequence $S = \{s_n\}$, $n \in N$, in X such that $s_n \to y$, but $f(S)$ does not converge to $f(y)$. Since $f(S)$ does not converge to $f(y)$, there must be a positive number p such that $N(f(y), p)$ excludes infinitely many of the $f(s_n)$. But since f is continuous, there is a positive number q such that $f(N(y, q)) \subset N(f(y), p)$. By assumption, $s_n \to y$; hence $N(y, q)$ contains all but finitely many of the s_n. This implies that $N(f(y), p)$ contains all but finitely many of the $f(s_n)$, a contradiction.

Conversely, suppose that given any sequence $S = \{s_n\}$, $n \in N$, in X such that $s_n \to y$, then $f(S)$ converges to $f(y)$; assume that f is not continuous. Then, since f is not continuous, there is a point $f(a)$ in Y and a positive number p for which there is no positive number q such that

$$f(N(a, q)) \subset N(f(a), p).$$

Consider the family $\{U_n\}$, $n \in N$, of neighborhoods of a, where $U_n = N(a, 1/n)$. For each n we can select $s_n \in U_n$ such that $f(s_n) \notin N(f(a), p)$; this selection is possible because f is not continuous. Then $s_n \to a$ (Exercise 1); but $\{f(s_n)\}$, $n \in N$, is a sequence in Y which does not converge to $f(a)$, since $N(f(a), p)$ by construction of $\{s_n\}$, $n \in N$, contains no $f(s_n)$ whatsoever. This contradicts our initial hypothesis that f preserves the limits of sequences; hence f could not be discontinuous. Therefore f is continuous.

A continuous function is thus seen to be one which in some sense preserves the convergence of sequences. (See Exercise 7 also.)

Example 17. The identity function i from R^2 with metric D onto R^2 with metric D_2 (Example 3) is not continuous. For the sequence defined by $s_n = (1, 1/n)$ converges to $(1, 0)$ with respect to metric D (Example 12), but the sequence $\{i(s_n)\} = \{s_n\}$, $n \in N$ does not converge to $i(1, 0) = (1, 0)$ with respect to D_2. Note, however, that

$$i^{-1}: R^2, D_2 \to R^2, D$$

is continuous. This follows from the fact that any sequence in R^2 which converges with respect to D_2 is essentially a constant sequence (Section 2.5, Exercise 4), and any constant sequence converges with respect to any metric whatsoever on R^2. We thus see that even the identity function from a set X with one metric onto the same set with a different metric may fail to be continuous.

EXERCISES

1. In the converse part of Proposition 10, prove $s_n \to a$.
2. Discuss the continuity of the function in each of the following:
   a) the function defined by $f(x) = 5x + 7$ from the space R, D onto itself, where R is the set of real numbers and D is the absolute value metric;
   b) the function defined by $f(x, y) = x + y$ from R^2, D_1 onto R, D [with R, D as in (a)];
   c) the function defined by $f(g) = g(0)$ from the space X, D of Example 5 onto the closed interval [0, 1] considered as a subspace of the space of real numbers with the absolute value metric;
   d) the identity function i from R^2, D_1 onto R^2, D_3.
3. Suppose f to be a function from a metric space X, D into the metric space Y, D' such that $D(x, x') \geq kD'(f(x), f(x'))$, where k is a constant positive real number. Prove that f is continuous.
4. Suppose that f is a continuous function from X, D into Y, D', and g a continuous function from Y, D' into Z, D''. Prove that $g \circ f$ is a continuous function from X, D into Z, D''.
5. Assume f to be a function from X, D onto a subspace W of Y, D'. Prove that f is continuous as a function from X, D into Y, D' if and only if f is continuous as a function from X, D onto W, $D' \mid W$.
6. Suppose W is a subset of Y, D. Prove that the function $i: W \to Y$ defined by $i(w) = w$ for each $w \in W$ is continuous as a function from W, $D \mid W$ into Y, D.
7. Prove that a function f from a space X, D into a space Y, D' is continuous if and only if given any convergent sequence S in X, $f(S)$ is a convergent sequence in Y. [*Hint:* It must be shown that if S converges to x in X, then $f(S)$ converges to $f(x)$.]
8. The concept of equivalent metrics was introduced in Section 2.3, Exercise 6. Prove that metrics D and D' on a set X are equivalent if and only if the identity map from both X, D onto X, D' and from X, D' onto X, D is continuous.

9. a) Suppose $f: R^2 \to R^2$ takes any circle in R^2 onto a circle. Need f be continuous if R^2 has the usual Pythagorean metric?
   b) Suppose $f: R^2 \to R^2$ takes collinear points into collinear points. Need f be continuous?
10. Prove that the set $f^{-1}(N(a, p)) \subset R^2$ of Example 15 is open with respect to the usual Pythagorean metric.

2.7 "DISTANCE" BETWEEN TWO SETS

Definition 7. Let X, D be any metric space. Suppose that $x \in X$ and $A \subset X$. Then define

$$D(x, A) = \text{greatest lower bound } \{D(x, a) \mid a \in A\}.$$

If A and B are subsets of X, define

$$D(A, B) = \text{greatest lower bound } \{D(a, b) \mid a \in A, b \in B\}.$$

The following equalities follow immediately from the definition:

$$D(x, A) = D(\{x\}, A)$$

and

$$D(A, B) = \text{glb } \{D(a, B) \mid a \in A\} = \text{glb } \{D(A, b) \mid b \in B\}.$$

Note, however, that D is not a metric for the set of all subsets of X. It is quite possible, for example, to have sets W and Y such that $D(W, Y) = 0$, but $W \cap Y = \phi$ (a contradiction to Definition 1iii) as we see from the following.

Example 18. Let R^2 be the plane with the Pythagorean metric D. Set

$$Y = \{(x, y) \mid x^2 + y^2 < 1\} \qquad \text{and} \qquad W = \{(1, 0)\}$$

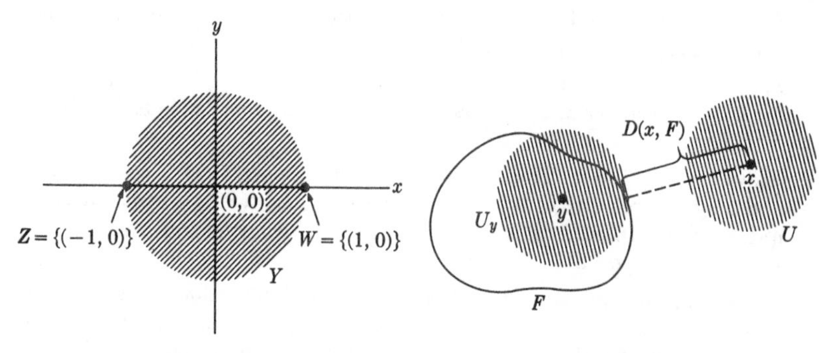

Figure 2.17 **Figure 2.18**

(Fig. 2.17). Then $D(Y, W) = 0$, since there are points of Y arbitrarily close to $(1, 0)$, but $(1, 0)$ is not a point of Y. If we let $Z = \{(-1, 0)\}$, then

$$D(Z, W) = 2 > D(W, Y) + D(Y, Z) = 0 + 0 = 0.$$

Therefore the "metric" D on the set of all subsets of R^2 does not even satisfy the triangle inequality.

Note too that the distance between a set and any of its nonempty subsets is always 0.

Even though the "metric" for the subsets of a metric space is not really a metric according to Definition 1, it is still of great use in helping us describe the properties of metric spaces.

Proposition 11. Let X, D be a metric space. A subset F of X is closed if and only if given any point x in $X - F$, $D(x, F) \neq 0$.

Proof. If F is a closed subset of X, then $X - F$ is open. Therefore, given any $x \in X - F$, there is a positive number p such that $N(x, p) \subset X - F$. But then $D(x, F) \geq p$; hence $D(x, F) \neq 0$.

Suppose, on the other hand, that given any point x in $X - F$, $D(x, F) \neq 0$. Then setting $p = D(x, F)$, we have $N(x, p) \subset X - F$. That is, for each $x \in X - F$, we have a positive number p such that

$$N(x, p) \subset X - F,$$

which is to say that $X - F$ is open. Therefore $F = X - (X - F)$ is closed.

Proposition 12. Let X, D be a metric space. Suppose F is a closed subset of X and $x \in X - F$. Then there are open sets U and V of X such that $x \in U$, $F \subset V$, and $U \cap V = \phi$.

Proof. Since F is closed and $x \in X - F$, $D(x, F) \neq 0$ (Fig. 2.18). For each $y \in F$, set

$$U_y = N(y, \tfrac{1}{2}D(x, F)).$$

Then $V = \bigcup_F U_y$ is an open set which contains F. Also $U = N(x, \frac{1}{2}D(x, F))$ is an open set which contains x. In order to complete the proof that U and V satisfy the terms of Proposition 12, we must show that $U \cap V = \phi$. Suppose $U \cap V \neq \phi$, and select w from $U \cap V$. Then $D(w, y) < \frac{1}{2}D(x, F)$ for some $y \in Y$, and $D(w, x) < \frac{1}{2}D(x, F)$. It then follows that

$$D(x, y) \leq D(w, x) + D(w, y) < \tfrac{1}{2}D(x, F) + \tfrac{1}{2}D(x, F) = D(x, F).$$

But $D(x, F) = \text{glb}\,\{D(x, z) \mid z \in F\}$, and $y \in F$; hence $D(x, y) \geq D(x, F)$, a contradiction. Since the assumption that $U \cap V \neq \phi$ led to a contradiction, it must be that $U \cap V = \phi$.

We now prove an even stronger result.

Proposition 13. Let X, D be a metric space. Then if F and F' are two closed subsets of X such that $F \cap F' = \phi$, there are open sets U and V of X such that

$$F \subset U, \qquad F' \subset V, \qquad \text{and} \qquad U \cap V = \phi.$$

(Note that this proposition contains Proposition 12 as a special case, since each one-element subset of X is a closed subset by Proposition 5.)

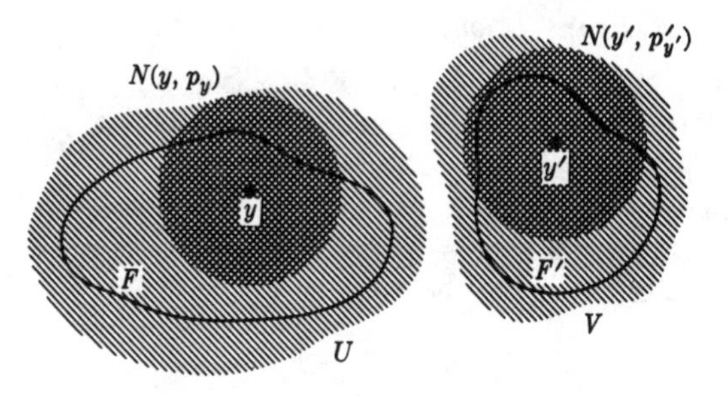

Figure 2.19

Proof. For each $y \in F$, set $p_y = \frac{1}{2}D(y, F')$, and for each $y' \in F'$, set $p'_{y'} = \frac{1}{2}D(y', F)$. Set $U = \bigcup_F N(y, p_y)$ and $V = \bigcup_{F'} N(y', p'_{y'})$ (Fig. 2.19). Since both U and V are the union of a family of open sets, both are open. Then, since $F \subset U$ and $F' \subset V$, it remains to show that $U \cap V = \phi$. Suppose we can find $w \in U \cap V$. It follows that $D(y, w) < p_y$ for some $y \in F$, and that $D(y', w) < p'_{y'}$ for some $y' \in F'$. We may suppose $p_y \geq p'_{y'}$. Then

$$D(y, y') \leq D(y, w) + D(y', w) < p_y + p'_{y'} \leq 2p_y = D(y, F').$$

But $D(y, y') \geq D(y, F')$, hence a contradiction. It follows then that $U \cap V = \phi$.

Propositions 12 and 13 represent what are called *separation* properties because they measure our ability to "separate" or distinguish disjoint closed subsets. Oddly enough, as the following example demonstrates, Proposition 13 does not imply that the "distance" between two disjoint nonempty closed subsets of a metric space is always greater than 0.

Example 19. Let R^2 be the coordinate plane with the Pythagorean metric D, and let

$$F = \{(x, y) \mid y = 1/x, x \neq 0\}$$

and

$$F' = \{(x, y) \mid y = 0\}$$

(Fig. 2.20). That is, F is the graph of the rectangular hyperbola $y = 1/x$, while F' is the x-axis. Both these sets are closed and $F \cap F' = \phi$. Since $y = 1/x$ has the x-axis for an asymptote, $D(F, F') = 0$. Nevertheless, F and F' can still be separated in the sense of Proposition 13.

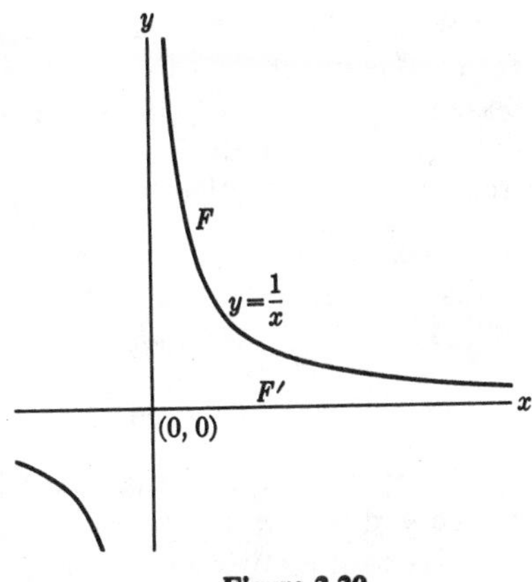

Figure 2.20

We saw in Proposition 11 that a subset F of a metric space X, D is closed if and only if each point of X which is 0 distance from F is an element of F. This inspires the following definition.

Definition 8. Let X, D be a metric space and $A \subset X$. We define the *closure of* A, denoted by Cl A, by

$$\text{Cl } A = \{x \in X \mid D(x, A) = 0\}.$$

Example 20. Let R be the space of real numbers with the absolute value metric. Set $A = \{1/n \mid n = 1, 2, 3, \ldots\}$. Since $1/n \to 0$, then $d(0, A) = 0$ (Exercise 5). Since $D(1/n, 1/n) = 0$ for each n, then $A \subset \text{Cl } A$. On the other hand, if y is any number other than 0 or an element of A, then it is readily verified that $D(y, A) > 0$. Therefore $\text{Cl } A = A \cup \{0\}$.

Proposition 14. Cl A as given in Definition 8 is a closed subset of X.

Proof. Suppose that Cl A is not closed. Then $X - \text{Cl } A$ is not open; therefore there is an element $x \in X - \text{Cl } A$ such that for any positive number p, $N(x, p) \cap \text{Cl } A \neq \phi$. Select $w \in N(x, p) \cap \text{Cl } A$. Then since $N(x, p)$ is open, there is a positive number q such that $N(w, q) \subset N(x, p)$. But since $w \in \text{Cl } A$, then $D(w, A) = 0$; therefore there is at least one element

$$a \in A \cap N(w, q) \subset N(x, p).$$

But this means that, for any positive number p, there is an element $a \in A$ such that $D(x, a) < p$. It follows then that glb $\{D(x, a) \mid a \in A\} = D(x, A) = 0$; thus $x \in \text{Cl } A$, a contradiction, since $x \in X - \text{Cl } A$. Cl A must therefore be a closed subset of X.

We see then that Cl A is a closed subset of X which contains A. In fact, if F is any closed subset of X which contains A, then F contains Cl A, because F contains all points 0 distance from itself (Proposition 11) and, consequently, contains as a subset all those points which are 0 distance from A, i.e., Cl A. We therefore have the following.

Proposition 15. Cl A is the smallest closed subset of X which contains A; that is, if F is any closed subset of X which contains A, then Cl $A \subset F$. In other words,

$$\text{Cl } A = \bigcap \{F \mid F \text{ is closed and } A \subset F\}.$$

There is much more that could be said about metric spaces, and more will be said later in the book. The primary purpose of this chapter, however, is one of motivation for what follows. Most of the concepts of point-set topology are generalized in some manner from the properties of metric spaces. We begin this process of generalization in the next chapter.

EXERCISES

1. Prove that Cl has the following properties.

   a) $A \subset \text{Cl } A$ b) $\text{Cl}(\text{Cl } A) = \text{Cl } A$ c) $\text{Cl}(A \cup B) = \text{Cl } A \cup \text{Cl } B$

2. Let R^2, the coordinate plane, have the Pythagorean metric. What are the closures of each of the following subsets of R^2?

   a) $\{(x, y) \mid |x| < 9\}$ b) $\{(x, y) \mid x^2 < y\}$ c) $\{(x, y) \mid |x| \leq 1, |y| < 1\}$

3. Prove that a subset A of a metric space X, D is closed if and only if $A = \text{Cl } A$.

4. Suppose A and B are subsets of a metric space X, D. Prove or disprove each of the following.

   a) If $\text{Cl } A \cap \text{Cl } B = \phi$, then $D(A, B) \neq 0$.
   b) $D(A, B) = 0$ if and only if some sequence of points in A converges to a point in B.
   c) If $A \cap B = \phi$, then there are disjoint open sets U and V such that $A \subset U$ and $B \subset V$ (even if A and B are not closed).
   d) If $D(A, B) \neq 0$, then there are disjoint open sets U and V such that $A \subset U$ and $B \subset V$.

5. Let $\{s_n\}$, $n \in N$ be a sequence in a metric space X, D and suppose $y \in X$ such that $y \neq s_n$ for any n. Prove that $s_n \to y \in X$ if and only if $D(\{s_n\}, y) = 0$.

6. If X, D is a metric space and $A \subset X$, we define the *frontier* of A, denoted by Fr A, by $\text{Fr } A = \{x \in X \mid D(x, A) = 0 \text{ and } D(x, X - A) = 0\}$. Find each of the following.

   a) $\text{Fr}(0, 1)$ in the set R of real numbers, with the absolute value metric
   b) $\text{Fr}\{(x, y) \mid x^2 + y^2 < 1\} \subset R^2$, with the Pythagorean metric

c) $\text{Fr}\{x \mid x \text{ is a rational number}\} \subset R$, as in (a)
d) $\text{Fr}\{(x, y) \mid x = 3\} \subset R^2$, with metric D_2 (Example 3)

7. Prove that a subset A of a metric space X, D is open if and only if $\text{Fr}A \cap A = \phi$. Prove that A is closed if and only if $\text{Fr}A \subset A$.

3
TOPOLOGIES

3.1 THE NOTION OF A TOPOLOGY

The fundamental properties of open subsets of a metric space are outlined in Proposition 2 of Chapter 2. Mathematicians have found from experience that families of subsets having these same properties arise in contexts other than those of metric spaces; hence it is reasonable to study these properties in their own right, abstracted from the limitations that metric spaces impose. In particular, the properties of open sets in metric spaces inspire the following definition.

Definition 1. Let X be any set. A collection τ of subsets of X is said to be a *topology* on X if the following axioms are satisfied:

i) X and ϕ are members of τ.

ii) The intersection of any two members of τ is a member of τ.

iii) The union of any family of members of τ is again in τ.

The members of τ are then said to be *τ-open* subsets of X, or merely *open* subsets of X if no confusion may result.

Example 1. If X, D is a metric space, then the D-open subsets of X form a topology on X. This topology is called the *metric topology* induced on X by D. It was, of course, this topology that we studied in Chapter 2.

Example 2. Let X be any set. Then the family of all subsets of X forms a topology on X. This topology consisting of all of the subsets of X is called the *discrete topology* on X. The discrete topology contains the maximum possible number of open sets since, relative to the discrete topology, every subset of X is open.

Example 3. If X is any set, then the collection $\{X, \phi\}$ of subsets of X also forms a topology on X. This topology is called the *trivial* (by some, the *indiscrete*) topology on X. It contains the fewest possible open sets compatible with having a topology on X.

The discrete and trivial topologies represent opposite extremes. Topologies which are of genuine interest usually lie somewhere in between; for example, the topology induced on the set of real numbers by the absolute value metric is neither the trivial nor the discrete topology. We now give an example of a topology which is neither discrete nor trivial, but also is not related to any metric.

Example 4. Let $X = \{a, b\}$. Define $\tau = \{X, \phi, \{a\}\}$. It is easily verified that τ is a topology on X. Suppose D to be any metric on X, and set $p = D(a, b)$. Then $N(b, p) = \{b\}$, and it follows that $\{b\}$ is a D-open set. But $\{b\}$ is not a τ-open set; hence τ could not be the topology induced on X by D.

Definition 2. A set X with topology τ is called a *topological space*. Just as X, D was used to denote a set X with metric D, so X, τ will be used to denote a set X with topology τ.

As in metric spaces, so in a topological space X, τ we say that a subset F of X is *τ-closed* (or merely *closed*) if $F = X - U$, where U is a τ-open set. (Compare this to Definition 4 of Chapter 2.)

Proposition 1. Let X, τ be a topological space. Then the closed subsets of X have the following properties.

a) X and ϕ are closed subsets of X.
b) The union of any two closed subsets of X is again a closed subset of X.
c) The intersection of any family of closed subsets of X is again a closed subset of X.

The proof is the same as the proof of Proposition 4, Chapter 2.

The next proposition shows that rather than defining a topology on a set by specifying the open subsets, we may equally well determine the topology by specifying the closed subsets.

Proposition 2. Let X be any set, and suppose that $\mathfrak{F}$ is a family of subsets of X such that

i′) X and ϕ are in $\mathfrak{F}$;

ii′) the union of any two members of $\mathfrak{F}$ is a member of $\mathfrak{F}$;

iii′) the intersection of any family of members of $\mathfrak{F}$ is a member of $\mathfrak{F}$.

If we now define a subset U of X to be open if and only if $U = X - F$, where F is some element of $\mathfrak{F}$, then the set τ of open sets thus formed is a topology on X with $\mathfrak{F}$ as the set of (τ-) closed subsets of X.

Proof. We first show that τ is a topology on X by verifying that τ satisfies Definition 1.

i) X and ϕ are in τ. Since X and ϕ are in $\mathfrak{F}$, and since $X = X - \phi$ and $\phi = X - X$, then X and ϕ are in τ.

ii) The intersection of any two members of τ is a member of τ. Suppose U and V are members of τ. Then

$$U = X - F_1 \qquad \text{and} \qquad V = X - F_2,$$

where F_1 and F_2 are in $\mathfrak{F}$. Therefore

$$U \cap V = (X - F_1) \cap (X - F_2) = X - (F_1 \cup F_2).$$

But, by (ii′), $F_1 \cup F_2 \in \mathfrak{F}$; hence $U \cap V \in \tau$.

iii) The union of any family of members of τ is a member of τ. Suppose $\{U_i\}$, $i \in I$, is a family of members of τ. It follows that for each $i \in I$, $U_i = X - F_i$, where $F_i \in \mathfrak{F}$. Then

$$\bigcup_I U_i = \bigcup_I (X - F_i) = X - \bigcap_I F_i.$$

But, by (iii′), $\bigcap_I F_i \in \mathfrak{F}$; hence $\bigcup_I U_i \in \tau$.

Therefore τ satisfies the definition of a topology on X.

It remains to be shown that $\mathfrak{F}$ is the set of closed sets for the topology τ. Suppose F is τ-closed. Then $F = X - U$, where $U \in \tau$. But $U = X - F'$, where $F' \in \mathfrak{F}$; therefore

$$F = X - (X - F') = F' \in \mathfrak{F}.$$

On the other hand, if $F \in \mathfrak{F}$, then $X - F \in \tau$. Then, since $F = X - (X - F)$, F is τ-closed. The members of $\mathfrak{F}$ are therefore precisely the τ-closed subsets of X.

Example 5. We define a family $\mathfrak{F}$ of subsets of R^2, the coordinate plane, as follows: Let $F \in \mathfrak{F}$ if and only if $F = R^2$, $F = \phi$, or F is a set consisting of finitely many points together with the union of finitely many straight lines. By hypothesis, X and ϕ are in $\mathfrak{F}$. It follows from the fact that two straight lines can only intersect in either a straight line (if they coincide), a point, or the empty set, that the intersection of any family of members of $\mathfrak{F}$ is again a member of $\mathfrak{F}$. Since the union of finitely many lines and points with finitely many more lines and points still consists of finitely many lines and points, the union of any two members of $\mathfrak{F}$ is also a member of $\mathfrak{F}$. Therefore, $\mathfrak{F}$ satisfies (i′) through (iii′) of Proposition 2, and hence determines a topology on R^2. The topology which $\mathfrak{F}$ determines is in fact the smallest topology in which lines and points are closed sets. It is not, however, the topology induced on R^2 by the Pythagorean metric D; for a subset of R^2 can exclude at most finitely many lines and still be open in the topology determined by $\mathfrak{F}$, but $\{(x, y) \mid x^2 + y^2 < 1\}$ excludes infinitely many lines and still is D-open.

EXERCISES

1. Prove that the intersection of finitely many open sets is open and that the union of finitely many closed sets is closed.
2. Suppose X, τ a topological space, and $A \subset X$. The *interior* of A, denoted by A°, is defined by $A^\circ = \cup\{U \in \tau \mid U \subset A\}$. Prove the following:
   a) $(A^\circ)^\circ = A^\circ$.
   b) $A^\circ \subset A$.
   c) $(A \cap B)^\circ = A^\circ \cap B^\circ$.
   d) A subset U of X is open if and only if $U = U^\circ$.
3. Let R^2 be the plane with the Pythagorean metric. Let $\mathfrak{N}$ be the set of all p-neighborhoods in R^2. Which properties of a topology for R^2 does $\mathfrak{N}$ fail to satisfy? Let τ be the set of all unions of elements of $\mathfrak{N}$. Prove that τ is a topology for R^2. What topology is this?
4. Find all possible topologies for the set $\{1, 2, 3\}$.
5. Suppose X a set with more than one element. Prove that there is no metric on X which induces the trivial topology on X. Find a metric for X which induces the discrete topology on X.
6. Let N be the set of positive integers. Define a subset F of N to be closed if F contains a finite number of positive integers, or $F = N$. Show that the closed subsets of N thus defined satisfy the conditions of Proposition 2, and hence can be used to define a topology on N. Prove that this topology is not induced by any metric. [*Hint:* Show that the topology does not satisfy Proposition 12 of Chapter 2.]
7. Prove that the topology defined on R^2 in Example 5 is really the smallest topology in which lines and points are closed sets. Would it be possible to have a topology on R^2 in which every line was a closed set, but every one-point subset was not? in which every one-point subset was closed, but every line was not? Try to find a topology satisfying the latter condition.
8. a) Define explicitly, that is, characterize completely the members of, the topology on R^2 which has the fewest members and relative to which each one point subset of R^2 is closed.
   b) Characterize the smallest topology on R^2 relative to which each straight line of R^2 is closed. Is this the same topology found in (a)? Does it contain the topology found in (a)?

3.2 BASES AND SUBBASES

Quite often it is impractical to explicitly specify all the open sets in order to define a topology on some set. Note that when we defined an open subset of a metric space, it was done in terms of p-neighborhoods and not by listing each open set separately. Proposition 2 has also shown us that we could equally well define a topology by giving the closed sets instead of the open ones. We now investigate other methods of determining a topology on a given set. In the first of these methods, a collection of sets is furnished which "generates" the topology (in much the same way that a basis of a vector space generates the vector space).

Definition 3. Suppose that X, τ is a topological space. A subset $\mathcal{B}$ of τ (i.e., a collection of open sets) is said to be a *basis* for the topology τ if each member of τ is the union of members of $\mathcal{B}$.

There is no analog of linear independence in Definition 3. Any topology has at least one basis, namely itself. Generally it is of no consequence whether or not a basis is in any sense minimal.

Example 6. Suppose R^2 to be the plane with the Pythagorean metric D. Then the p-neighborhoods of R^2 form a basis for the topology induced by D (see Section 3.1, Exercise 3). In fact, if X, D is any metric space, then the p-neighborhoods of X form a basis for the topology induced by D.

Note that if X, D is any metric space $x \in X$ and $p > 0$, then there is a rational number q, $0 < q < p$, with $N(x, q) \subset N(x, p)$. This fact can be used to prove that

$$\{N(x, q) \mid x \in X, q \text{ a positive rational number}\}$$

is also a basis for the topology induced by D (Exercise 6). It can also be proved that

$$\{N(x, 1/n) \mid x \in X \text{ and } n \text{ a positive integer}\}$$

is a basis for the metric topology (Exercise 6). Thus we see that bases for a topology can be quite diverse.

Proposition 3. Suppose that X, τ is a topological space and that $\mathcal{B}$ is a basis for τ. Then the intersection of any two members of $\mathcal{B}$ is the union of members of $\mathcal{B}$, and X itself is the union of members of $\mathcal{B}$.

Proof. Since both X and the intersection of any two members of $\mathcal{B}$ are members of τ, such sets must be the union of members of $\mathcal{B}$.

The case often occurs when rather than being given a topology for a set X, we are merely given a collection of subsets of X. For example, in our study of metric spaces, it was the p-neighborhoods which arose most naturally; the open sets were defined after the p-neighborhoods had been introduced. We might, then, reasonably ask, When is a collection of subsets of X the basis for a topology on X? The following proposition answers this question.

Proposition 4. Let X be any set. Assume $\mathcal{B}$ to be a family of subsets of X such that

i′) X is the union of members of $\mathcal{B}$;

ii′) the intersection of any two members of $\mathcal{B}$ is the union of members of $\mathcal{B}$.

Define $\tau = \{U \subset X \mid U \text{ is the union of members of } \mathcal{B}\}$. Then τ is a topology on X and $\mathcal{B}$ is a basis for τ. (The topology τ for which $\mathcal{B}$ is a basis is, in fact, unique; see Exercise 3.)

Proof. We must verify that τ satisfies Definition 1.

i) X is the union of members of $\mathcal{B}$, by (i′), and ϕ is the union of the empty subfamily of $\mathcal{B}$. Therefore X and ϕ are members of τ.

ii) Suppose that U and V are in τ. Then $U = \bigcup_I B_i$ and $V = \bigcup_J B_j$, where I and J are appropriate index sets and where B_i and B_j are members of $\mathcal{B}$ for each $i \in I$ and $j \in J$. Then

$$U \cap V = \bigcup_{I,J} (B_i \cap B_j).$$

Since each $B_i \cap B_j$ is the union of members of $\mathcal{B}$ by (ii′), $U \cap V$ is the union of members of $\mathcal{B}$, and hence is in τ. The intersection of any two members of τ is again a member of τ.

iii) If $\{U_k\}$, $k \in K$, is any family of members of τ, then U_k is the union of members of $\mathcal{B}$ for each $k \in K$. Therefore $\bigcup_K U_k$ is the union of members of $\mathcal{B}$, and hence is in τ. That is, the union of any family of members of τ is again a member of τ. Therefore τ satisfies the definition of a topology on X. Since each member of τ is by definition the union of members of $\mathcal{B}$, $\mathcal{B}$ is a basis for τ.

Example 7. Let R be the set of real numbers. Clearly R is the union of open intervals. Since the intersection of any two open intervals in R is either empty or again an open interval, condition (ii′) of Proposition 4 is satisfied by the collection of open intervals in R. The family of open intervals in R thus forms the basis for a topology on R. This topology is the same as the topology induced on R by the absolute value metric (Exercise 1).

Suppose X to be any set, and $\mathcal{S}$ any collection of subsets of X. Combining Propositions 3 and 4, we see that $\mathcal{S}$ is the basis for a topology on X if and only if $\mathcal{S}$ satisfies conditions (i′) and (ii′) of Proposition 4; but not every family of subsets of X satisfies these conditions. We may ask, therefore, in what topologies on X the given sets are open. There is, however, generally no unique topology on X for which the given sets are open.

For example, the p-neighborhoods in R^2 with the Pythagorean metric D are open in the topology induced by D, but they are also open with respect to the discrete topology. We may therefore rephrase our question by asking, What is the "smallest" topology τ on X, that is, the topology having the "fewest" open sets, for which $\mathcal{S} \subset \tau$? This question is answered in the following proposition.

Proposition 5. Let X be any set and suppose $\mathcal{S}$ to be a collection of subsets of X. Set

$\mathcal{B} = \{B \mid B \text{ is the intersection of finitely many sets in } \mathcal{S} \text{ or } B = X\}.$
Then $\mathcal{B}$ is the basis for a topology τ on X defined by

$$\tau = \{U \mid U \text{ is the union of members of } \mathcal{B}\}.$$

Moreover, $\mathcal{S} \subset \tau$, and if τ' is any topology on X such that $\mathcal{S} \subset \tau'$, then $\tau \subset \tau'$; that is, τ is the smallest topology on X for which $\mathcal{S}$ is a collection of open sets.

Proof. We first show that $\mathcal{B}$ satisfies (i′) and (ii′) of Proposition 4.

i′) X is itself the union of members of $\mathcal{B}$, since $X \in \mathcal{B}$ by hypothesis.

ii′) If B_1 and B_2 are both in $\mathcal{B}$, then both B_1 and B_2 are the intersection of finitely many members of $\mathcal{S}$. Therefore $B_1 \cap B_2$ is itself the intersection of finitely many members of $\mathcal{S}$, and is hence in $\mathcal{B}$. The intersection of any two members of $\mathcal{B}$ is thus again a member of $\mathcal{B}$. By Proposition 4, then, $\mathcal{B}$ is the basis for a topology, the topology τ as defined above, on X. Clearly $\mathcal{S} \subset \tau$.

If $\mathcal{S}$ is to be a collection of open subsets in any topology on X, then all finite intersections of members of $\mathcal{S}$ must also be open sets (Section 3.1, Exercise 1), and hence any union of a family of these intersections must also be open. Since τ is the smallest topology on X in which these conditions are fulfilled, it is therefore the smallest topology on X for which $\mathcal{S}$ is a family of open sets.

Proposition 5 inspires the following definition.

Definition 4. Let X, τ be a topological space. A subset $\mathcal{S}$ of τ is said to be a *subbasis* for τ if the set

$$\mathcal{B} = \{B \mid B \text{ is the intersection of finitely many members of } S\}$$

is a basis for τ.

Proposition 5 tells us that any collection of subsets of X whose union is X is the subbasis for a unique topology on X.

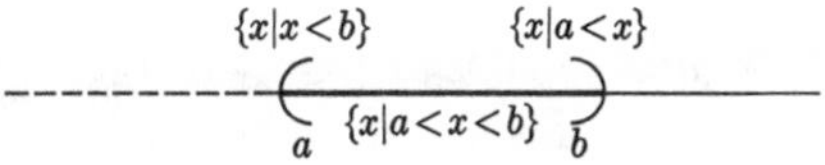

$\{x|x<b\}$ $\{x|a<x\}$

a $\{x|a<x<b\}$ b

Figure 3.1

Example 8. We saw in Example 7 that the family of open intervals in the set R of real numbers is the basis for a topology on R. Each open interval (Fig. 3.1) is, however, the intersection of two half-lines (rays) without endpoints, and each such half-line is open in the topology determined by the open intervals. Thus the set of all half-lines without endpoints forms a subbasis for the open-interval topology.

Example 9. Let R^2 be the coordinate plane with metric D_3 as in Chapter 2, Examples 3 and 6. As in any metric space, the D_3-p-neighborhoods in R^2 form a basis for the topology induced on R^2 by D_3. Note that each D_3-p neighborhood of R^2 is the intersection of finitely many open half-planes (Fig. 3.2), and that each of these half-planes is open in the topology induced by D_3. The collection of open half-planes is therefore a *subbasis* for this topology.

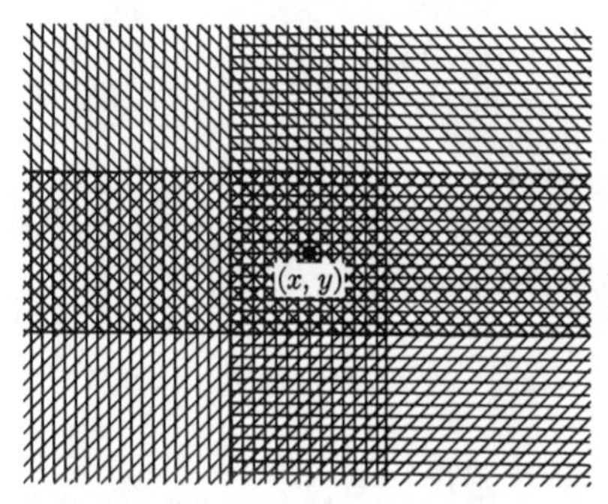

Figure 3.2

Thus far we have four means of specifying a topology on a set X: (1) by explicitly giving the open sets, that is, the members of the topology; (2) by explicitly giving the closed sets; (3) by giving a basis for the topology; or (4) by giving a subbasis for the topology.

EXERCISES

1. Prove that the topology on R, the set of real numbers, for which the collection of open intervals is a basis is the same as the topology induced on R by the absolute value metric.
2. Let R^2 be the coordinate plane and let D, D_1, D_2, and D_3 be the metrics described in Chapter 2, Examples 2 and 3.

a) Prove that the collection of open half-planes is a subbasis for the topologies induced by D and D_1.
b) Prove that the collection of closed half-planes (that is, a half-plane and its bounding line) is a subbasis for the topology induced by D_2.
c) Suppose that X, τ is a topological space and $\mathcal{S}$ a subbasis for τ. Prove that τ is the only possible topology on X for which $\mathcal{S}$ is a subbasis; that is, if τ' is a topology on X for which $\mathcal{S}$ is also a subbasis, then $\tau = \tau'$.
d) Using (c), prove that the topologies induced by D, D_1 and D_3 on R^2 are equal.

3. Let X be any set. Suppose a collection $\mathcal{B}$ of subsets of X is the basis for topologies τ and τ' on X. Prove $\tau = \tau'$. Thus any collection of subsets of X which satisfies (i′) and (ii′) of Proposition 4 is a basis for one and only one topology on X.

4. Prove that each of the following are bases for topologies on the prescribed sets.
   a) the set of intervals of the form $[a, b)$ in the set of real numbers
   b) $X = \{f \mid f \text{ is a function from } [0, 1] \text{ into } [0, 1]\}$ and the collection of subsets of X of the form
   $$B_S = \{f \in X \mid f(x) = 0 \text{ for } x \in S\},$$
   where S is some subset of $[0, 1]$
   c) $X = \{p \mid p \text{ is a polynomial with real coefficients}\}$ and the collection of subsets of X of the form
   $$B_n = \{p \in X \mid \text{degree of } p = n\},$$
   where n is a nonnegative integer

5. Prove that
$$\{N(x, q) \mid q \text{ is a rational number, } x \in X\}$$
and
$$\{N(x, 1/n) \mid x \in X \text{ and } n \text{ a positive integer}\}$$
are both bases for the topology induced on X by D as is claimed in Example 6.

6. Let N be the set of positive integers. Find explicitly all the open sets in the smallest topologies on N for which each of the following is a collection of open sets.
   a) N and ϕ b) N, $\{1, 2\}$, $\{3, 4, 5\}$ c) N, $\{1, 2\}$, $\{3, 4, 5\}$, $\{1, 4, 7\}$

7. Let X, D be a metric space. For each $x, y \in X$, define $H_1(x, y)$ to be $\{w \in X \mid D(x, w) > D(y, w)\}$ and $H_2(x, y) = \{w \in X \mid D(x, w) < D(y, w)\}$.
   a) Prove that $H_1(x, y)$ and $H_2(x, y)$ are open with respect to D.
   b) Describe these sets relative to two distinct points of R^2 with the Pythagorean metric.
   c) Prove or disprove: $\{H_1(x, y) \mid (x, y) \in X \times X, x \neq y\}$ is a subbasis for the topology induced on X by D.

3.3 OPEN NEIGHBORHOOD SYSTEMS

Although we already have four ways to specify a topology on a set, we have not as yet formally introduced one of the most widely used manners of determining a topology. Actually, however, we have already encountered this method since it is nothing more than the generalization of the p-neighborhoods of a point in a metric space.

Definition 5. Suppose X, τ to be a topological space, and suppose that for each point $x \in X$ we have a collection $\mathfrak{N}_x$ of open sets having the following properties:

i) $\mathfrak{N}_x \neq \phi$.

ii) $x \in N$ for each $N \in \mathfrak{N}_x$.

iii) If N_1 and N_2 are in $\mathfrak{N}_x$, then there is $N_3 \in \mathfrak{N}_x$ such that $N_3 \subset N_1 \cap N_2$.

iv) Given $N \in \mathfrak{N}_x$ and any $y \in N$, there is $N' \in \mathfrak{N}_y$ such that $N' \subset N$.

v) A subset U of X is open if and only if for each $x \in U$, there is $N \in \mathfrak{N}_x$ such that $N \subset U$.

Then the collection of families of members of τ (one for each $x \in X$) is called an *open neighborhood system* for τ.

Example 10. Suppose X, D is a metric space. Set $\mathfrak{N}_x = \{N(x, p) \mid p > 0\}$, for each $x \in X$. We will verify that the collection of $\mathfrak{N}_x$ forms an open neighborhood system for the topology induced on X by D.

i) Since $N(x, 1) \in \mathfrak{N}_x$ for each $x \in X$, $\mathfrak{N}_x \neq \phi$ for each $x \in X$.

ii) $D(x, x) = 0$ for any $x \in X$ implies that $x \in N(x, p)$ for any $p > 0$. Therefore $x \in N$ for any $N \in \mathfrak{N}_x$.

iii) Suppose N_1 snd N_2 are in $\mathfrak{N}_x$. Then $N_1 = N(x, p_1)$ and $N_2 = N(x, p_2)$, where p_1 and p_2 are positive numbers. We may suppose $p_1 \geq p_2$. Then

$$N_1 \cap N_2 = N(x, p_2) \in \mathfrak{N}_x.$$

iv) This is essentially Proposition 1, Chapter 2.

v) This is the definition of open set in the topology induced by D.

The following proposition relates open neighborhood systems and bases for a topology.

Proposition 6. Suppose that X, τ is a topological space. Then if $\mathfrak{N}$ is any open neighborhood system for τ, the collection of subsets of X contained in $\mathfrak{N}$ forms a basis for τ. On the other hand, if $\mathfrak{B}$ is any basis for τ, then, setting $\mathfrak{N}_x = \{B \in \mathfrak{B} \mid x \in B\}$ for each $x \in X$, we obtain an open neighborhood system for τ.

Proof. We first show that all the subsets of X in an open neighborhood system $\mathfrak{N}$ for τ form a basis for τ. Assume U to be any open subset of X. Then for each $x \in U$, we can choose $N_x \in \mathfrak{N}_x$ such that $N_x \subset U$ (Definition 5v). It follows that $\bigcup_U N_x \subset U$, since each $N_x \subset U$, but since every $x \in U$ is also in N_x, $U \subset \bigcup_U N_x$. Therefore $U = \bigcup_U N_x$. Thus U is the union of members of $\mathfrak{N}$, and hence the members of $\mathfrak{N}$ form a basis for τ.

Suppose $\mathcal{B}$ is any basis for τ. We now show that the collection of $\mathfrak{N}_x$, one for each $x \in X$, as defined above is an open neighborhood system for τ.

i) Since X is the union of members of $\mathcal{B}$, given any $x \in X$, $x \in B$ for some $B \in \mathcal{B}$. Therefore for each $x \in X$, $\mathfrak{N}_x \neq \phi$.

ii) By definition of $\mathfrak{N}_x$, $x \in N$ for any $N \in \mathfrak{N}_x$.

iii) Suppose N_1 and N_2 are in $\mathfrak{N}_x$. Then $N_1 \cap N_2$ is open, and hence is the union of members of $\mathcal{B}$. But $x \in N_1 \cap N_2$; hence there is $N_3 \in \mathcal{B}$ such that $x \in N_3 \subset N_1 \cap N_2$.

iv) Given $N \in \mathfrak{N}_x$, N is an open set, and thus N is the union of members of $\mathcal{B}$. Therefore if $y \in N$, there is $N' \in \mathcal{B}$ such that $y \in N' \subset N$; but then $N' \in \mathfrak{N}_y$.

The proof of (v) is left as an exercise (Exercise 1).

Recall that when metric spaces were being discussed, it was not the topology induced by the metric (that is, the open sets) which was considered first, but rather the p-neighborhoods. We can in fact define a topology on a set by providing a collection of sets which will serve as an open neighborhood system. This is brought out in the next proposition.

Proposition 7. Suppose that X is any set, and that for each $x \in X$ we have a collection of subsets of X which satisfy (i) through (iv) of Definition 5. Then if we define open subsets of X by means of (v) in Definition 5, the set τ of open subsets of X is a topology on X for which the collection of $\mathfrak{N}_x$ is an open neighborhood system.

Proof. If we can show that τ is a topology on X, it is clear that the collection of $\mathfrak{N}_x$ will form an open neighborhood system for τ. For by hypothesis, the collection of $\mathfrak{N}_x$ satisfies (i) through (iv) of Definition 5, and we are defining τ so that (v) holds as well; by (iv) each $\mathfrak{N}_x \subset \tau$. We therefore verify that τ satisfies (i) through (iii) of Definition 1.

i) If $x \in X$, then $\mathfrak{N}_x \neq \phi$. Consequently, for each $x \in X$, there is $N \in \mathfrak{N}_x$ such that $N \subset X$. Therefore X is in τ, and ϕ is vacuously an open set.

ii) Suppose U and V are any two open sets and $x \in U \cap V$. Then there is $N_1 \in \mathfrak{N}_x$ such that $N_1 \subset U$, and $N_2 \in \mathfrak{N}_x$ such that

$N_2 \subset V$. Therefore

$$x \in N_1 \cap N_2 \subset U \cap V.$$

By Definition 5(iii), there is $N_3 \in \mathfrak{N}_x$ such that

$$x \in N_3 \subset N_1 \cap N_2 \subset U \cap V;$$

hence $U \cap V$ is also an open set. The intersection of any two open sets is again an open set.

iii) Suppose $\{U_i\}$, $i \in I$, is a family of open sets, and $x \in \bigcup_I U_i$. Then $x \in U_i$ for some i; hence there is $N \in \mathfrak{N}_x$ such that $x \in N \subset U_i \subset \bigcup_I U_i$. The union of any family of open sets is thus again an open set. Therefore τ is a topology on X.

The reader should compare the proof of Proposition 7 with the proof of Proposition 2, Chapter 2. Why might one expect to see many similarities?

Example 11. Let R be the set of real numbers. For each $x \in R$, let $\mathfrak{N}_x$ be the set of all half-open intervals having x as a left-hand endpoint; that is,

$$\mathfrak{N}_x = \{[x, a) \mid a \in R, x < a\}.$$

The reader should verify at once (Exercise 3) that the collection of $\mathfrak{N}_x$ satisfies (i) through (iv) in Definition 5. In accordance with Proposition 7, then the collection of $\mathfrak{N}_x$ determines a topology on X. Exercise 2 shows that this topology is unique.

Example 12. Let R^2 be the coordinate plane. For each $x \in R^2$, let $\mathfrak{N}_x$ be the set of interiors of all triangles which contain x in their interior. Then the collection of $\mathfrak{N}_x$ forms an open neighborhood system for a topology on X.

Example 13. The following topology has applications in algebraic geometry. Let S be a ring. For each $s \in S$, define

$$\mathfrak{N}_s = \{s + A \mid A \text{ is a nonzero ideal of } S\},$$

that is, $\mathfrak{N}_s$ is the set of cosets of s. It can be verified that the collection of $\mathfrak{N}_s$ satisfies (i) through (iv) of Definition 5 and hence forms an open neighborhood system for a topology on S (Exercise 5).

The reader should note that even though the definition of an open neighborhood system seems more cumbersome than other methods of specifying a topology, in actual practice it is often the easiest and most natural way.

EXERCISES

1. Prove (v) in Proposition 6.
2. Suppose that X is a set and that for each $x \in X$ we have a collection $\mathfrak{N}_x$ of subsets of X satisfying (i) through (iv) of Definition 5. Assume that the $\mathfrak{N}_x$ are used to specify a topology τ on X in accordance with Proposition 7. Prove that if τ' is a topology on X for which the collection of $\mathfrak{N}_x$ is an open neighborhood system, then $\tau = \tau'$.
3. Let R be the set of real numbers. Assume τ to be the topology induced on R by the absolute value metric and τ' to be the topology defined on R by means of the open neighborhood system given in Example 11.
   a) Prove that the collection of $\mathfrak{N}_x$ described in Example 11 actually satisfies (i) through (iv) of Definition 5.
   b) Prove that every τ-open set is also τ'-open.
   c) Prove that there are τ'-open sets which are not τ-open.
   d) Prove that τ' is not the discrete topology.
4. Prove that the collection of $\mathfrak{N}_x$ given in Example 12 really do satisfy (i) through (iv) of Definition 5. Prove that the topology which the $\mathfrak{N}_x$ determine is the same as the topology induced on R^2 by the Pythagorean metric.
5. Verify that the collection of $\mathfrak{N}_s$ in Example 13 forms an open neighborhood system for a topology on the ring S. If the reader is not sufficiently familiar with rings to carry out the proof, he might restrict his attention to the ring Z of integers (with ordinary addition and multiplication as operations). An ideal of the integers is any subset of Z of the form $mZ = \{mz \mid z \in Z\}$, where m is any integer.
6. Suppose X, D is a metric space and $\{a_n\}$, $n \in N$, is a sequence of positive real numbers which converges to 0. For each $x \in X$, set
$$\mathfrak{N}_x = \{N(x, a_n) \mid n \text{ a positive integer}\}.$$
   Prove that the collection of $\mathfrak{N}_x$ thus formed forms an open neighborhood system for the topology induced by D. We thus see that if X, D is a metric space, we can find an open neighborhood system for the topology induced by D such that $\mathfrak{N}_x$ is countable for each $x \in X$.
7. Prove or disprove: Let $\mathfrak{N}_x$ be a collection of subsets of a set X satisfying (i) through (iv) of Definition 5 and with the property that each member of each $\mathfrak{N}_x$ is finite. Then the topology τ specified on X by the family of $\mathfrak{N}_x$ is necessarily the discrete topology.

3.4 FINER AND COARSER TOPOLOGIES

We have already encountered the problem of determining if two topologies on a set X are equal, that is, if they contain precisely the same open sets. This section is devoted to a more complete discussion of this topic.

Definition 6. Let X be any set, and suppose that τ and τ' are two topologies on X. Then τ is said to be *finer* than τ' if $\tau' \subset \tau$, that is, any τ'-open set is also a τ-open set. τ is said to be *strictly finer* than τ' if τ is finer than τ', but $\tau \neq \tau'$. If τ is finer than τ', then we may say that τ' is *coarser* than τ.

Proposition 8. Let X be any set. Then two topologies τ and τ' on X are equal if and only if τ is finer than τ' and τ' is finer than τ.

Proof. τ finer than τ' means $\tau' \subset \tau$. τ' finer than τ means $\tau \subset \tau'$. Therefore $\tau = \tau'$.

Example 14. Let X be any set. Then the discrete topology on X is finer than any topology on X and is strictly finer than any other topology on X. The trivial topology on X is coarser than any topology on X.

Example 15. Let τ and τ' be any two topologies on some set X. Let

$$\mathcal{S} = \tau \cap \tau' \qquad \text{and} \qquad \mathcal{S}' = \tau \cup \tau'.$$

Then $\mathcal{S}$ and $\mathcal{S}'$ are subbases for unique topologies τ_1 and τ_2, respectively, on X. (By Proposition 5, any collection of subsets of X is a subbasis for a unique topology on X. The reader should be certain that he understands that $\tau \cap \tau'$ is the family of all subsets of X which are both τ-open and τ'-open, e.g. X and ϕ, and *not* intersections of τ-open and τ'-open sets.) τ_1 is coarser than both τ and τ', since any τ_1-open set is both τ-open and τ'-open. τ_2, on the other hand, is finer than both τ and τ'. The reader should also see Exercise 5.

Proposition 9. Let τ and τ' be two topologies on some set X. Suppose that $\mathfrak{N}$ and $\mathfrak{N}'$ are open neighborhood systems for τ and τ', respectively. Then $\tau \subset \tau'$ if and only if for each $x \in X$ and $N \in \mathfrak{N}_x$, there is $N' \in \mathfrak{N}'_x$ such that $N' \subset N$.

Proof. Assume U to be any τ-open set and $x \in U$. Then there is $N \in \mathfrak{N}_x$ such that $x \in N \subset U$. Now if there is $N' \in \mathfrak{N}'_x$ such that $x \in N' \subset N$, then $x \in N' \subset U$; hence U is also τ'-open. Therefore $\tau \subset \tau'$. On the other hand, if $\tau \subset \tau'$, then since $N \in \tau$, $N \in \tau'$; hence there is $N' \in \mathfrak{N}'_x$ such that $N' \subset N$. (Note that the reason for virtually every step in this proof is Definition 5v.)

Corollary 1. Let τ and τ' be two topologies on the set X. Suppose $\mathfrak{N}$ and $\mathfrak{N}'$ are open neighborhood systems for τ and τ', respectively. Then $\tau = \tau'$ if and only if for each $x \in X$ and $N \in \mathfrak{N}_x$, there is $N' \in \mathfrak{N}'_x$ such that $N' \subset N$, and for each $N' \in \mathfrak{N}'_x$, there is $N \in \mathfrak{N}_x$ such that $N \subset N'$.

Proof. Applying Proposition 9, we see that this corollary merely states that $\tau = \tau'$ if and only if $\tau \subset \tau'$ and $\tau' \subset \tau$.

Corollary 2. Suppose X to be a set and D and D' possible metrics on X. Then the topology induced by D is the same as the topology induced by D' if and only if for each $x \in X$ and positive number p, there are positive numbers p_1 and p_2 such that the D-p_1-neighborhood of x is a subset of the D'-p-neighborhood of x, and the D'-p_2-neighborhood of x is a subset of the D-p-neighborhood of x.

Proof. The p-neighborhoods of points in a metric space form an open neighborhood system for the metric topology (see Example 10). Corollary 2 then is merely a restatement of Corollary 1 applied to metric spaces.

Example 16. Suppose that R^2, the coordinate plane, is given either the Pythagorean metric D, or the metric D_1 of Chapter 2, Example 3. It was shown in Chapter 2, Example 6 that the hypotheses of Corollary 2 apply; hence the topologies induced by D and D_1 on R^2 are the same.

Example 17. Let τ be the topology on R^2 which is defined by the open neighborhood system described in Example 12. It is easily verified (Fig. 3.3) that τ is the same topology as that induced on R^2 by the Pythagorean metric (Exercise 7).

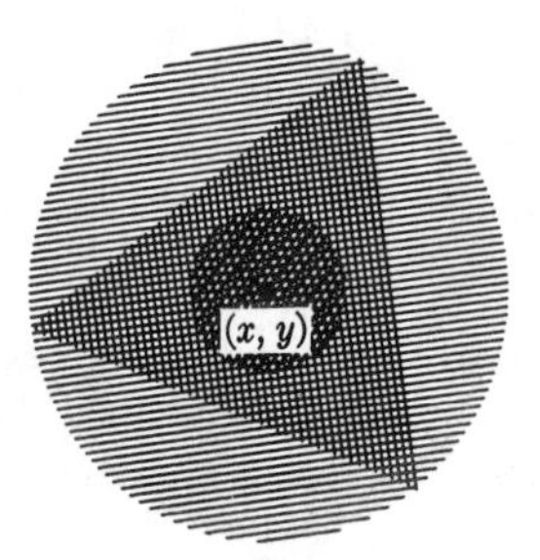

Figure 3.3

The reader might be tempted to conjecture that two topologies τ and τ' on some set X are equal if and only if given bases $\mathcal{B}$ and $\mathcal{B}'$ for τ and τ', respectively, for each $B \in \mathcal{B}$ there is $B' \in \mathcal{B}'$ with $B' \subset B$, and for each $B' \in \mathcal{B}'$, there is $B \in \mathcal{B}$ such that $B \subset B'$. This conjecture is, however, false, as is shown by Example 11 and Section 3.3, Exercise 3. Each interval of the form $[x, a)$ contains some interval of the form (p, q) [though of course (p, q) could not contain x], and each interval of the form (p, q) contains some interval of the form $[x, a)$. The set of all half-open intervals of the form $[x, a)$ forms a basis for a topology τ', and the set of open intervals also forms a basis for a topology τ. If the conjecture were correct, these two topologies should be equal, which they are not.

EXERCISES

1. In Example 15, prove as asserted that τ_1 is coarser than both τ and τ', and that τ_2 is finer than both τ and τ'.
2. Prove that the topology induced on R, the set of real numbers, by the absolute value metric is the same as the topology for which the set of all open intervals is a basis.
3. Prove that the topologies induced on R^2 by the metrics D_1 and D_3, Chapter 2, Example 3, are equal.
4. Define a metric D' on the set R of real numbers by $D'(x, y) = 3|x - y|$. How does the topology induced on R by D' compare with the topology induced on R by the absolute value metric? Answer this same question with D' replaced by D'', where D'' is defined by $D''(x, y) = |x - y|^2$.
5. In Example 15, prove that τ_1 is the finest topology which is coarser than both τ and τ', and that τ_2 is the coarsest topology which is finer than τ and τ'.
6. Find all possible topologies on $\{x, y, z\}$. Order these topologies as to fineness and coarseness. Construct a diagram which illustrates the relationships between the topologies.
7. In Example 17, prove that τ is the same as the topology induced on R^2 by the Pythagorean metric.
8. Suppose $\mathcal{B}$ and $\mathcal{B}'$ are bases for topologies τ and τ', respectively, on a set X. Suppose that each member of $\mathcal{B}'$ contains a member of $\mathcal{B}$. Are τ and τ' necessarily comparable? If so, in what way?

3.5 DERIVED SETS

Let X, τ be a topological space. Then associated with any subset A of X, there are a number of sets which are topologically related to or "derived" from A. We have already encountered some of these sets in the discussion of metric spaces.

Definition 7. If $A \subset X$, where X, τ is a topological space, then we define

a) the *closure* of A, denoted by Cl A, to be the intersection of all closed sets which contain A (cf. Chapter 2, Definition 8 and Proposition 15);
b) the *interior* of A, denoted by A°, to be the union of all open sets which are contained in A (Section 3.1, Exercise 2);
c) the *frontier* of A, denoted by Fr A, to be

$$\{x \mid \text{each open set which contains } x \text{ contains points of both } A \text{ and } X - A\}$$

(Section 2.7, Exercise 6);

d) the *exterior* of A, denoted by Ext A, to be $X - \text{Cl } A$; and
e) the *derived set* of A (sometimes called the *weak derived set*), denoted by A', to be

$$\{x \mid \text{if } x \in U,\ U \text{ an open set, then } A \cap (U - \{x\}) \neq \phi;$$
$$\text{that is, if } x \in U, \text{ then } U - \{x\} \text{ contains some point of } A\}.$$

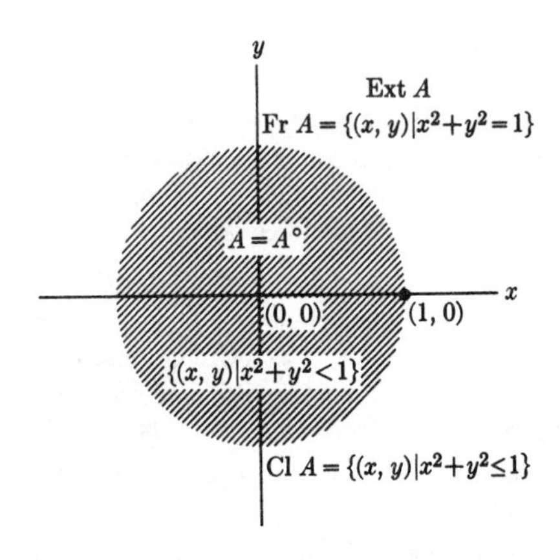

Figure 3.4

Example 18. Let R^2 be the plane with the topology induced by the Pythagorean metric D. Let

$$A = \{(x, y) \mid x^2 + y^2 < 1\}$$

(Fig. 3.4). Then

$$\text{Cl } A = \{(x, y) \mid x^2 + y^2 \leq 1\}, \qquad A^\circ = A,$$
$$\text{Fr } A = \{(x, y) \mid x^2 + y^2 = 1\},$$
$$\text{Ext } A = \{(x, y) \mid x^2 + y^2 > 1\}, \qquad \text{and} \qquad A' = \text{Cl } A.$$

The reader is not expected to see all of these equalities until more information has been obtained about these topologically derived sets, but he should verify as many as possible. He should also examine this example for possible relations that might hold between the sets topologically associated with A. These reflections also hold for the following example.

Example 19. Let R be the set of real numbers with the topology induced by the absolute value metric, and let A be the set of rational numbers. Then $\text{Cl } A = R$, $A^\circ = \phi$, $\text{Fr } A = R$, $\text{Ext } A = \phi$ (note that both A° and Ext A can simultaneously be empty), and $A' = R$. If R is given the trivial topology, then these sets topologically associated with A are exactly the same as in the metric topology. We can conclude then that what the topologically derived sets happen to be for any one subset of R

does not give much information about the topology. If, however, we know, say, Cl A for *every* $A \subset X$, then the topology is completely determined, as we shall see from Proposition 11.

Proposition 10. Suppose X, τ is a topological space and A and B are any subsets of X. Then

i) $A \subset \text{Cl}\, A$; ii) $\text{Cl}(\text{Cl}\, A) = \text{Cl}\, A$;

iii) $\text{Cl}(A \cup B) = \text{Cl}\, A \cup \text{Cl}\, B$; iv) $\text{Cl}\, \phi = \phi$;

v) A is closed if and only if $A = \text{Cl}\, A$.

Proof. Cl A is the intersection of a family of sets each of which contains A; therefore $A \subset \text{Cl}\, A$, and (i) is proved. Since Cl A is the intersection of a family of closed sets, Cl A is closed. If A is already closed, then A is one of the closed sets which contains A; hence $\text{Cl}\, A \subset A$. Since $A \subset \text{Cl}\, A$ by (i), $A = \text{Cl}\, A$. We have therefore proved (v), (ii), and (iv).

It still remains to prove (iii). Since $A \subset \text{Cl}(A \cup B)$ and $B \subset \text{Cl}(A \cup B)$, we have

$$\text{Cl}\, A \subset \text{Cl}(\text{Cl}(A \cup B)) = \text{Cl}(A \cup B) \qquad \text{and} \qquad \text{Cl}\, B \subset \text{Cl}(A \cup B).$$

Therefore $\text{Cl}\, A \cup \text{Cl}\, B \subset \text{Cl}(A \cup B)$. On the other hand, since $\text{Cl}\, A \cup \text{Cl}\, B$ is the union of two closed sets, it is closed. Thus $\text{Cl}\, A \cup \text{Cl}\, B$ is a closed set which contains $A \cup B$; consequently, $\text{Cl}(A \cup B) \subset \text{Cl}\, A \cup \text{Cl}\, B$. Therefore $\text{Cl}(A \cup B) = \text{Cl}\, A \cup \text{Cl}\, B$.

Proposition 11. Let X be any set, and suppose that Cl is a function from the set of subsets of X into the set of subsets of X such that Cl satisfies (i) through (iv) of Proposition 10. Then if we define a subset of X to be closed in accordance with (v), the collection $\mathfrak{F}$ of closed subsets thus obtained satisfies (i′) through (iii′) of Proposition 2 and hence determines a topology τ on X. Moreover, Cl A is the closure of A with respect to τ for each subset A of X.

Proof. We must verify (i′) through (iii′) of Proposition 2.

i′) Since $X \subset \text{Cl}\, X \subset X$, $X = \text{Cl}\, X$; therefore X is closed. Then $\text{Cl}\, \phi = \phi$ by (iv).

ii′) Let F and F' be any two closed sets. Then $F = \text{Cl}\, A$ and $F' = \text{Cl}\, B$, where A and B are subsets of X. Hence

$$F \cup F' = \text{Cl}\, A \cup \text{Cl}\, B = \text{Cl}(A \cup B),$$

and it follows that $F \cup F'$ is also a closed subset.

iii′) Let $\{F_i\}$, $i \in I$, be any family of closed subsets of X. Then $F_i = \text{Cl}\, F_i$ for each i. Now $\bigcap_I F_i \subset F_i$ for each i; it follows therefore that

$$\mathrm{Cl}\left(\bigcap_I F_i\right) \subset \mathrm{Cl}\, F_i = F_i.$$

[For $\bigcap_I F_i \subset F_i$ implies $(\bigcap_I F_i) \cup F_i = F_i$, and thus

$$\begin{aligned}\mathrm{Cl}\left(\left(\bigcap_I F_i\right) \cup F_i\right) &= \mathrm{Cl}\left(\bigcap_I F_i\right) \cup \mathrm{Cl}\, F_i \\ &= \mathrm{Cl}\, F_i = \mathrm{Cl}\left(\bigcap_I F_i\right) \cup F_i = F_i.]\end{aligned}$$

Therefore

$$\mathrm{Cl}\left(\bigcap_I F_i\right) \subset \bigcap_I F_i.$$

By (i), however, $\bigcap_I F_i \subset \mathrm{Cl}(\bigcap_I F_i)$; hence

$$\bigcap_I F_i = \mathrm{Cl}(\bigcap_I F_i).$$

We have shown then that $\bigcap_I F_i$ is closed. The family $\mathfrak{F}$ of closed subsets of X therefore satisfies (i′) through (iii′) of Proposition 2 and hence defines a topology on X.

Now if A is any subset of X, then since $\mathrm{Cl}(\mathrm{Cl}\, A) = \mathrm{Cl}\, A$, $\mathrm{Cl}\, A$ is closed with respect to τ. But $A \subset \mathrm{Cl}\, A$ by (i); hence the τ-closure of A is a subset of $\mathrm{Cl}\, A$. If F is any set such that $\mathrm{Cl}\, F = F$ and $A \subset F$, then $\mathrm{Cl}\, A \subset \mathrm{Cl}\, F = F$. Therefore the intersection of all such F, the τ-closure of A, contains $\mathrm{Cl}\, A$; that is, $\mathrm{Cl}\, A \subset \tau$-closure of A. Hence $\mathrm{Cl}\, A$ is the same as the τ-closure of A.

EXERCISES

1. Suppose X, τ a topological space. If A and B are any two subsets of X, show that it is not true in general that $\mathrm{Cl}(A \cap B) = \mathrm{Cl}\, A \cap \mathrm{Cl}\, B$.
2. Let X, τ be any topological space. Compute $\mathrm{Cl}\, X$, X°, $\mathrm{Fr}\, X$, $\mathrm{Ext}\, X$, X', and the corresponding sets for ϕ.
3. Suppose X is any set and $^\circ$ is a function from the set of subsets of X to the set of subsets of X with the following properties:

   i) $A^\circ \subset A$, ii) $(A^\circ)^\circ = A^\circ$,

   iii) $(A \cap B)^\circ = A^\circ \cap B^\circ$, and iv) $X^\circ = X$, where A and B are any subsets of X.

   Define a subset U of X to be open if and only if $U^\circ = U$. Prove that the set of open sets thus defined gives a topology on X.
4. Suppose that X is any set and that τ and τ' are topologies on X with τ finer than τ'. If $A \subset X$, we will denote the closure of A with respect to τ by $\mathrm{Cl}\, A$ and the closure of A with respect to τ' by $\mathrm{Cl}'\, A$. Analogous notation will be used with regard to the other sets topologically derived from A. Prove

that

a) $\text{Cl } A \subset \text{Cl}' A$; b) $A^{\circ\prime} \subset A^{\circ}$.

c) Find relations between the corresponding other sets topologically derived from A.

5. Try to find a method for specifying a topology on a set X by specifying $\text{Fr } A$ for each $A \subset X$. Do likewise for Ext.
6. Suppose that X is a set with the discrete topology and that $A \subset X$. Find sets topologically associated with A.
7. Is it possible for two distinct subsets of a topological space to have exactly the same topologically derived sets? Support your assertion.
8. Suppose τ and τ' are topologies on a set X. Determine if each of the following conditions implies either $\tau \subset \tau'$ or $\tau' \subset \tau$. In the following, A stands for any subset of X; we use $'$ to indicate that a derived set is being taken relative to τ'.

a) $\text{Fr } A \subset \text{Fr}' A$ b) $\text{Cl } A \subset \text{Cl}' A$ c) $\text{Ext } A \subset \text{Ext}' A$

3.6 MORE ABOUT TOPOLOGICALLY DERIVED SETS

In this section we continue the discussion begun in Section 3.5. Throughout this section X, τ will be assumed to be a topological space.

Proposition 12. If $A \subset X$, then

a) $\text{Cl } A = A \cup A'$; b) $\text{Cl } A = A^{\circ} \cup \text{Fr } A$;

c) $\text{Fr } A = \text{Fr}(X - A)$; d) $\text{Cl } A - \text{Fr } A = A^{\circ}$.

Proof

a) Assume that $x \in \text{Cl } A$, but that $x \notin A$. We prove first that for each open set U which contains x, $U \cap A \neq \phi$. If $A \cap U = \phi$, then $X - U$ is a closed set which contains A; hence $\text{Cl } A \subset X - U$. But since $x \in U$, x could not be in $\text{Cl } A$, a contradiction. We have then that for each open set U which contains x, $A \cap (U - \{x\}) \neq \phi$. But then $x \in A'$. Therefore

$$\text{Cl } A \subset A \cup A'.$$

Now suppose that $y \in A \cup A'$. If $y \in A$, then $y \in \text{Cl } A$, since $A \subset \text{Cl } A$. Suppose further that $y \in A'$, and that F is a closed subset of X which contains A, but not y. Then $X - F$ is open; hence $X - F$ is an open subset which contains y such that $((X - F) - \{y\}) \cap A = \phi$. Therefore y could not be in A', a contradiction. Thus y is contained in any closed set which contains A, and hence $y \in \text{Cl } A$. This gives $A \cup A' \subset \text{Cl } A$; it follows that

$$\text{Cl } A = A \cup A'.$$

b) Suppose $x \in \text{Cl } A$, but $x \notin \text{Fr } A$. Since $x \notin \text{Fr } A$, there is some open set U which contains x such that either $U \subset A$, or $U \subset X - A$. If $U \subset X - A$, then $X - U$ is a closed set which contains A; therefore $\text{Cl } A \subset X - U$, contradicting the assumption that $x \in \text{Cl } A$. It must be then that $U \subset A$, and hence $x \in A°$. Therefore

$$\text{Cl } A \subset A° \cup \text{Fr } A.$$

Assume that $y \in A° \cup \text{Fr } A$, but that $y \notin \text{Cl } A$. Since $A° \subset A \subset \text{Cl } A$, y must be in $\text{Fr } A$. Since $y \notin \text{Cl } A$, there is a closed set F which contains A, but not y. Then $X - F$ is an open set which contains y, but does not intersect A; hence y could not be in $\text{Fr } A$, a contradiction. Therefore y must be in $\text{Cl } A$. It follows that $A° \cup \text{Fr } A \subset \text{Cl } A$; hence

$$\text{Cl } A = A° \cup \text{Fr } A.$$

c) Suppose $x \in \text{Fr } A$. Then any open set U which contains x meets both A and $X - A$. Then U meets $X - A$ and $X - (X - A) = A$, and hence $x \in \text{Fr}(X - A)$. Thus $x \in \text{Fr}(X - A)$. Similarly, if $x \in \text{Fr}(X - A)$, $x \in \text{Fr } A$.

d) Proposition 12(d) will follow from (b) if we show that

$$A° \cap \text{Fr } A = \phi.$$

Suppose $A° \cap \text{Fr } A \neq \phi$, and choose x in this intersection. Then since $x \in \text{Fr } A$, every open set which contains x meets $X - A$. Since $x \in A°$, however, there is an open set U which contains x such that $U \subset A$. Clearly these two possibilities mutually exclude one another; thus it is impossible to have x in both $A°$ and $\text{Fr } A$.

The following terminology is introduced as an aid in making certain statements about topological spaces.

Definition 8. Let X, τ be a topological space. If $x \in X$, then any open set which contains x is said to be a *neighborhood* of x. (Some texts define a neighborhood of x to be any set which contains x in its interior, and refer to what we have defined to be a neighborhood as an *open neighborhood*. Such variations in terminology should be expected, however, in topology, since topology is still a rather young branch of mathematics and much terminology still has not become universally accepted.)

The following proposition relates neighborhoods and the topologically derived sets.

Proposition 13. Suppose that X, τ is any topological space and $A \subset X$.

Then

a) $x \in \text{Cl } A$ if and only if every neighborhood of x meets A;
b) $x \in A^\circ$ if and only if some neighborhood of x is contained in A;
c) $x \in \text{Ext } A$ if and only if x has some neighborhood disjoint from A;
d) $x \in \text{Fr } A$ if and only if every neighborhood of x meets both A and $X - A$;
e) $x \in A'$ if and only if every neighborhood of x meets A in some point other than x.

Proof.

a) By Proposition 12(a), $\text{Cl } A = A \cup A'$. If $x \in \text{Cl } A$ and $x \in A$, then every neighborhood of x meets A (in x). If $x \in A'$, then every neighborhood of x meets A also. On the other hand, if every neighborhood of x meets A, then either $x \in A$, or $x \in A'$; hence

$$x \in A \cup A' = \text{Cl } A.$$

b) If $x \in A^\circ$, then there is a neighborhood (open set) U of x such that $U \subset A$ by definition of A°. Conversely, if there is a neighborhood U of x with $U \subset A$, then $x \in A^\circ$.
c) If $x \in \text{Ext } A$, then since $\text{Ext } A = X - \text{Cl } A$ is open, $\text{Ext } A$ is a neighborhood of x disjoint from A. On the other hand, if there is a neighborhood of x disjoint from A, then $x \in X - \text{Cl } A = \text{Ext } A$.
d) and e) are merely the definitions of $\text{Fr } A$ and A' stated in terms of neighborhoods.

Example 20. Let R be the set of real numbers with the topology induced by the absolute value metric D. Let $A = (0, 1]$. Given any $a \in A$, $a \neq 1$, it is possible to find a neighborhood of a which lies entirely in A [since $A - \{1\} = (0, 1)$ is open]. On the other hand, no neighborhood of 1 lies entirely in A; hence $A^\circ = (0, 1)$. There are only two points, 0 and 1, with the property that every neighborhood of each of these points meets both A and $R - A$. Therefore $\text{Fr } A = \{0, 1\}$. Since $\text{Cl } A = A^\circ \cup \text{Fr } A$, $\text{Cl } A = [0, 1]$. If U is a neighborhood of any point x in $[0, 1]$, then U intersects A in some point other than x; hence $A' = [0, 1]$. Note that although $A^\circ \cap \text{Fr } A = \phi$, it is not true in general that $A \cap A' = \phi$. We also have (see Fig. 3.5).

$$\text{Ext } A = R - \text{Cl } A = \{x \in R \mid x > 1, \text{ or } x < 0\}$$

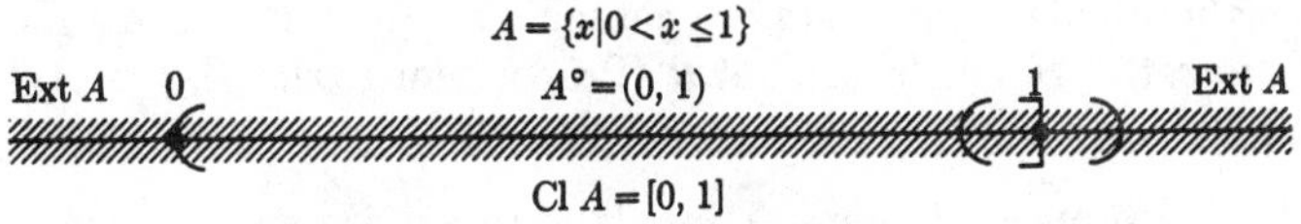

$A = \{x | 0 < x \leq 1\}$

Ext A 0 $A^\circ = (0, 1)$ 1 Ext A

Cl $A = [0, 1]$

Figure 3.5

Example 21. Let N be the set of positive integers, and define a subset U of N to be open if U contains all but finitely many positive integers. The set τ of open subsets of N forms a topology for N (Section 3.1, Exercise 6). Let A be the set of even positive integers. Then $\mathrm{Cl}\, A = N$, since any open set which contains any integer must contain at least one (in fact an infinite number) of even integers. For if the open set excluded all even integers, it would exclude infinitely many positive integers and hence would not be open. It is true that $A^\circ = \phi$, since any subset of A excludes infinitely many positive integers and hence could not be open. Also,

$$\mathrm{Fr}\, A = \mathrm{Cl}\, A - A^\circ = N - \phi = N.$$

Note that $\mathrm{Fr}\, A$ can be larger than A. Then $\mathrm{Ext}\, A = N - \mathrm{Cl}\, A = \phi$. Finally, $A' = N$, since any neighborhood of any integer contains both even and odd integers.

The notion of *denseness* is important in topology. Although we will not develop the concept in this chapter, this is an appropriate place to define it.

Definition 9. Let X, τ be a topological space. A subset A of X is said to be *somewhere dense* if

$$(\mathrm{Cl}\, A)^\circ \neq \phi,$$

that is, if the closure of A contains some open set. A is said to be *nowhere dense* if A is not somewhere dense. A is said to be *dense* if

$$\mathrm{Cl}\, A = X.$$

If A is any subset of a topological space such that $A^\circ \neq \phi$, then A is somewhere dense, since $A^\circ \subset A \subset \mathrm{Cl}\, A$.

Example 22. Let R be the set of real numbers with the topology induced by the absolute value metric. The set $A = [0, 1)$ is somewhere dense, since $A^\circ = (0, 1) \neq \phi$. The set of integers Z is nowhere dense, since Z is closed [$R - Z$ is the union of open sets of the form $(n - 1, n)$, n an integer, and hence is open] and no neighborhood of any integer contains only integers; that is,

$$\mathrm{Cl}\, Z = Z \qquad \text{and} \qquad Z^\circ = \phi.$$

Since any neighborhood of any number contains a rational number, the closure of Q, the set of rationals, is all of R (Proposition 13a). Therefore Q is dense in R.

The following proposition gives a simple criterion for determining if any given set is dense.

Proposition 14. Let X, τ be a topological space. A subset A of X is dense if and only if every open subset of X contains some point of A. The proof is left as an exercise.

EXERCISES

1. Prove Proposition 14.
2. Suppose X is any set with the trivial topology. Prove that any nonempty subset of X is dense. Then suppose instead that X has the discrete topology, and show that the only dense subset of X is X itself.
3. Let R be the set of real numbers and $A = \{x \mid x \text{ is rational}\}$. Find Cl A, A°, Ext A, A', and Fr A with respect to each of the following topologies on R. Also decide if A is dense, somewhere dense, or nowhere dense.
   a) the discrete topology
   b) the trivial topology
   c) the topology defined on R in Example 11
   d) the topology formed by defining a set to be open if it contains all but at most countably many numbers
4. Let R^2 be the coordinate plane with the topology induced by the Pythagorean metric. Let $A = \{(x, y) \mid x \text{ and } y \text{ are rational and } x^2 + y^2 < 1\}$. Find
   a) Fr(Cl A) and Cl(Fr A);
   b) Fr(A') and (Fr A)$'$;
   c) (Cl A)$^\circ$ and Cl(A°).

   Do the same for $\{(x, y) \mid y = 0\}$.
5. Prove that A' need not be closed for every subset A of a topological space. Try to find some condition on a space X which will make A' closed for each $A \subset X$.
6. Prove or disprove:
   a) $(A')' = A'$;
   b) Ext(Ext A) $= A^\circ$;
   c) Fr(Fr A) $=$ Fr A;
   d) Fr A is always closed;
   e) Fr($A \cup B$) $\subset$ Fr $A \cup$ Fr B.
7. We define the *order* of a topological operator O to be n if n is the smallest positive integer greater than 1 for which $O^n = O$; thus, the order of Cl is 2 since Cl(ClA) = ClA. Find the order of Ext, Fr, $'$, and $^\circ$. If there is no $n \geq 2$ for which $O^n = O$ for any one of these operators, write *infinite* for that operator.

4
DERIVED TOPOLOGICAL SPACES. CONTINUITY

4.1 SUBSPACES

We have already noted that if X, D is a metric space and $Y \subset X$, then $Y, D \mid Y$ is also a metric space (Section 2.1, Example 4). Recall that a subset W of Y is $D \mid Y$-open if and only if $W = Y \cap U$, where U is a D-open subset of X (Section 2.4, Exercise 2). However, Y is not merely a subset of X, but is a *subspace* of X, and the topology which $D \mid Y$ induces on Y can be defined by means of the topology which D induces on X. Suppose Y is a subset of a topological space X, τ. It is reasonable then to inquire whether there is any topology on Y which is "induced" by τ. Using what we learned about metric spaces, the following definition seems in order.

Definition 1. Let X, τ be a topological space and $Y \subset X$. A subset W of Y is said to be *open in* Y if

$$W = Y \cap U, \qquad \text{where} \qquad U \in \tau.$$

Proposition 1. If X, τ is a topological space and $Y \subset X$, then the set of all subsets of Y which are open in Y forms a topology for Y.

Proof. We shall show that the set of all subsets of Y which are open in Y satisfies the definition of a topology on Y (Chapter 3, Definition 1).

i) X and ϕ are members of τ. Therefore $Y \cap X = Y$ and $Y \cap \phi = \phi$ are open in Y.

ii) Suppose U and V are open in Y. Then $U = Y \cap U'$ and $V = Y \cap V'$, where U' and V' are open subsets of X. Then

$$U \cap V = (Y \cap U') \cap (Y \cap V') = Y \cap (U' \cap V').$$

But $U' \cap V'$ is an open subset of X; hence $U \cap V$ is an open subset of Y.

iii) Suppose $\{U_i\}$, $i \in I$, is a family of open subsets of Y. Then $U_i = Y \cap U_i'$, where U_i' is an open subset of X for each $i \in I$.

Then

$$\bigcup_I U_i = \bigcup_I (Y \cap U_i') = Y \cap \left(\bigcup_I U_i'\right).$$

Since $\bigcup_I U_i'$ is the union of open sets, it is open, and therefore $\bigcup_I U_i$ is open in Y. The set of subsets of Y which are open in Y therefore forms a topology on Y.

Proposition 1 enables us to make the following definition.

Definition 2. The topology on Y described in Definition 1 and Proposition 1 is called the *subspace topology on* Y. Y with this topology is said to be a *subspace of* X. If X, τ is a topological space and $Y \subset X$, then Y will be assumed to have the subspace topology when considered as a topological space, unless explicitly stated otherwise.

Example 1. If X, D is a metric space and $Y \subset X$, then the topology induced on Y by $D \mid Y$ is the same as the subspace topology on Y induced by the metric topology on X. It was in fact this example which inspired us to define the subspace topology as we did. (See the remarks opening this chapter.)

Example 2. Let N be the set of positive integers with the topology defined by declaring a set to be open if it contains all but at most finitely many elements of N. Let $Y = \{1, 2, 3, 4, 5\}$. We now show that the subspace topology on Y is the discrete topology. Suppose $n \in Y$. Then

$$U(n) = \{n\} \cup (N - Y)$$

is an open subset of N, since it excludes only four positive integers. Therefore $U(n) \cap Y = \{n\}$ is open in Y. Every one-point subset of Y is therefore open in Y, and hence every subset of Y (being the union of one-point subsets) is open in Y. Consequently Y has the discrete topology. We thus see that it is quite possible for a subspace to have the discrete topology even when the space itself does not have the discrete topology.

Example 3. If a set X has the discrete topology, then every subspace of X has the discrete topology. If X has the trivial topology, then every subspace of X has the trivial topology. The proof of these assertions is left as an exercise.

Proposition 2. If Y is a subspace of X and W is a subspace of Y, then W is a subspace of X. That is, if Y is given the subspace topology from X, and then a subset W of Y is given the subspace topology considered as a subset of the topological space Y, then W would be given the same topology as though W were considered as a subset of X and were given the subspace topology.

Proof. Let τ_Y be the topology on W considered as a subspace of Y, and let τ_X be the topology on W considered as a subspace of X. We must show that $\tau_X = \tau_Y$. Suppose $U \in \tau_X$. Then $U = W \cap U'$, where U' is open in X. But then

$$U = W \cap U' = (W \cap Y) \cap U' = W \cap (Y \cap U'),$$

since $W \subset Y$; hence $U \in \tau_Y$. On the other hand, if $U \in \tau_Y$, then $U = W \cap U'$, where U' is open in Y. But since U' is open in Y, $U' = Y \cap U''$, where U'' is open in X. It follows that

$$U = W \cap U' = W \cap (Y \cap U'') = (W \cap Y) \cap U'' = W \cap U'',$$

and hence $U \in \tau_X$. Therefore $\tau_X = \tau_Y$.

EXERCISES

1. a) Prove that every subspace of a topological space with the discrete topology has the discrete topology.
   b) Prove that every subspace of a space with the trivial topology has the trivial topology.
2. Suppose that X, τ is a topological space and that F is a closed subset of X. Prove that a subset W of F is closed in F if and only if W is a closed subset of X. Make and prove the corresponding statement about open subsets of X.
3. Assume X, τ a topological space and $\mathcal{B}$ a basis for τ. If $Y \subset X$, define

$$\mathcal{B}_Y = \{B \cap Y \mid B \in \mathcal{B}\}.$$

   Prove that $\mathcal{B}_Y$ is a basis for the subspace topology on Y.
4. Suppose R is the space of real numbers with the topology induced by the absolute value metric. Prove that the following subsets of R do not have the discrete subspace topology.
   a) the set of rational numbers
   b) $\{x \mid x = 0, \text{ or } x = 1/n, \text{ where } n \text{ is a positive integer}\}$
   c) $\{x \mid x = q\pi, \text{ where } q \text{ is a rational number}\}$
   d) any subset of R which is somewhere dense
5. Let X, D be any metric space. Prove that if Y is a finite subset of X, then the subspace Y has the discrete topology. Is this true if we replace *finite* by *countable*?
6. Prove or disprove: A subset A of a topological space X, τ is nowhere dense if and only if the subspace A has the discrete topology.
7. Suppose X, τ is a topological space with the property that every two-point subspace of X has the trivial topology. Prove that X has the trivial topology. Show that the corresponding statement about the discrete topology is not true.

8. Find the coarsest topology on R^2 for which each finite subspace of R^2 has the discrete topology. Find the finest topology on R^2 for which each finite subspace of R^2 has the trivial topology.

4.2 THE TOPOLOGICALLY DERIVED SETS IN SUBSPACES

Suppose that X, τ is a topological space and that Y is a subspace of X. If A is a subset of Y, then A is also a subset of X. We may wish to know the sets topologically associated with A either with respect to the topology on X or with respect to the subspace topology on Y. As we shall see, the corresponding sets are not always equal, nor should we expect them to be, since it is not at all true that a set open in Y is necessarily open in X. Nevertheless, the subspace topology on Y is defined in terms of the topology on X; there should therefore be some relationships between the corresponding sets. It is the purpose of this section to investigate these relationships.

Example 4. Assume that R is the set of real numbers with the topology induced by the absolute value metric. Let

$$Y = (0, 1) \qquad \text{and} \qquad A = \{x \mid 0 < x < 1 \text{ and } x \text{ is rational}\}.$$

Then the closure of A in Y is $(0, 1)$, while the closure of A in R is $[0, 1]$. Fr Y in Y is ϕ, since no subset of Y, open or otherwise, contains any elements of $Y - Y = \phi$. But Fr Y in R is $\{0, 1\}$.

Example 5. Let the plane R^2 have the topology described in Example 5 of Chapter 3. Let

$$Y = \{(x, y) \mid x^2 + y^2 \leq 1\} \qquad \text{and} \qquad A = \{(x, y) \mid x^2 + y^2 < 1\}.$$

There is no closed subset of R^2, except R^2, which contains A; that is, no union of finitely many lines and points contains all of A. Therefore the closure of A in R^2 is all of R^2. Suppose F is a subset of Y which is closed in Y and which contains A. Applying Proposition 3 below, then $F = Y \cap F'$, where F' is a closed subset of R^2. But F' contains F; hence $A \subset F'$. Since R^2 is the only closed subset of R^2 which contains A, it follows that $F' = R^2$. Thus $F = Y \cap R^2 = Y$. We have then that the closure of A in Y is Y.

Proposition 3. If X, τ is a topological space and Y is a subspace of X, then a subset F of Y is closed in Y if and only if $F = Y \cap F'$, where F' is a closed subset of X.

Proof. Suppose F is closed in Y. Then $Y - F$ is open in Y; therefore

$$Y - F = Y \cap U,$$

where U is an open subset of X. But then $Y - (Y - F) = F = Y \cap (X - U)$. Since U is open, $X - U$ is closed; hence F is of the desired form.

Suppose further that $F = Y \cap F'$, where F' is a closed subset of X. Then

$$Y - F = Y \cap (X - F').$$

But $X - F'$ is open; hence $Y - F = Y \cap (X - F')$ is open in Y. Therefore F is closed in Y.

Note that this proposition tells us that the subspace topology on Y could equally well have been defined by defining the subsets of Y which are closed in Y in the same way that the subsets of Y which are open in Y are defined (substituting closed for open, of course) in Definition 1.

Proposition 4. Let X, τ be a topological space, and let Y be a subspace of X. If $A \subset Y$, then

$$\text{Cl } A \quad \text{in } Y = Y \cap \text{Cl } A \text{ (in } X).$$

Proof. Cl A is closed in X and $A \subset \text{Cl } A$; hence $Y \cap \text{Cl } A$ is a closed (in Y) subset of Y (Proposition 3) which contains A. Therefore Cl A in $Y \subset Y \cap \text{Cl } A$. Now Cl A in Y is a closed (in Y) subset of Y, and thus, again by Proposition 3, there is a closed subset F of X such that Cl A in $Y = Y \cap F$. But $A \subset F$; hence $\text{Cl } A \subset F$. It follows that

$$Y \cap \text{Cl } A \subset Y \cap F = \text{Cl } A \text{ in } Y.$$

We therefore have Cl A in $Y = Y \cap \text{Cl } A$.

The reader might be tempted to conjecture that A° in $Y = Y \cap A^\circ$. This is not true, as is shown by the next example.

Example 6. Let R^2 be the coordinate plane with the topology induced by the Pythagorean metric, $Y = \{(x, y) \mid y = 0\}$, that is, Y is the x-axis, and $A = Y$. Then A° in $Y = Y$, while

$$Y \cap A^\circ = Y \cap \phi = \phi,$$

since Y contains no open subset of R^2. We might also note that Fr A in $Y = \phi$, while Fr $A = A$; hence it is also false that Fr A in $Y = Y \cap \text{Fr } A$.

Proposition 5. Suppose that X, τ is a topological space and that for each $x \in X$, we have a collection $\mathfrak{N}_x$ of subsets of X such that the $\mathfrak{N}_x$ form an open neighborhood system for X. Let $Y \subset X$. Then setting

$$\mathfrak{N}'_y = \{Y \cap N \mid N \in \mathfrak{N}_y\}$$

for each $y \in Y$, we obtain an open neighborhood system for the subspace topology of Y.

Proof. We must show that the $\mathfrak{N}'_y$ satisfy Definition 5 of Chapter 3.

i) Since there is at least one $N \in \mathfrak{N}_y$ for each $y \in Y \subset X$, there is $N \cap Y \in \mathfrak{N}'_y$; hence $\mathfrak{N}'_y \neq \phi$.

ii) Definition 5(ii) follows at once from the fact that $y \in N$ for each $N \in \mathfrak{N}_y$.

iii) Suppose N'_1 and N'_2 are in $\mathfrak{N}'_y$. Then $N'_1 = Y \cap N_1$ and $N'_2 = Y \cap N_2$ for some N_1 and N_2 in $\mathfrak{N}_y$. There is, however, $N_3 \in \mathfrak{N}_y$ such that $N_3 \subset N_1 \cap N_2$. Therefore

$$N'_3 = Y \cap N_3 \subset N'_1 \cap N'_2 \quad \text{and} \quad N'_3 \in \mathfrak{N}'_y.$$

The proofs of (iv) and (v) are left as exercises.

Since $N \in \mathfrak{N}_y$ is an open subset of X, $N' = Y \cap N$ is an open (in Y) subset of Y. Therefore we do have an open neighborhood system for the subspace topology on Y.

EXERCISES

1. Prove (iv) and (v) in Proposition 5.
2. Assume U to be an open subset of a topological space X, τ and $A \subset U$. Is it true that Fr A in U = Fr $A \cap U$? Does this equality hold if U is a closed subset of X rather than an open subset?
3. Why is it true that Cl A in Y = Cl $A \cap Y$, but that $A°$ in $Y \neq A° \cap Y$? (See Proposition 4 and Example 6.) Try to reproduce the proof of Proposition 4 for $A°$ instead of Cl A and see where the proof fails.
4. Let R^2 be the coordinate plane with the usual Pythagorean metric. Let
$$Y = \{(x, y) \mid x^2 + y^2 < 1, \text{ or } x = 0 \text{ or } 1 \text{ and } y = 0 \text{ or } 1\}.$$
For each of the following subsets of Y compare the sets topologically derived from these sets in Y with those topologically derived in X. That is, compute Cl A in Y and compare it with Cl A in X, etc.
   a) $\{(x, y) \mid x = 0 \text{ and } y = 1/n, n \text{ a positive integer, or } y = 0\}$
   b) $\{(x, y) \mid x^2 + y^2 < 1\}$
   c) $\{(x, y) \mid \text{either } x \text{ or } y \text{ is irrational}\}$
5. Suppose that Y is a subspace of X, τ and that $A \subset Y$. Prove
   a) Fr A in $Y \subset$ Fr $A \cap Y$;
   b) $A° \subset A°$ in Y.
6. Compute Ext A, A', $A°$, and Fr A in Y, for A and Y in Example 5.
7. Find a necessary and sufficient condition for each subset A of a subspace W of a space X, τ to have the same frontier relative to W as A has relative to X. Find such a condition on W in order to have A' in W equal to A' (in X) for each subset A of W.

4.3 CONTINUITY

We now come to one of the central notions in all of topology: continuity. We have already encountered continuity in connection with metric spaces. Then a continuous function was a "nearness-preserving" function. Neighborhoods of a point in a general topological space are in a sense measures of nearness, just as the term "neighborhood" implies. As we have seen, however, many topological spaces are not metric spaces, nor can they be made into metric spaces by defining an appropriate metric. We therefore need a definition of continuity which will reduce to the metric definition of continuity when we are dealing with a metric space, generalize appropriately the idea of a "nearness-preserving" function, and not depend on metrics (or anything else which is not common to all topological spaces) for its definition. In Chapter 2 we found at least one criterion for the continuity of a function from one metric space to another which does not include any mention of the metrics in its statement (Proposition 8, Chapter 2). We will therefore use this proposition for a generalized definition of continuity.

Definition 3. Let X, τ and Y, τ' be topological spaces. Then a function f from X to Y is said to be *continuous* if given any open subset U of Y, then $f^{-1}(U)$ is an open subset of X.

Example 7. If X is any space with the discrete topology and Y is any topological space, then any function f from X into Y is continuous. For if U is any subset (open or not) of Y, then $f^{-1}(U)$ is an open subset of X, since every subset of X is open.

Example 8. If X is any space with the trivial topology and f is any function from X onto a space Y, then f is continuous if and only if Y has the trivial topology. For if Y has the trivial topology, then Y and ϕ are the only open subsets of Y; hence $f^{-1}(U)$ is open in X (being either ϕ for $U = \phi$, or X for $U = Y$) for any open subset U of Y. On the other hand, if Y does not have the trivial topology, then there is an open subset U of Y which is neither Y nor ϕ. Then $f^{-1}(U)$ is neither X nor ϕ, and hence is not an open subset of X. Therefore f could not be continuous.

The following proposition is the generalized version of Definition 6 of Chapter 2.

Proposition 6. A function f from a topological space X, τ to a space Y, τ' is continuous if and only if given any $f(x) \in Y$ and any neighborhood V of $f(x)$, there is a neighborhood U of x such that

$$f(U) \subset V.$$

Proof. Suppose f is continuous. Then if V is any neighborhood of $f(x)$, V is an open subset of Y. Therefore $f^{-1}(V)$ is an open subset of X which contains x; that is, $f^{-1}(V)$ is a neighborhood of x. Setting $U = f^{-1}(V)$, we have the desired result.

Suppose that given any $f(x) \in Y$ and any neighborhood V of $f(x)$, there is a neighborhood U of x such that $f(U) \subset V$. Let W be any open subset of Y; we must show that $f^{-1}(W)$ is an open subset of X. Suppose $z \in f^{-1}(W)$. Then $f(z) \in W$, that is, W is a neighborhood of $f(z)$. Then there is a neighborhood U of z such that $f(U) \subset W$. But then $U \subset f^{-1}(W)$. We therefore have that for each $z \in f^{-1}(W)$, $z \in U \subset f^{-1}(W)$, where U is an open subset of X. Hence $f^{-1}(W)$ is the union of open subsets of X, and thus is an open subset of X. Therefore f is continuous.

Propositions 7 and 8 give further criteria for the continuity of a function.

Proposition 7. Suppose that $X, \mathcal{T}$ and $Y, \mathcal{T}'$ are topological spaces, and that f is a function from X to Y. Let $\mathcal{B}$ be any basis for $\mathcal{T}'$. Then f is continuous if and only if for each $B \in \mathcal{B}$, $f^{-1}(B)$ is an open subset of X. (Compare this with Proposition 9, Chapter 2.)

Proof. Assume f continuous. Then since each $B \in \mathcal{B}$ is an open subset of Y, $f^{-1}(B)$ is an open subset of X. Suppose instead that $f^{-1}(B)$ is an open subset of X for each $B \in \mathcal{B}$. Let V be any open subset of Y. Then $V = \bigcup_I B_i$, where each B_i is a member of $\mathcal{B}$ and I is a suitable index set. It follows that

$$f^{-1}(V) = f^{-1}\left(\bigcup_I B_i\right) = \bigcup_I f^{-1}(B_i).$$

But $f^{-1}(B_i)$ is open in X for each $i \in I$; hence $f^{-1}(V)$ is the union of a family of open sets and is therefore open. Consequently, f is continuous.

Corollary. Suppose that f is a function from $X, \mathcal{T}$ into $Y, \mathcal{T}'$ and that $\{\mathfrak{N}_y\}$, $y \in Y$, is an open neighborhood system for $\mathcal{T}'$. Then f is continuous if and only if given any $N \in \mathfrak{N}_y$ for any $y \in Y$, $f^{-1}(N)$ is an open subset of X.

Proof. The collection of all N contained in some $\mathfrak{N}_y$ forms a basis for $\mathcal{T}'$ by Proposition 6 of Chapter 3. The corollary then follows at once from Proposition 7.

Example 9. Let R^2 be the coordinate plane with the usual metric topology. A "rotation" of R^2 is best described using polar coordinates. If A_0 is an angle measured in radians, define

$$R_{A_0}(r, A) = (r, A + A_0)$$

for any point (r, A) (expressed in polar coordinates) of R^2. The reader may recall from analytic geometry that R_{A_0} is a rotation through angle A_0. Then $R_{A_0}^{-1} = R_{(-A_0)}$, the rotation through angle $-A_0$. Any rotation preserves congruences; in particular, if U is the interior of some triangle or square, then $R_{A_0}^{-1}(U)$ is also the interior of a triangle or square. Since the family of interiors of triangles, or the family of interiors of squares, forms a basis for the standard topology on R^2, any rotation is continuous. The inverse of any rotation, also being a rotation, is continuous.

Proposition 8. Let X, τ and Y, τ' be topological spaces. Then a function f from X to Y is continuous if and only if given any closed subset F of Y, $f^{-1}(F)$ is a closed subset of X.

The proof is left as an exercise.

Proposition 9. Suppose that X is any set and that τ and τ' are topologies for X. Then τ is finer than τ' if and only if the identity function i from X to X defined by

$$i(x) = x \qquad \text{for all} \quad x \in X$$

is continuous from the topological space X, τ to the topological space X, τ'.

Proof. Assume i continuous. If $U \in \tau'$, then $i^{-1}(U) = i(U) = U$ is in τ. Therefore $\tau' \subset \tau$, that is, τ is finer than τ'. Suppose τ is finer than τ'. Then if $U \in \tau'$,

$$i^{-1}(U) = U \in \tau$$

(since any τ'-open set is τ-open). Therefore i is continuous.

The following proposition shows that the composition of continuous functions is continuous.

Proposition 10. If f is a continuous function from the space X, τ to the space Y, τ' and if g is a continuous function from Y, τ' to Z, τ'', then $g \circ f$ is a continuous function from X, τ to Z, τ''.

Proof. Suppose U is an open subset of Z. Then $g^{-1}(U)$ is an open subset of Y, since g is continuous. But then since f is continuous,

$$f^{-1}(g^{-1}(U)) = (g \circ f)^{-1}(U)$$

is an open subset of X. Therefore $g \circ f$ is continuous.

Propositions 11 and 12 pertain to continuous functions as they are related to subspaces. Proposition 11 deals with a function which is known to be continuous on certain subspaces of a space X.

Proposition 11. If f is a function from X, τ to Y, τ', $X = A \cup B$, and $f \mid A$ and $f \mid B$ are both continuous (where A and B are considered as subspaces of X), then if A and B are both open or both closed, f is continuous.

Proof. We will prove Proposition 11 for the case when A and B are both closed. The case when A and B are both open is left as an exercise. We use Proposition 8. Let F be any closed subset of Y. We must show that $f^{-1}(F)$ is a closed subset of X. $(f \mid A)^{-1}(F)$ is closed in A and $(f \mid B)^{-1}(F)$ is closed in B, since $f \mid A$ and $f \mid B$ are both assumed to be continuous. But since A and B are closed, $(f \mid A)^{-1}(F)$ and $(f \mid B)^{-1}(F)$ are closed subsets of X (see Section 4.1, Exercise 2). Since $A \cup B = X$,

$$f^{-1}(F) = (f \mid A)^{-1}(F) \cup (f \mid B)^{-1}(F),$$

which is a closed subset of X since it is the union of two closed subsets of X. Therefore f is continuous.

Example 10. Let R be the set of real numbers with the topology induced by the absolute value metric. Define $f: R \to R$ by

$$f(x) = \begin{cases} x, & \text{if } x \geq 0, \\ 0, & \text{if } x \leq 0. \end{cases}$$

Set $A = \{x \mid x \geq 0\}$ and $B = \{x \mid x \leq 0\}$. Then $A \cup B = R$, A and B are closed, and $f \mid A$ and $f \mid B$ are easily seen to be continuous. Therefore, by Proposition 11, f is continuous.

Note that A and B must either *both* be closed, or both be open. One cannot be closed and the other open. For if we continue to let R be the space of real numbers with the absolute value topology and set

$$A = \{x \mid x \geq 0\},$$

$$B = \{x \mid x < 0\},$$

and define $g: R \to R$ by

$$g(x) = \begin{cases} 3, & \text{if } x \in A, \\ 0, & \text{if } x \in B, \end{cases}$$

then A is closed, B is open; but g is not continuous.

The next proposition answers the following questions:

a) Suppose that f is a continuous function from X, τ onto a subspace Y of Z, τ''. Is f then continuous as a function from X to Z?

b) Suppose that f is a continuous function from X, τ to Y, τ' and that W is a subspace of X. Is $f \mid W$ continuous?

Proposition 12. Suppose f is a continuous function from X, τ to Y, τ'.

a) If Y is a subspace of Z, τ'', then f is a continuous function from X to Z.

b) If W is a subspace of X, then $f \mid W$ is a continuous function from W to Y.

Proof

a) Suppose U is an open subset of Z. Then $Y \cap U$ is open in Y; hence $f^{-1}(Y \cap U)$ is an open subset of X. But $f(x) \in Y$ for every $x \in X$, and thus

$$f^{-1}(U) = f^{-1}(Y \cap U)$$

is an open subset of X. Therefore f is continuous as a function from X to Z.

The proof of (b) is left as an exercise.

Example 11. Let W be a subspace of an X, τ. The identity function restricted to W, $i \mid W$, is sometimes called the *inclusion mapping* of W into X. If τ_W denotes the subspace topology on W, then

$$i \mid W \colon W, \tau_W \to W, \tau_W$$

is continuous; hence, applying Proposition 12(a), $i \mid W \colon W, \tau_W \to X, \tau$ is continuous. Looking at it another way, $i \colon X, \tau \to X, \tau$ is continuous, and therefore, by Proposition 12(b), $i \mid W$ is also continuous.

Note that Proposition 12(a) implies that we never would lose any generality by assuming that a continuous function was onto.

EXERCISES

1. Prove Proposition 8.
2. Prove Proposition 12(b).
3. Suppose that f is a function from a space X, τ to a set Y. Define a subset U of Y to be open if $f^{-1}(U)$ is an open subset of X. Prove that the set of open subsets of Y then forms a topology τ'. Further, prove that f is continuous from X, τ to Y, τ'. Show that τ' is the finest topology for which f is continuous.
4. Let f be a function from a set X to a topological space Y, τ'. Define a subset U of X to be open if $U = f^{-1}(V)$ for some open subset V of Y. Prove that the family of open subsets of X thus obtained forms a topology τ on X. Prove that $f \colon X, \tau \to Y, \tau'$ is continuous. Prove that τ is the coarsest topology for which f is continuous.
5. In Example 11, show that the subspace topology is the coarsest topology for which $i \mid W$ is continuous.

6. Let R be the space of real numbers with the absolute value topology. Discuss the continuity of each of the following functions from R to R.

   a) $f(x) = \sin x$, for all $x \in R$
   b) $f(x) = \sin x^2$, for all $x \in R$
   c) $f(x) = \begin{cases} \sin(1/x), & x \in R - \{0\} \\ 0, & \text{if } x = 0 \end{cases}$

7. Let $X = \{1, 2, 3, 4, 5\}$ with topology $\{\phi, X, \{1\}, \{3, 4\}, \{1, 3, 4\}\}$, and let $Y = \{A, B\}$ with topology $\{\phi, Y, \{A\}\}$. Find all continuous functions from X to Y. How many functions from X to Y are there altogether?

8. Let f be a function from X, τ to Y, τ'. Suppose $\mathcal{S}$ is a subbasis for τ'. Prove that f is continuous if and only if given any $S \in \mathcal{S}, f^{-1}(S)$ is an open set in X.

9. Suppose f to be a function from X, τ to Y, τ'. Prove that f is continuous if and only if $\text{Cl}\, f(A)$ in $Y \supset f(\text{Cl}\, A$ in $X)$ for each $A \subset X$.

10. Prove Proposition 11 for the case when A and B are both open. Show that Proposition 11 is not true without the condition that $X = A \cup B$.

11. Prove or disprove each of the following.

    a) $f: X, \tau \to Y, \tau'$ is continuous if given any $A \subset X, f(A^\circ) = (f(A))^\circ$.
    b) $f: X, \tau \to Y, \tau'$ is continuous if given any $A \subset X, f(A') \subset (f(A))'$.

4.4 HOMEOMORPHISMS

If a function f from a set X onto a set Y is one-one, then f^{-1} is a function from Y to X. If X and Y also happen to be topological spaces and if f is continuous, it does not follow, however, that f^{-1} is continuous. It may be that f^{-1} is continuous (see Example 9), but f^{-1} may fail to be continuous (see Example 17, Chapter 2). We are therefore justified in making the following definition.

> **Definition 4.** Let f be a function from a topological space X, τ to a space Y, τ'. The function f is said to be a *homeomorphism* if f is one-one, onto, and continuous, and if f^{-1} is continuous. If, given spaces X, τ and Y, τ', we can find a homeomorphism from X onto Y, then X and Y are said to be *homeomorphic*. A function g from X into Y is said to be a *homeomorphism of X into Y*, or an *embedding of X in Y*, if g is a homeomorphism from X onto a subspace of Y.

Example 12. Any two closed line segments in the plane R^2 with the usual Pythagorean metric topology are homeomorphic. We may position the line segments as shown in Fig. 4.1. Choosing an appropriate point P, we can project the first segment into the second in the obvious fashion. This projection gives a one-one, onto function, which we will denote by j_P, from one segment to the other. Both j_P and j_P^{-1} are continuous, since both preserve open subsets of each interval; in particular, a basis for the topology of the first segment is carried onto a basis for the topology of the second segment. Therefore j_P is a homeomorphism.

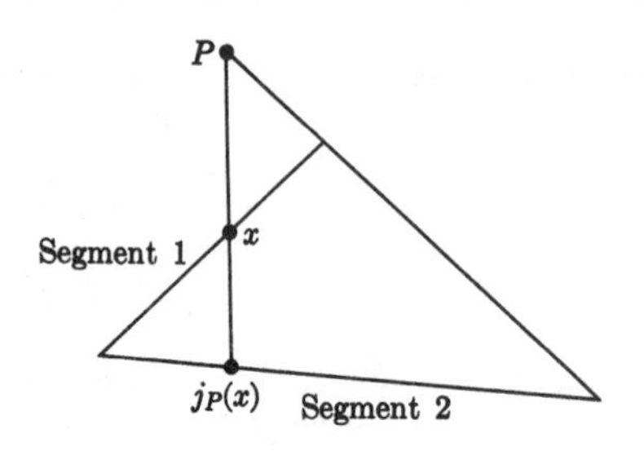

Figure 4.1

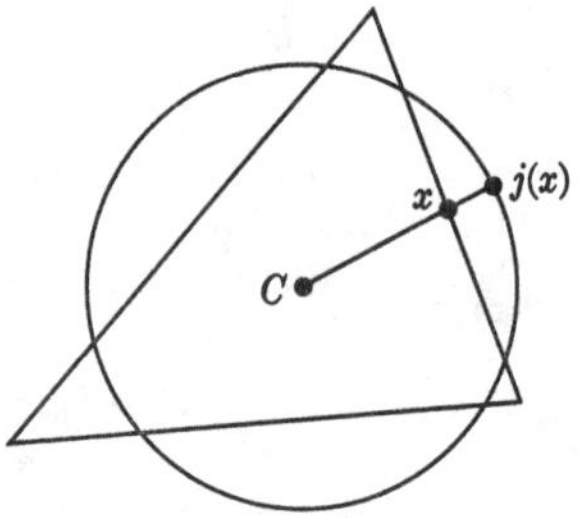

Figure 4.2

Example 13. Let R^2 again be the plane with the Pythagorean metric topology. Then any triangle in R^2 is homeomorphic to any circle. We can prove this by positioning the triangle and circle so that the center of the circle lies inside the triangle. We can then "project" the circle onto the triangle as shown in Fig. 4.2. An argument similar to that of Example 12 shows that this projection is a homeomorphism.

Proposition 13. Let f be a one-one function from a space X, τ onto a space Y, τ'. Then the following statements are equivalent:

a) f is a homeomorphism.
b) A subset U of Y is open if and only if $f^{-1}(U)$ is open in X.
c) A subset F of Y is closed if and only if $f^{-1}(F)$ is closed in X.
d) If $\mathcal{B}$ is a basis for τ, then $f(\mathcal{B}) = \{f(B) \mid B \in \mathcal{B}\}$ is a basis for τ'.

Proof. Statement (a) implies statement (b): If f is a homeomorphism, then both f and f^{-1} are continuous. Therefore if U is any open subset of Y, $f^{-1}(U)$ is open in X. Suppose U is a subset of Y such that $f^{-1}(U)$ is open in X. Since f is onto, $f(f^{-1}(U)) = U$. But $f = (f^{-1})^{-1}$, and since f^{-1} is continuous and $f^{-1}(U)$ is open in X,

$$(f^{-1})^{-1}(f^{-1}(U)) = U$$

is open in Y. Therefore U is open in Y if and only if $f^{-1}(U)$ is open in X.

Statement (b) implies statement (c): Suppose F is a closed subset of Y. Then $Y - F$ is an open subset of Y; hence $f^{-1}(Y - F)$ is open in X. But $f^{-1}(Y - F) = X - f^{-1}(F)$, and thus $f^{-1}(F)$ is a closed subset of X. Suppose that F is a subset of Y such that $f^{-1}(F)$ is a closed subset of X. Then

$$f^{-1}(Y - F) = X - f^{-1}(F)$$

is an open subset of X; hence $Y - F$ is an open subset of Y; hence F is closed. Therefore (b) implies (c).

Statement (c) implies statement (a): By Proposition 8, (c) states that f and f^{-1} are continuous; hence f is a homeomorphism.

Statement (a) implies statement (d): Suppose U is any open subset of Y. Since f is continuous, $f^{-1}(U)$ is an open subset of X. Therefore $f^{-1}(U) = \bigcup_I B_i$, where each $B_i \in \mathcal{B}$ and I is a suitable index set. Now

$$f(f^{-1}(U)) = U = f\left(\bigcup_I B_i\right) = \bigcup_I f(B_i).$$

But each B_i is open in X, and it has been shown that (a), (b), and (c) are equivalent (a implies b implies c implies a); hence, by (b), $f(B_i)$ is an open subset of Y. Then U is the union of members of $f(\mathcal{B})$, and each member of $f(\mathcal{B})$ is an open subset of Y. Therefore $f(\mathcal{B})$ is a basis for τ'.

Statement (d) implies statement (a): Suppose U is an open subset of Y. Then $U = \bigcup_I f(B_i)$, where $B_i \in \mathcal{B}$ and I is again a suitable index set. It follows that

$$f^{-1}(U) = f^{-1}\left(\bigcup_I f(B_i)\right) = \bigcup_I f^{-1}(f(B_i)) = \bigcup_I B_i,$$

which is an open subset of X. Therefore f is continuous. On the other hand, if V is any open subset of X, then $V = \bigcup_J B_j$, where each $B_j \in \mathcal{B}$. Thus

$$f(V) = f\left(\bigcup_J B_j\right) = \bigcup_J f(B_j)$$

is the union of open subsets of Y and is therefore open. But $f = (f^{-1})^{-1}$; hence $(f^{-1})^{-1}(V)$ is open in Y if V is open in X. Therefore f^{-1} is continuous, and f is a homeomorphism.

We see from Proposition 13 that homeomorphic spaces are essentially equivalent from a topological point of view. If two spaces are homeomorphic, there is not only a one-one function from one space to the other, but also a natural one-one correspondence between their open sets (Proposition 13b). Put another way, if X, τ and Y, τ' are homeomorphic spaces, then by suitably relabeling the points of Y, we obtain X, and τ' becomes τ.

We must keep in mind, however, that homeomorphic spaces can appear quite different from other points of view than the topological. We have seen, for example, that a circle and a triangle are homeomorphic. From a geometric point of view, a circle and a triangle are quite different, even though from a topological point of view they are indistinguishable. Recall that Euclidean geometry is primarily concerned with the properties of objects which are preserved under rigid motions, that is, in Euclidean geometry, we are interested in studying properties common to all objects which are congruent. Almost all geometric studies can be classified according to the type of properties they study; in particular, these types of properties are those which are preserved by certain kinds of functions. Topology, considered as a branch of geometry, studies properties preserved by a very special type of function, the homeomorphism.

Proposition 14. Let $\mathcal{T}$ be the class of all topological spaces. Then the relation R defined on $\mathcal{T}$ by "is homeomorphic to" is an equivalence relation on $\mathcal{T}$.

Proof. If X, τ is any topological space, then X, τ is homeomorphic to X, τ by the identity function. Therefore X, τ is R-equivalent to X, τ.

Suppose X, τ is homeomorphic to Y, τ' by some homeomorphism f. Then f^{-1} is a homeomorphism from Y, τ' to X, τ. Hence if X, τ is R-equivalent to Y, τ', then Y, τ' is R-equivalent to X, τ.

Now suppose that f is a homeomorphism from X, τ to Y, τ', and that g is a homeomorphism from Y, τ' to Z, τ''. Since both f and g are one-one and onto, $g \circ f: X \to Z$ is one-one and onto. Since f, g, f^{-1}, and g^{-1} are all continuous, $g \circ f$ and $(g \circ f)^{-1} = f^{-1} \circ g^{-1}$ are also continuous (Proposition 10). Therefore

$$g \circ f: X, \tau \to Y, \tau'$$

is a homeomorphism. Hence the relation R is transitive, and, consequently, R is an equivalence relation on $\mathcal{T}$.

From a topological point of view then, any two homeomorphic spaces are equivalent, or interchangeable. The question of determining whether or not two given spaces are homeomorphic is often extremely difficult. As a matter of fact, it is usually very difficult to determine whether there is even a continuous function from one space onto another. In most instances, this problem has not been solved.

Proposition 15. Suppose that X is any set and that τ and τ' are two topologies for X. Then $\tau = \tau'$ if and only if the identity function i on X is a homeomorphism from X, τ to X, τ'.

The proof is left as an exercise.

EXERCISES

1. Prove Proposition 15.
2. Let R be the set of real numbers with the absolute value topology.
   a) Prove that any open interval (a, b) is homeomorphic to the interval $(0, 1)$. [*Hint:* Use $f(x) = (x - a)/(b - a)$.]
   b) Prove that the ray (a, ∞) is homeomorphic to $(1, \infty)$.
   c) Prove that (a, ∞) is homeomorphic to $(-\infty, -a)$.
   d) Prove that R is homeomorphic to $(-\pi/2, \pi/2)$. [*Hint:* Use $f(x) = \tan^{-1} x$.]
   e) Prove that $(1, \infty)$ is homeomorphic to $(0, 1)$. [*Hint:* Use $g(x) = 1/x$.]

   We thus conclude that any two open intervals of the real line are homeomorphic.

   f) Prove that any two closed intervals of the real line are homeomorphic.

3. Homeomorphic spaces have essentially the same topological properties, that is, properties related exclusively to their topologies. Although the reader has yet encountered very few topological properties, he should be able to make an intelligent conjecture about whether the spaces in each of the following pairs are homeomorphic to one another. If the spaces are homeomorphic, try to describe a homeomorphism. If they are not homeomorphic, try to find a topological property which one of the spaces has, but which the other space does not have.
   a) the open interval $(0, 1)$ and the closed interval $[0, 1]$ considered as subspaces of the real numbers with the absolute value topology
   b) $(0, 1)$ and $[0, 1]$ considered as subspaces of the real numbers with the discrete topology
   c) a circle C considered as a subspace of the plane R^2 with the usual topology and the interval $(0, 1)$ from (a)
   d) $\{x \mid x \text{ is a rational number}\}$ and $\{n \mid n \text{ is an integer}\}$, both considered as subspaces of the real numbers with the absolute value topology
4. Let X be the space of functions described in Example 5, Chapter 2, with the topology induced by the metric D.
   a) Prove that X cannot be homeomorphic to the space R of real numbers with the absolute value topology. [*Hint:* Prove that X has greater cardinality than $[0, 1]$, which has the same cardinality as R. Do this by assuming that there is a one-one correspondence between the elements of $[0, 1]$ and the elements of X and then constructing a function which does not correspond to any element of $[0, 1]$. More particularly, if $x \leftrightarrow f$, set $g(x) \neq f(x)$, for each $x \in [0, 1]$.]
   b) Find an embedding of R as a subspace of X.
5. Let N be the set of positive integers. Define a subset F of N to be closed if F contains a finite number of positive integers, or $F = N$ (cf. Exercise 6 of Section 3.1). Prove that any infinite subspace of N is homeomorphic to N. Is it possible to find a nontrivial topology on the set R of real numbers such that every uncountable subspace of R is homeomorphic to R?

4.5 IDENTIFICATION SPACES

If X is any set and R is an equivalence relation on X, then R determines a partition of X into R-equivalence classes. We will denote the set of equivalence classes by X/R. If X also has a topology τ, we might inquire if τ can be used in a natural way to give a topology on X/R.

We note that there is a natural function f from X to X/R defined by $f(x) = \bar{x}$, where x is any element of X and $\bar{x}$ is the R-equivalence class of x. If X is a topological space, it is reasonable to want a topology on X/R which would at least make f continuous. Of course, if X/R is given the trivial topology, then f is continuous. But the trivial topology is pretty much what its name implies, trivial. Furthermore, the trivial topology on X/R is not necessarily related to τ, and we are looking for a topology which is derived from τ. We know that the function f will be continuous if and only if given any open set U of X/R, $f^{-1}(U)$ is open in X. We will use this fact to define a topology on X/R; that is, we will say that a subset U of X/R will be open if $f^{-1}(U)$ is open in X.

Definition 5. Suppose X, τ is a topological space and R is an equivalence relation on X. Let X/R denote the set of R-equivalence classes. Define the function f from X to X/R by $f(x) = \bar{x}$, where x is any element of X and $\bar{x}$ is the R-equivalence class of x. Then f is called the *identification mapping* from X to X/R. Define a subset U of X/R to be open if $f^{-1}(U)$ is open in X. The topology thus obtained on X/R (Proposition 16) is called the *identification topology* on X/R. (Some topologists refer to this topology as the *quotient topology*, and of X/R as a *quotient space*.)

Proposition 16. The collection of open sets of X/R actually forms a topology for X/R, that is, the identification topology is really a topology.

Proof. We verify that the collection of open sets in X/R satisfies Definition 1, Chapter 3.

i) $f^{-1}(\phi) = \phi$ and $f^{-1}(X/R) = X$. Since X and ϕ are both open subsets of X, ϕ and X/R are open subsets of X/R.

ii) Let U and V be open subsets of X/R. Then $f^{-1}(U)$ and $f^{-1}(V)$ are open subsets of X. Now $f^{-1}(U) \cap f^{-1}(V)$ is also an open subset of X. But

$$f^{-1}(U) \cap f^{-1}(V) = f^{-1}(U \cap V).$$

Therefore $U \cap V$ is also an open subset of X/R.

iii) Suppose $\{U_i\}$, $i \in I$, is a family of open subsets of X/R. Then $f^{-1}(U_i)$ is open in X for each $i \in I$, and thus $\bigcup_I f^{-1}(U_i)$ is an open subset of X. But since

$$\bigcup_I f^{-1}(U_i) = f^{-1}\left(\bigcup_I U_i\right),$$

$\bigcup_I U_i$ is an open subset of X/R. Therefore the collection of open subsets of X/R forms a topology on X/R.

Proposition 17. Let X, τ be a topological space, let R be an equivalence relation on X, and suppose that X/R has the identification topology τ'. Then the identification map

$$f: X, \tau \rightarrow X/R, \tau'$$

is continuous. Furthermore, τ' is the finest topology on X/R for which f is continuous.

Proof. By definition of the identification topology, $f^{-1}(U)$ is an open subset of X whenever U is an open subset of X/R. The identification topology has been specifically defined so as to make f continuous. Suppose τ'' is a topology on X/R which is strictly finer than τ'. Then there is $U \in \tau''$ such that $U \notin \tau'$. Then $f^{-1}(U)$ is not an open subset of X (if it were, U would also be in τ'); hence f is not a continuous function from X, τ onto $X/R, \tau''$.

The identification topology is so called because it may be viewed in the following manner: Let X, τ be a space, and let R be an equivalence relation on X. Then we obtain X/R by identifying all R-equivalent elements of X with one another, that is, we make an equivalence class a point of a new set. We put a topology on X/R by defining a subset U of X/R to be open if all elements of X contained in all the members of U form an open subset of X.

Example 14. Let the closed interval [0, 1] have the usual (absolute value) topology. An equivalence relation on any set can be specified either by giving the equivalence classes, that is, a partition of the set, or by defining the relation. For [0, 1], we will let 0 be equivalent to 1, and every other element of [0, 1] be equivalent only to itself. Then the equivalence classes are $\{0, 1\}$ and $\{x\}$, for $x \neq 0, 1$. By identifying 0 and 1, we obtain a circle (Fig. 4.3). We have joined the endpoints of [0, 1] by making the endpoints a single point of a new topological space, which is a simple closed curve. We therefore have a continuous function from [0, 1] onto a circle.

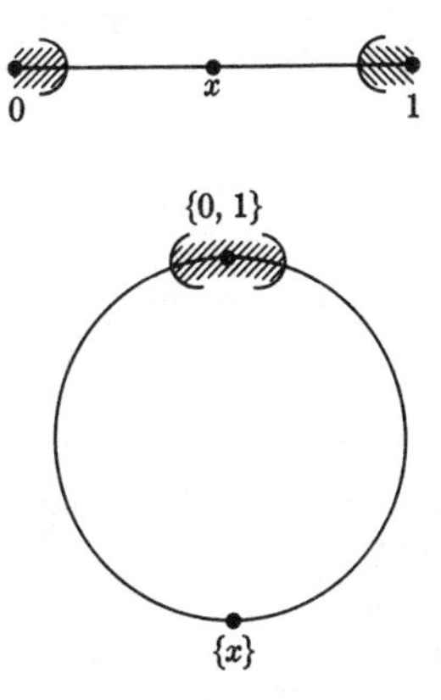

Figure 4.3

Example 15. Suppose X is any set and Y, τ' is a topological space. Let g be a function from X to Y. Suppose we want to find a topology on X which will make g continuous. Of course, the discrete topology will do, but as with the trivial topology, this possibility is not of much interest. Since g will be continuous if and only if given any open set U of Y, $g^{-1}(U)$ is open in X, we will define a subset V of X to be open if there is an open subset U of Y such that $V = g^{-1}(U)$. Then the family of open subsets of X forms a topology τ on X; moreover, this topology is the coarsest topology which makes g continuous (Section 4.3, Exercise 4).

Define an equivalence relation R on X by letting $x \; R \; x'$ if $g(x) = g(x')$ for any x and x' in X. If $\bar{x}$ is the R-equivalence class of x, then there is a natural function $\bar{g}$ from X/R into Y defined by $\bar{g}(\bar{x}) = g(x)$. The function $\bar{g}$ is well-defined, for if $\bar{x} = \bar{x}'$, then

$$\bar{g}(\bar{x}) = g(x) = g(x') = \bar{g}(\bar{x}').$$

Let f be the identification mapping from X to X/R. Then if g is onto and X/R is given the identification topology τ'', $\bar{g}$ is a homeomorphism from $X/R, \tau''$ onto Y, τ'. This is proved as follows: Since g is onto, $\bar{g}$ is onto. Suppose $\bar{g}(\bar{x}_1) = \bar{g}(\bar{x}_2)$. Then $g(x_1) = g(x_2)$. Therefore $\bar{x}_1 = \bar{x}_2$, and hence $\bar{g}$ is one-one.

It remains to show that $\bar{g}$ and $\bar{g}^{-1}$ are continuous. Suppose U is any open subset of Y. Then

$$f^{-1}(\bar{g}^{-1}(U)) = g^{-1}(U),$$

which by definition is an open subset of X. Since $f^{-1}(\bar{g}^{-1}(U))$ is an open subset of X, and since X/R has the identification topology, $\bar{g}^{-1}(U)$ is open in X/R. Therefore $\bar{g}$ is continuous. Assume that V is any open subset of X/R. Then $f^{-1}(V)$ is an open subset of X. Hence

$$f^{-1}(V) = g^{-1}(U),$$

where U is some open subset of Y, and then

$$\bar{g}(V) = g(g^{-1}(U)) = U$$

is an open subset of Y. Since $\bar{g} = (\bar{g}^{-1})^{-1}$, $(\bar{g}^{-1})^{-1}(V)$ is open in Y; hence $\bar{g}^{-1}$ is continuous. Therefore $\bar{g}$ is a homeomorphism.

If the reader is familiar with some group theory, he may find it instructive to recall the relationship between quotient groups and homomorphisms. If G and G' are groups and h is a homomorphism from G onto G', then G' is isomorphic to the quotient group G/K, where K is the kernel of h. The quotient group G/K is nothing but the set of equivalence classes of the relation R, defined by $g\ R\ g'$ if $h(g) = h(g')$. There is a function $\bar{h}$ from the set of R-equivalence classes G/K onto G', defined by $\bar{h}(\bar{g}) = h(g)$ where $\bar{g}$ is the equivalence class of $g \in G$. If h is onto, then $\bar{h}$ is one-one and onto. An operation is then defined on G/K by means of the operation on G such that $\bar{h}$ and $\bar{h}^{-1}$ are homomorphisms; hence $\bar{h}$ is an isomorphism.

This procedure is quite important in mathematics. That is, starting with a function g onto a structured set Y from an unstructured set X, we might wish to find a structure on X so that g becomes a structure-preserving function with the structure on X derived from the structure on Y in a natural way. It might be that X already has a structure and that g is structure preserving, hence making it unnecessary to define another structure on X. This is the case for example, if X and Y are groups and g is a homomorphism. In any event, taking the equivalance classes determined by g [that is, x is equivalent to x' if $g(x) = g(x')$], we have a function $\bar{g}$ from the set of equivalence classes onto Y which is one-one. We also have the identification mapping from X onto the set of equivalence classes. We then find a structure on the set of equivalence classes so that both the identification mapping, $\bar{g}$, and $\bar{g}^{-1}$ are structure preserving. Note the close parallel in this respect between quotient groups and homomorphisms in group theory, and identification spaces and continuous functions in topology (also see Exercise 2).

EXERCISES

1. Verify that the identification space obtained from $[0, 1]$ in Example 14 is really homeomorphic to a circle. An actual homeomorphism might be given by "wrapping" $[0, 1]$ around a circle in R^2 of radius $1/2\pi$.
2. A fundamental theorem of group homomorphisms states: There is a homomorphism from the group G onto the group G' if and only if there is a normal subgroup K of G such that G' is isomorphic to the quotient group G/K. Provide an example to show that the following analogous statement about topological spaces is not true: There is a continuous function g from the space X, τ onto the space Y, τ' if and only if there is an equivalence relation R on X such that the identification space X/R is homeomorphic to Y, τ'. Explain why this statement fails to be true. Is it true if the phrase "if and" is omitted? Is it true if the phrase "and only if" is omitted?

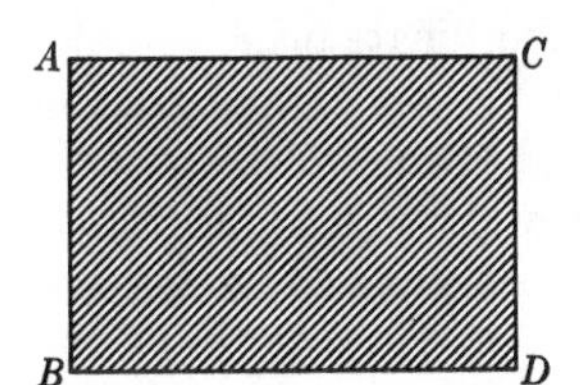

Fig. 4.4 $\{\{x \mid x \in \overline{AB} \cup \overline{CD}\}\} \cup \{\{x \mid x = x\} \mid x \notin \overline{AB} \cup \overline{CD}\}\}$

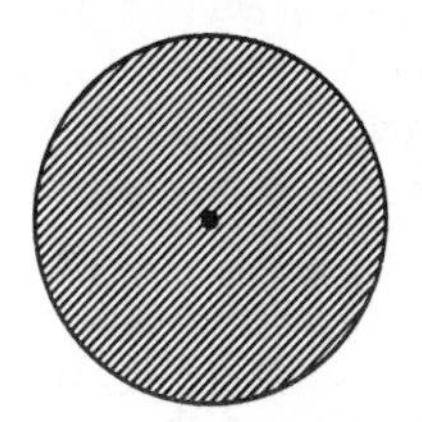

Fig. 4.5 $\{\overline{P} = \{w \mid w$ is diagonally opposite P, or $w = P\}$, if P is on the circumference$\} \cup \{\{P \mid P = P\}$, if P is not on the circumference$\}$

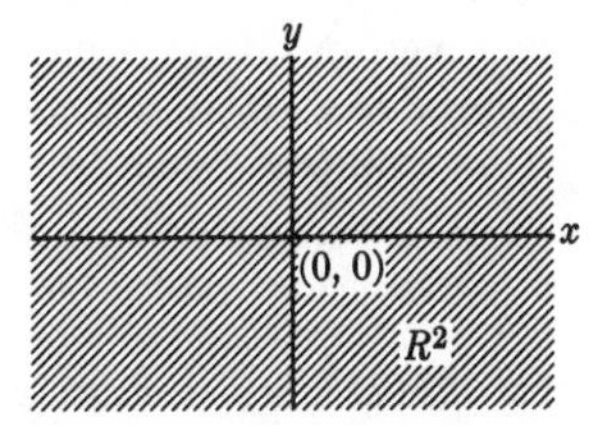

Fig. 4.6 $\{\{(x, y) \mid x = y + m$, where m is an integer$\}\}$

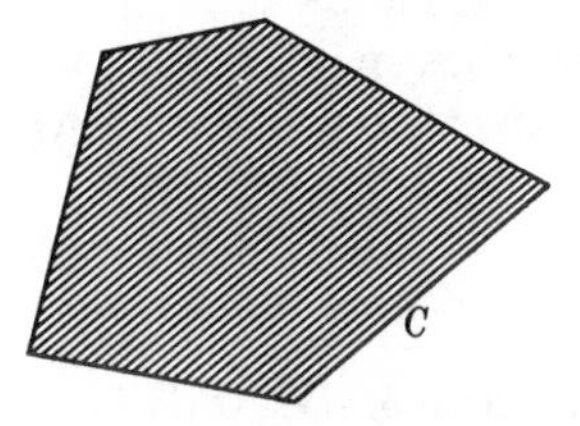

Fig. 4.7 $\{\{z \mid z$ is on the perimeter of $C\}\} \cup \{\{z \mid z = z\} \mid z$ is not on the perimeter of $C\}\}$

3. Each of Figs. 4.4 through 4.7 is to be considered as a subspace of the plane R^2 with the usual Pythagorean metric topology. Under each figure is given a partition of the set which the figure represents. Draw a picture of the identification space determined by each partition.

4. Let g be a function from a space X, τ onto a set Y. Define a subset U of Y to be open if $g^{-1}(U)$ is an open subset of X. It was shown in Section 4.3, Exercise 3 that the open subsets of Y then form a topology τ' on Y and that $g: X, \tau \to Y, \tau'$ is continuous. Prove that Y, τ' is homeomorphic to the identification space X/R, where R is the equivalence relation associated with the function g.
5. Find a quotient space of R^2 homeomorphic to each of the following.
   a) a rectangle with its interior
   b) a sphere
   c) a straight line

4.6 PRODUCT SPACES

The reader has undoubtedly encountered the concept of the Cartesian product of finitely many sets in previous studies. If $S_1, S_2, \ldots, S_n$ are sets, then the *Cartesian product* of these sets $\times_{i=1}^{n} S_i$ is defined by

$$\mathop{\times}_{i=1}^{n} S_i = \{(s_1, s_2, \ldots, s_n) \mid s_i \in S_i,\ i = 1, \ldots, n\}.$$

That is, the *Cartesian product*, or simply the *product*, of the S_i is the set of ordered n-tuples of elements of the S_i. The coordinate plane is nothing but the Cartesian product $R \times R$, where R is the set of real numbers. The ith place in an ordered n-tuple is usually called the ith *coordinate*.

If, however, the reader has already adjusted to the n-tuple definition of the product of n sets, then he may find it somewhat hard to begin the study of product topological spaces by having to learn a new and more general definition of the product of a family of sets—one which allows us to take the product of infinitely many sets as well as finitely many. In the body of this text, we will extend the definition of the product so that we can deal with the product of countably many sets. The Appendix gives the definition and some properties of more general products. Wherever possible, proofs about product spaces will be given in a form which easily adapts to the more general definition of a product. It should be kept in mind, however, that not all statements about finite or countable products are true when applied to the product of an arbitrary family of sets or topological spaces.

Definition 6. Let $\{S_i\}, i \in I$, be a family of sets, where I is a countable index set (either finite or infinite). We may then choose I either to be $\{1, 2, \ldots, n\}$, where n is an appropriate positive integer, or to be the set of positive integers. By $\times_I S_i$ we will mean the set of ordered tuples of the form $(s_1, s_2, \ldots, s_i, \ldots)$, where $s_i \in S_i$ for each $i \in I$. If

$$s = (s_1, \ldots, s_i, \ldots) \in \mathop{\times}_{I} S_i,$$

s_i is called the ith *coordinate* of s, and S_i is called the ith *component* of $\times_I S_i$. The set $\times_I S_i$ is itself called the *product* of the sets $\{S_i\}$, $i \in I$. If $s \in \times_I S_i$, then s_i will generally be used to denote the ith coordinate of s. For each $s \in \times_I S_i$, define $p_i(s) = s_i$. Then p_i is a function from $\times_I S_i$ onto S_i. The function p_i is called the *projection into the ith component.*

Example 16. If each S_i is the set of real numbers and I is the set of positive integers, then $\times_I S_i$ is precisely the set of sequences in R.

Suppose now that $\{S_i\}$, $i \in I$, is not only a countable family of sets, but that each set S_i has a topology τ_i. If we take the product set $\times_I S_i$, we might ask if a topology could be put on this product set which is related in a natural way to the topologies on the S_i.

Recall that in the discussion of identification spaces, a topology was defined on X/R so as to make the identification mapping continuous. The identification mapping was a function naturally associated with X/R, so it was reasonable to want it to be continuous. There is, however, a family of functions naturally associated with $\times_I S_i$; namely, for each $i \in I$, we have the projection mapping $p_i\colon \times_I S_i \to S_i$. It is reasonable then to want a topology on $\times_I S_i$ which makes each p_i continuous.

Suppose U_j is an open subset of S_j. Then $p_j^{-1}(U_j) = \times_I W_i$, where $W_i = S_i$, if $i \neq j$, and $W_j = U_j$. If p_j is to be continuous, then any such subset of $\times_I S_i$ must be open, no matter what element j is of I. That is, if each p_j, $j \in I$, is to be continuous, then every set of the form $\times_I W_i$, where W_i is an open subset of S_i for each i and $W_i = S_i$ for all but at most one $i \in I$, must be an open subset of $\times_I S_i$.

The collection of open subsets of $\times_I S_i$ thus formed do not, however, form a topology for the product set. Although $\times_I S_i$ is, of course, such a subset, the union or intersection of even finitely many such subsets is not necessarily such a subset. However, any collection of subsets whose union is the set forms a subbasis for a topology on the set. We therefore make the following definition.

Definition 7. Suppose $\{S_i, \tau_i\}$, $i \in I$, is a countable family of topological spaces. Set

$$\mathcal{S} = \{\times_I W_i \mid W_i = S_i \text{ for all but at most one } i, \\ \text{and } W_i \text{ is an open subset of } S_i \text{ for every } i \in I\}.$$

Then the topology τ on $\times_I S_i$ for which $\mathcal{S}$ is a subbasis is called the *product topology*, and $\times_I S_i, \tau$ is said to be the *product space* of the spaces $\{S_i, \tau_i\}$, $i \in I$.

Proposition 18. Using the notation of Definitions 6 and 7, each projection

$$p_i\colon \underset{I}{\times} S_i, \tau \to S_i, \tau_i$$

is continuous. Moreover, τ is the coarsest topology on $\times_I S_i$ for which each p_i is continuous.

Proof. The product topology has been specifically defined so as to make each projection continuous. If any topology on $\times_I S_i$ were strictly coarser than the product topology, then some member of $\mathcal{S}$, the subbasis for τ, would not be open; hence at least one of the projections could not be continuous.

Example 17. It is not true that the product space of countably many discrete spaces necessarily has the discrete topology, although the product topology will be discrete if only finitely many spaces are involved. Let I be the set of positive integers. For each $i \in I$, let $S_i = \{1, 2\}$ with the discrete topology. Set $U = \times_I W_i$, where $W_i = \{1\}$ for each $i \in I$. Then U is the product of open subsets of the S_i (specifically, $U = \{(1, 1, \ldots, 1, \ldots)\}$), but U is not open. The proof that U is not open is left as an exercise. It is, however, a rather easy corollary of Proposition 19.

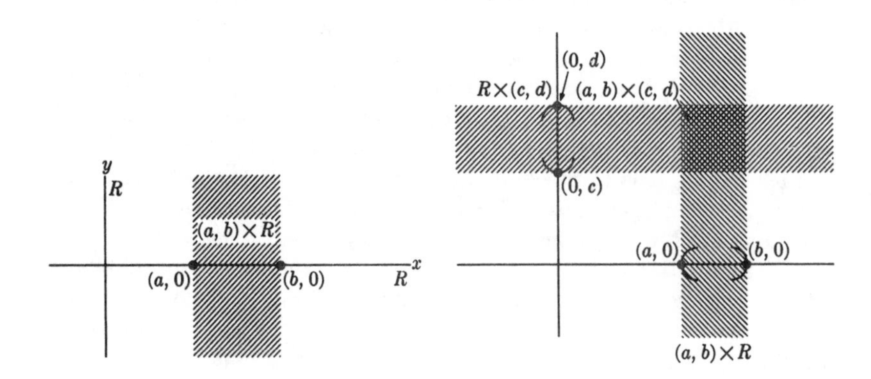

Figure 4.8 **Figure 4.9**

Example 18. Let R be the space of real numbers with the absolute value topology. We will show that the product topology on the plane $R^2 = R \times R$ is the same as the topology on R^2 induced by the Pythagorean metric D. The topology induced by D is the same as the topology induced by the metric D_3 of Chapter 2, Example 3 (Section 3.2, Exercise 2). A typical subbasis element for the product topology on R^2 is shown in Fig. 4.8. This means that a typical basis element for the product topology is given by Fig. 4.9 (see Proposition 19). Each element of the basis for the product topology is D_3-open, and each D_3-p-neighborhood of any point of R^2 is exactly a basis element of the product topology. We can easily verify that a basis for the product topology is also a basis for the topology induced by D_3; hence the topologies are the same.

Note that not every open set of the product topology is of the form $U \times V$, where U and V are open subsets of R. For example,

$$\{(x, y) \mid x^2 + y^2 < 1\}$$

is open, but is not a product set.

Proposition 19. Let $\times_I S_i, \tau$ be the product space of the countable family of spaces $\{S_i, \tau_i\}$, $i \in I$. Set

$$\mathcal{B} = \{\times_I V_i \mid V_i \text{ is open in } S_i, \text{ and } V_i = S_i \text{ for all but at most finitely many } i\} \cdot$$

Then $\mathcal{B}$ is a basis for τ.

Proof. Let $\mathcal{S}$ be the subbasis for τ described in Definition 7. Then a basis for τ is obtained by taking all finite intersections of members of $\mathcal{S}$ (Chapter 3, Definition 4). Suppose $U_1, \ldots, U_n$ are elements of $\mathcal{S}$, with $U_k = \times_I W_i$, where $W_i = S_i$ for all i, except possibly $i = i_k$, $k = 1, \ldots, n$. Then

$$U_1 \cap U_2 \cap \cdots \cap U_n = \underset{I}{\times} V_i,$$

where $V_i = S_i$, except possibly for $i_1, i_2, \ldots, i_k$. This means that each element of the basis derived from $\mathcal{S}$ is also a member of $\mathcal{B}$; but each member of $\mathcal{B}$ is the intersection of finitely many members of $\mathcal{S}$. Therefore $\mathcal{B}$ is a basis for τ.

Corollary. If I is finite, then a basis for τ consists of all sets of the form $\times_I V_i$, where V_i is open in S_i.

Proposition 20. Suppose $\times_I S_i, \tau$ is the product space of the nonempty spaces $\{S_i, \tau_i\}$, $i \in I$. Then S_i, τ_i is homeomorphic to a subspace of $\times_I S_i, \tau$ for each $i \in I$.

Proof. We lose no generality in proving this proposition for S_1, τ_1, since the same proof could be used for any $i \in I$. Let $y_2, \ldots, y_i, \ldots$ be fixed points of $S_2, \ldots, S_i, \ldots$, respectively. Define the function q_1 from S_1 into $\times_I S_i$ by

$$q_1(x) = (x, y_2, \ldots, y_i, \ldots)$$

for each $x \in S_1$. Let Y be the subspace of $\times_I S_i$, defined by

$$Y = \{(x, y_2, \ldots, y_i, \ldots) \mid x \in S_1\}.$$

Then q_1 takes S_1 onto Y; moreover, q_1 is one-one. It remains to show that q_1 and q_1^{-1} are continuous.

Assume U an open subset of Y. Then $U = Y \cap U'$, where U' is an open subset of $\times_I S_i$. If p_1 is the projection into the first component, then

$p_1(U')$ is an open subset of S_1 (Exercise 2). But it is readily seen that $q_1^{-1}(U) = p_1(U')$. Therefore $q_1^{-1}(U)$ is an open subset of S_1, and thus q_1 is continuous. On the other hand, suppose V is an open subset of S_1. Then

$$q_1(V) = (q_1^{-1})^{-1}(V) = Y \cap \underset{I}{\times} W_i,$$

where $W_i = S_i$, $i \geq 2$, and $W_1 = V$. It follows that $(q_1^{-1})^{-1}(V)$ is an open subset of Y; hence q_1^{-1} is continuous. Therefore q_1 is a homeomorphism.

Example 19. Let R be the set of real numbers with the absolute value topology, and let R^2 be the plane with the product topology. Figure 4.10 illustrates one possible embedding of R as a subspace of R^2. One generally thinks of the x-axis as being the real line, whereas, strictly speaking, it is a space which is homeomorphic to the real line. Note that even when restricting oneself to the procedure of Proposition 20, there are uncountably many subspaces of R^2 homeomorphic to R.

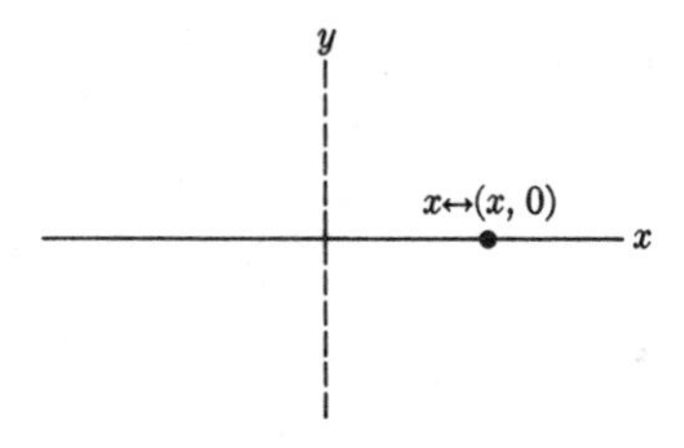

Figure 4.10

Proposition 21. Suppose f is a function from a space X, τ into the product space $\times_I S_i, \tau'$. Define $f_i: X, \tau \to S_i, \tau_i$ by $f_i(x) = p_i \circ f(x)$ for each $x \in X$, where p_i is the projection into the ith component. Then f is continuous if and only if f_i is continuous for each $i \in I$.

Proof. If f is continuous, then $f_i = p_i \circ f$ is the composition of two continuous functions and therefore is continuous.

Assume now that f_i is continuous for each $i \in I$. We will use Propositions 7 and 19. We first note that

$$f(x) = (f_1(x), f_2(x), \ldots, f_i(x), \ldots)$$

for each $x \in X$. Suppose $\times_I V_i$ is any member of the basis $\mathcal{B}$ for τ' described in Proposition 19, where $V_i = S_i$ for each $i \in I$, except $i_1, \ldots, i_m$. Now $f^{-1}(\times_I V_i)$ is the set of all points x of X such that $f(x) \in \times_I V_i$. But this is easily seen to be $\bigcap_I f_i^{-1}(V_i)$. For every i, except $i_1, \ldots, i_m$,

$$f_i^{-1}(V_i) = X \qquad \text{(because } V_i = S_i\text{)}.$$

Since f_i is continuous for each $i \in I$, $f_{i_j}^{-1}(V_{i_j})$ is open in X for $j = 1, \ldots, m$.

Therefore

$$f^{-1}\left(\bigtimes_I V_i\right) = f_{i_1}^{-1}(V_{i_1}) \cap \cdots \cap f_{i_m}^{-1}(V_{i_m}),$$

which is open in X since it is the intersection of finitely many open sets. Hence, by Proposition 7, f is continuous.

Proposition 21 is extremely important in the study of product spaces. Note that its proof would not have gone through if we had defined a subbasis of the product topology to consist of sets of the form $\bigtimes_I V_i$, where V_i is open in S_i, since we could not have been sure then that $\bigcap_I f_i^{-1}(V_i)$ was an open subset of X. (Note, however, with this topology that each projection is still continuous.) Proposition 21 is, in fact, another good reason why the product topology was defined as it was.

Example 20. Let R be the space of real numbers with the absolute value topology, and let R^2 be the plane with product topology from R. Define $f\colon R \to R^2$ by

$$f(x) = (\sin x, 3x + 1)$$

for each $x \in R$. Then $f_1(x) = \sin x$ and $f_2(x) = 3x + 1$ for each $x \in R$. Since f_1 and f_2 are both continuous functions from R into R, f is continuous.

EXERCISES

1. Suppose that $\{S_i\}$, $i \in I$, is any countable family of sets, and that $S_i = \phi$, for some i. Prove $\bigtimes_I S_i = \phi$.
2. A function f from a space X, τ to a space Y, τ' is said to be *open* if $f(V)$ is open in Y whenever V is an open subset of X. Prove that the projection p_i from the product space $\bigtimes_I S_i, \tau$ into S_i, τ_i is open for each $i \in I$.
3. Prove that the product space of a countable family of spaces, each with the trivial topology, has the trivial topology.
4. Each of the sets involved in the following is to be considered as a subspace of the space R of real numbers with the absolute value topology. Sketch each of the following spaces.

   a) $[0, 1] \times [0, 1]$ b) $\{0, 1\} \times [0, 1]$ c) $\{0, 1\} \times R$
   d) $\{x \mid x > 0\} \times \{x \mid x \text{ is an integer greater than } 1\}$
5. Let $C = \{(x, y) \mid x^2 + y^2 = 1\} \subset R^2$ with the Pythagorean topology. Describe

   a) $C \times C$; b) $C \times [0, 1]$; c) $C \times (0, 1)$.
6. Prove that the set U in Example 17 is not an element of the product topology.
7. Let X, τ be any space, and let $X \times X$ have the product topology. The *diagonal* Δ of $X \times X$ is defined by $\Delta = \{(x, x) \mid x \in X\}$. Prove that X is homeomorphic to Δ.
8. Let $\{S_i, \tau_i\}$, $i \in I$, be a countable family of spaces, and let P be a permutation of I. Prove that the product spaces $\bigtimes_I S_i$ and $\bigtimes_I S_{P(i)}$ are homeomorphic.

That is, the order in which the components are used in the product does not affect the topological character of the product space.

9. Consider the space N described in Exercise 6 of Section 3.1 and Exercise 5 of Section 4.4. Prove or disprove: The product space $N \times N$ is homeomorphic to N.

5
THE SEPARATION AXIOMS

5.1 T_0- AND T_1-SPACES

Propositions 12 and 13 of Chapter 2, and Section 2.3, Exercise 1 furnish us with examples of "separation" properties for metric spaces. A "separation" property really does imply separation in the following sense: Given any two nonintersecting subsets A and B of a topological space X, where A and B are subsets of a certain type, there are other nonintersecting subsets U and V of X, generally open sets, such that $A \subset U$ and $B \subset V$ (Fig. 5.1). In other words, being separated in a topological space is a bit stronger than merely being disjoint. There are various degrees of separation. As we saw in Chapter 2, it is possible to separate disjoint closed subsets of a metric space in a rather strong way. But not all topological spaces are metric spaces; hence not all topological spaces have strong separation properties. Although most important topological spaces are at least T_2 (see below), many are not.

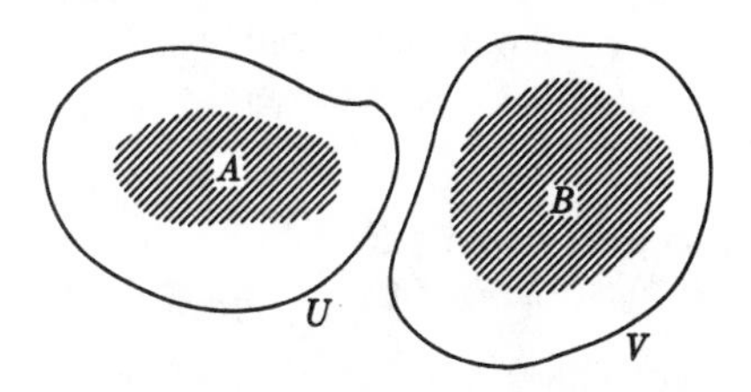

Figure 5.1

Many topologists will not even consider a topological space which is not T_2, and some won't touch anything which is not at least *normal*. But since it is hoped that the reader has not yet developed personal prejudices, at least in the area of topology, and since, too, the reader has not yet begun to specialize to the point where he feels justified in throwing out whatever does not fall in his sphere of interest, we shall even study some of the lesser separation axioms. The first separation axiom follows.

Definition 1. A topological space X, τ is said to be T_0 if given any two distinct points x and y of X, there is a neighborhood of at least one which does not contain the other.

Example 1. Let $X = \{0, 1\}$. If X is given the trivial topology, then X is not T_0, since there is then no way to separate 0 from 1 or 1 from 0. If X is given the topology $\{\phi, X, \{0\}\}$, then X becomes a T_0-space, since there is then a neighborhood $\{0\}$ of 0 which does not contain 1. Note, however, that there is no neighborhood of 1 which does not contain 0.

The following proposition enables us to obtain another criterion for a space to be T_0.

Proposition 1. If x and y are distinct points of a topological space X, τ, then every open set which contains either x or y contains both if and only if

$$\text{Cl}\{x\} = \text{Cl}\{y\}.$$

Proof. Suppose $\text{Cl}\{x\} = \text{Cl}\{y\}$. Since $x \in \text{Cl}\{x\}$, $x \in \text{Cl}\{y\}$; similarly, $y \in \text{Cl}\{x\}$. Suppose U is an open set which contains x, but not y. Then $X - U$ is a closed set which contains y, but not x. But, since $\{y\} \subset X - U$, and since $X - U$ is closed, $\text{Cl}\{y\} \subset X - U$; hence $x \notin \text{Cl}\{y\}$, a contradiction. There is therefore no open set which contains x, but not y; similarly, there is no open set which contains y, but not x.

Suppose now that every open set which contains either x or y contains both x and y. Let $z \in \text{Cl}\{x\}$. Then every open set which contains z contains x. For if U is an open set which contains z, but not x, then $X - U$ is a closed set which contains x, but not z. Thus z could not be in $\text{Cl}\{x\}$. Since every open set which contains x contains y, and every open set which contains z contains x, we have that every open set which contains z contains y. It therefore follows easily that $z \in \text{Cl}\{y\}$. Hence, $\text{Cl}\{x\} \subset \text{Cl}\{y\}$. But a similar argument (merely interchanging x and y) shows that $\text{Cl}\{y\} \subset \text{Cl}\{x\}$. Therefore $\text{Cl}\{x\} = \text{Cl}\{y\}$.

Corollary. A space X, τ is T_0 if and only if given any two distinct points x and y of X, either $x \notin \text{Cl}\{y\}$ or $y \notin \text{Cl}\{x\}$.

The proof of this corollary is left as an exercise.

Example 2. Suppose X is any set. Then a *pseudometric* on X is a function from $X \times X$ into the set of nonnegative real numbers which has all the properties of a metric on X (Definition 1, Chapter 2), except that we do not require the "distance" between two distinct points to be nonzero. It is thus possible to have points which are 0 distance apart, even though they are noncoincident.

Assume D a pseudometric on X. We may define D-p-neighborhoods and open sets exactly as we did for metric spaces; in fact, such sets will have essentially the same properties as in metric spaces. One difference, however, between a metric space and a pseudometric space is that in a metric space each one-point subset is closed, whereas this is not necessarily true in a pseudometric space. In fact, it is definitely false with a pseudometric space unless the pseudometric is a metric. For if x and y are distinct points such that $D(x, y) = 0$, then every open set which contains x contains y, and every open set which contains y contains x; hence x and y cannot be separated. Therefore $\{x, y\}$ is a subset of $\text{Cl}\{x\}$ and $\text{Cl}\{y\}$, and it follows that $\text{Cl}\{x\} \neq \{x\}$ and $\text{Cl}\{y\} \neq \{y\}$. A pseudometric space, then, is generally not even T_0.

A slightly stronger separation property than T_0 is given by the following.

Definition 2. A space X, τ is said to be T_1 if for any two distinct points x and y of X, there is a neighborhood of x which does not contain y *and* a neighborhood of y which does not contain x.

Proposition 2. A space X, τ is T_1 if and only if for each $x \in X$,

$$\text{Cl}\{x\} = \{x\}.$$

Proof. Suppose X is T_1, and suppose there is $x \in X$ such that $z \in \text{Cl}\{x\}$, $z \neq x$. Then every neighborhood of z must contain x. Hence there is no neighborhood of z which excludes x, a contradiction since X is assumed to be T_1. Therefore $\text{Cl}\{x\} = \{x\}$ for all $x \in X$.

Now assume that $\text{Cl}\{x\} = \{x\}$ for each $x \in X$ and that y and z are distinct points of X. If every neighborhood of y contains z, then

$$z \in \text{Cl}\{y\} = \{y\};$$

hence $y = z$, a contradiction. Therefore there is a neighborhood of y which excludes z. Similarly, there is a neighborhood of z which does not contain y; hence X is T_1.

Corollary. A space X, τ is T_1 if and only if every one-point subset of X is closed.

The proof is left as an exercise.

Example 3. Every metric space is T_1, since every one-point subset of a metric space is closed (Chapter 2, Proposition 5).

Example 4. Let N be the set of positive integers with the topology defined by making every finite subset of N closed (Section 3.1, Exercise 6). Then every one-element subset of N is closed; hence N is a T_1-space. Suppose x and y are two distinct positive integers. Then any neighborhood of either x or y contains all but finitely many positive integers. Therefore, while it is possible to find a neighborhood of x which excludes y and a neighborhood of y which excludes x, it is not possible to find a neighborhood U of x and a neighborhood V of y such that $U \cap V = \phi$.

It should be clear from the definitions that any T_1-space is also a T_0-space. In Example 1 the reader can find an example of a space which is T_0 but is not T_1.

EXERCISES

1. Prove the corollaries to Propositions 1 and 2.
2. Suppose X is any finite set. Prove that the only topology on X which makes X into a T_1-space is the discrete topology. If X is a set of n elements, what is the fewest number of members a topology can have which makes X into a T_0-space?
3. Let X be any set and D be a pseudometric on X. Define a relation R on X by $x\ R\ y$ if $D(x, y) = 0$, for any $x, y \in X$.
   a) Prove that R is an equivalence relation on X.
   b) Let X have the topology induced by D (defining open sets as if D were a metric). For each $x \in X$, let $\bar{x}$ denote the R-equivalence class of x. Define $\bar{D}(\bar{x}, \bar{y}) = D(x, y)$, for any $x, y \in X$.
      i) Prove that $\bar{D}$ is a metric for X/R.
      ii) Prove that the topology induced on X/R, considered merely as the set of equivalence classes, is the same as the identification topology on X/R.
      iii) Find a natural one-one correspondence between the open sets of X, D and the open sets of X/R, $\bar{D}$.
   c) Suppose X, τ is any topological space. Find an equivalence relation R on X such that the identification space X/R is T_0 and there is a natural one-one correspondence between the open sets of X and the open sets of X/R.
4. Let $X = \{1, 2, 3\}$. Find all topologies on X which are either T_0 or T_1.
5. Suppose a space X with topology τ is T_0 or T_1. Prove that if τ' is any topology on X which is finer than τ, then the space X, τ' is also T_0 or T_1.
6. a) Prove that every subspace of a T_1-space is T_1.
   b) Prove that every subspace of a T_0-space is T_0.
   c) Prove that the product space of a countable family of nonempty T_0-spaces is T_0 if and only if each component space is T_0. Prove the corresponding statement for T_1-spaces.
7. Prove that if a space X is homeomorphic to a space Y and X is $T_0(T_1)$, then Y is also.
8. Prove that a space X, τ is T_0 if and only if distinct one-point subsets of X have distinct closures.

9. Prove or disprove: A space X is T_0 if and only if every proper subspace of X is T_0. Does this statement become true if X contains more than two points? Prove or disprove the corresponding statement with T_1 substituted for T_0 and with the added assumption that X contains at least three points.
10. Prove or disprove: A space X is T_1 if and only if $\{x\}' = \phi$ for each $x \in X$.

5.2 T_2-SPACES

A still stronger separation property than being either T_0 or T_1 is the following.

Definition 3. A space X, τ is said to be T_2 if given any two distinct points x and y of X, there are open sets U and V such that $x \in U$, $y \in V$, and $U \cap V = \phi$. A T_2-space is often called a *Hausdorff space.*

Example 5. Every metric space is a T_2-space (Section 2.3, Exercise 1).

Example 6. Let X be any set which is totally ordered by a relation $<$. Let $\mathcal{S}$ be the family of all subsets of X of the form $\{x \mid x < a\}$ or $\{x \mid a < x\}$, for all $a \in X$. Then $\mathcal{S}$ is a subbasis for a topology on X called the *order topology induced by* $<$. The set X with the order topology is always T_2. For suppose a and b are distinct points of X. Since X is totally ordered, we may assume $a < b$. If there is $c \in X$ such that $a < c < b$, then

$$\{x \mid x < c\} \qquad \text{and} \qquad \{y \mid c < y\}$$

are disjoint neighborhoods of a and b, respectively. If there is no $c \in X$ such that $a < c < b$, then

$$\{x \mid x < b\} \qquad \text{and} \qquad \{y \mid a < y\}$$

are disjoint neighborhoods of a and b, respectively.

Note that for the set R of real numbers, the order topology (for which the family of open intervals forms a basis) and the absolute value topology are the same.

Proposition 3

a) Any subspace of a T_2-space is T_2.
b) Let $Y = \times_I X_i, \tau$ be the product space of the countable family of nonempty spaces $\{X_i, \tau_i\}$, $i \in I$. Then Y is T_2 if and only if each X_i is T_2.

Proof

a) Suppose W is a subspace of the T_2-space X, and let x and y be distinct points of W. Then there are open sets U and V in X such that $x \in U$, $y \in V$, and $U \cap V = \phi$. But $x \in U \cap W$, $y \in V \cap W$, and

$$(U \cap W) \cap (V \cap W) = (U \cap V) \cap W = \phi \cap W = \phi.$$

Therefore $U \cap W$ and $V \cap W$ are disjoint neighborhoods in W of x and y, respectively. Hence W is T_2.

b) Assume that each X_i, τ_i is T_2, and let x and y be distinct points of Y. We will use x_i and y_i to denote the ith coordinate of x and y, respectively. Since $x \neq y$, $x_i \neq y_i$ for at least one $i \in I$, say for i'. Therefore there are open sets $U_{i'}$ and $V_{i'}$ in $X_{i'}$ such that

$$x_{i'} \in U_{i'}, \; y_{i'} \in V_{i'}, \qquad \text{and} \qquad U_{i'} \cap V_{i'} = \phi.$$

Set $U = \times_I H_i$, where $H_i = X_i$, $i \neq i'$, and $H_{i'} = U_{i'}$; and set $V = \times_I G_i$, where $G_i = X_i$, $i \neq i'$, and $G_{i'} = V_{i'}$. Then U and V are neighborhoods of x and y, respectively. Since any point of U differs from any point of V at least in the i'th coordinate, $U \cap V = \phi$. Therefore Y is T_2.

By Proposition 20, Chapter 4, each X_i, τ_i is homeomorphic to a subspace of Y (regardless of whether Y is T_2). If Y is T_2, then every subspace of Y is T_2, by (a). Therefore if Y is T_2, each X_i, τ_i is homeomorphic to a T_2-space, and hence is T_2 (Exercise 1).

Example 7. The reader might conjecture from Proposition 3 that if X, τ is a T_2-space and if R is an equivalence relation on X, then the identification space X/R is also T_2. This is not true. For example, let R^2 be the plane with the Pythagorean topology. Let E be the equivalence relation defined by the partition

$$\{\{(x, y) \mid y < 0\}, \; \{(x, y) \mid y \geq 0\}\}.$$

Here $\{(x, y) \mid y < 0\}$ is open, whereas $\{(x, y) \mid y \geq 0\}$ is not open. Therefore R^2/E is homeomorphic to the set $X = \{0, 1\}$ with the topology $\{X, \phi, \{0\}\}$, which is not T_2. Since the identification mapping from R^2 onto R^2/E is continuous, we have also shown that the property of being T_2 is not preserved by continuous functions (although, of course, it is preserved by homeomorphisms).

EXERCISES

1. Suppose that the space X, τ is homeomorphic to the space Y, τ' and that X is T_2. Prove that Y is T_2.
2. Prove that a space X, τ is T_2 if and only if, given any two distinct points x and y of X, there is a neighborhood U of x such that $y \notin \text{Cl}\, U$.

3. Prove that a space X, τ is T_2 if and only if the diagonal $\Delta = \{(x, x) \mid x \in X\}$ is a closed subset of the product space $X \times X$.
4. Suppose that X, τ is a T_2-space and that $A \subset X$. Prove that $x \in \text{Cl } A$ if and only if $x \in A$, or each neighborhood of x contains infinitely many points of A.
5. Assume that f is a function from a set X onto a T_2-space Y, τ'. Assume further that X is given the topology τ, defined by taking a subset U of X to be open if $U = f^{-1}(V)$, where V is an open subset of Y. Is X with this topology necessarily T_2?
6. Suppose f is a function from a T_2-space X, τ onto a space Y, τ' such that f is one-one and f^{-1} is continuous. Prove that Y is a T_2-space.
7. Let X be a set partially ordered by $\leq$. Define $\mathcal{S}$ to be the collection of subsets of X of the form $\{x \mid x < a\}$ or $\{y \mid a < y\}$, for all $a \in X$. Is $\mathcal{S}$ necessarily the subbasis for a topology on X? If it is a subbasis for a topology τ on X, is X, τ a T_2-space?
8. Suppose X is a T_2-space. Is it necessarily true that given any subbasis $\mathcal{S}$ for the topology on X and any two distinct points x and y of X, there are disjoint members U and V of $\mathcal{S}$ with $x \in U$ and $y \in V$? Answer this question with *subbasis* replaced by *basis*.

5.3 T_3- AND REGULAR SPACES

Definition 4. A space X, τ is said to be T_3 if given any closed subset F of X and any point x of X which is not in F, there are open sets U and V such that $x \in U$, $F \subset V$, and $U \cap V = \phi$ (Fig. 5.2). A space X, τ is said to be *regular* if X is both T_3 and T_1. (The author is quite aware of the lack of uniformity in the literature about what constitutes a T_3- or a regular space. In some places, T_3 and regular are synonymous. In others, T_1 is assumed as part of T_3, and in still others, T_3 does not imply T_1. The author has therefore felt justified in making the definition to suit himself.)

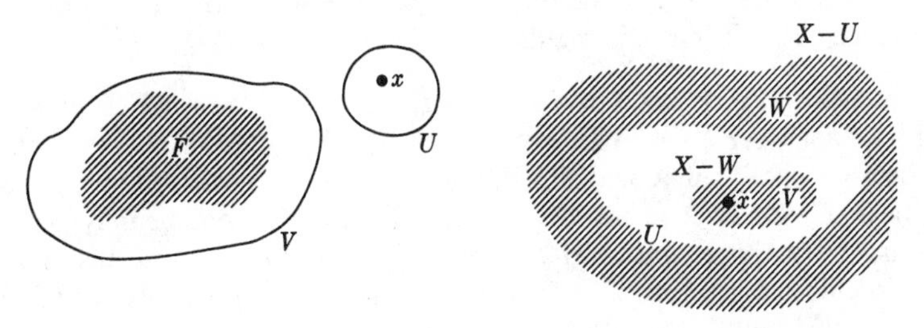

Figure 5.2

Figure 5.3

We first state and prove a very important criterion for being T_3.

Proposition 4. A space X, τ is T_3 if and only if given any $x \in X$ and any neighborhood U of x, there is a neighborhood V of x such that $\text{Cl } V \subset U$ (Fig. 5.3).

Proof. Suppose that X is T_3 and that U is a neighborhood of x. Then $X - U$ is a closed subset of X which does not contain x. Therefore there are open sets W and V such that $X - U \subset W$, $x \in V$, and $W \cap V = \phi$. Since $X - U \subset W$, $X - W \subset U$. Moreover, since $W \cap V = \phi$, we have $V \subset X - W \subset U$. But then $X - W$ is a closed set which contains V; hence

$$V \subset \text{Cl } V \subset X - W \subset U.$$

Suppose instead that given any $x \in X$ and any neighborhood U of x, there is a neighborhood V of x such that $\text{Cl } V \subset U$. Let $x \in X$, and let F be any closed subset of X which does not contain x. Then $X - F$ is a neighborhood of x; hence there is a neighborhood V of x such that $\text{Cl } V \subset X - F$. Then $X - \text{Cl } V$ is an open set which contains F, and V is an open set which contains x. Since $V \subset \text{Cl } V$,

$$(X - \text{Cl } V) \cap V = \phi.$$

Therefore $X - \text{Cl } V$ and V are suitable open sets for "separating" F and x; hence X is T_3.

Example 8. Any metric space X, D is regular. Although this has been proved previously, we may prove it again using Proposition 4. For if $x \in X$, and if U is any neighborhood of x, then U contains a D-p-neighborhood of x for some positive number p. Choose a number q such that $0 < q < p$. Then the D-q-neighborhod of x is a subset of the D-p-neighborhood of x, and

$$\text{Cl } N(x, q) \subset \{y \mid D(y, x) \leq q\} \subset N(x, p) = \{w \mid D(w, x) < p\} \subset U.$$

Therefore X, D is T_3. We have already seen that any metric space is T_1; hence any metric space is regular.

Example 9. If X is any set of more than one point, then if X is given the trivial topology, X is T_3 in a vacuous sort of way. For the only closed nonempty subset of X is X itself, and it follows that there is no point of X in $X - X$. It is to avoid such cases as this that one usually requires a space to be regular rather than merely T_3. We also see from this example that T_3 does not imply T_2. However, if every one-point subset of X is a closed subset of X, then X is T_2 if X is T_3.

We now give an example of a space which is T_2 but not T_3, or regular. This demonstrates that regularity is a stronger separation property than merely being T_2.

Example 10. Let R be the set of real numbers. We will define a topology on R by giving an open neighborhood system. If x is any real number other than 0, let $\mathfrak{N}_x$ be the family of all open intervals which contain x. If $x = 0$, we will let $\mathfrak{N}_0$ be the family of all sets of the form

$$(-p, p) - \{1/n \mid n \text{ is a positive integer}\},$$

where $0 < p$. The collection of $\mathfrak{N}_x$ for all $x \in R$ gives an open neighborhood system for a topology τ on R (Chapter 3, Proposition 7). It is easily verified that R, τ is T_2 (Exercise 1).

We now show that X is not T_3. Take $x = 0$ and $F = \{1/n \mid n$ is a positive integer$\}$. F is a closed subset of R, τ, since there is no point y of R such that each neighborhood of y contains a point of F, except those points in F itself. (Note that in the usual topology for R, each neighborhood of 0 would contain points of F; thus 0 would be in $\operatorname{Cl} F$. We have, however, purposely excluded the points of F from the neighborhoods of 0.) Suppose V is a neighborhood of 0 of the form

$$(-p, p) - \{1/n \mid n \text{ is a positive integer}\}$$

(any neighborhood of 0 contains such a neighborhood because of the manner in which the $\mathfrak{N}_x$ define the topology τ). Then $(-p, p)$ contains infinitely many of the $1/n$. Hence any open set U which contains F would have to overlap V (Fig. 5.4); thus we could not find an open set U which contains F such that $U \cap V = \phi$. Therefore R, τ is not T_3.

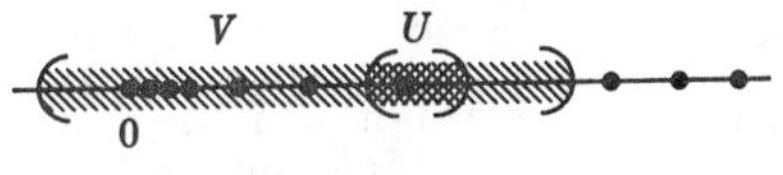

Figure 5.4

We now investigate how the property of being regular or T_3 behaves with respect to the derived topological spaces.

Proposition 5

a) Every subspace of a regular space is regular.
b) Suppose $Y = \times_I X_i$ is the product space of the (countable) family of nonempty spaces $\{X_i, \tau_i\}$, $i \in I$. Then Y is regular if and only if each X_i, τ_i is a regular space.

Proof

a) Suppose W is a subspace of a regular space X, τ. Let F be closed in W and $x \in W - F$. Since F is closed in W, $F = W \cap F'$, where F' is a closed subset of X. Then $x \in X - F'$. Since X is T_3, there are open sets U and V in X such that

$$x \in U, \qquad F' \subset V, \qquad \text{and} \qquad U \cap V = \phi.$$

Then $W \cap U$ and $W \cap V$ are disjoint subsets of W which are open in W such that $x \in W \cap U$ and $F \subset W \cap V$. W is therefore T_3. By Section 5.1, Exercise 6, W is also T_1, since X is T_1. Therefore W is regular.

b) Suppose each X_i, τ_i is a regular space. By Exercise 6 of Section 5.1, Y is T_1. Suppose that $x \in Y$ and that U is a neighborhood of x. We lose no generality in assuming that U is a basis element for the product topology, since any neighborhood of x contains a neighborhood of x which is a basis element. Then $U = \times_I U_i$, where each U_i is open in X_i. Each U_i is therefore a neighborhood in X_i of x_i, the ith coordinate of x and $U_i = X_i$ except for $i_1, \ldots, i_m$. Since each X_i is T_3, there is an open neighborhood V_{i_j} of x_{i_j}, $j = 1, \ldots, m$, such that

$$x_{i_j} \in V_{i_j} \subset \text{Cl}\, V_{i_j} \subset U_{i_j}.$$

For each $i \in I$, set $V_i = X_i$, $i \neq i_j$ and let the V_{i_j} be as above, $j = 1, \ldots, m$. Then

$$x \in \underset{I}{\times} V_i \subset \text{Cl}\left(\underset{I}{\times} V_i\right) \subset \underset{I}{\times} \text{Cl}\, V_i \subset U.$$

By Proposition 4, then, Y is T_3. Hence Y is regular.

If Y is regular, then since each X_i, τ_i is homeomorphic to a subspace of Y, and each subspace of Y is regular, each X_i is regular.

As with T_2-spaces, it is not true that if X, τ is regular and R is an equivalence relation on X, then the identification space X/R is regular (Exercise 2). However, the following is true.

Proposition 6. If X, τ is a regular space and F is a closed subset of X, then if R is the equivalence relation defined by the partition

$$\{\{x \mid x \in F\}\} \cup \{\{y \mid y = y\} \mid y \notin F\},$$

the identification space X/R is T_2. (Note that this identification has the effect of shrinking F to a point. See Figs. 5.5 and 5.6.)

Proof. Suppose $\bar{x}$ and $\bar{y}$ are distinct points of X/R, where $\bar{x}$ and $\bar{y}$ denote the equivalence classes of x and y, respectively. If x and y are not in F, then since X is regular and hence also T_2, there are open sets U and V in $X - F$ such that

$$x \in U, \qquad y \in V, \qquad \text{and} \qquad U \cap V = \phi.$$

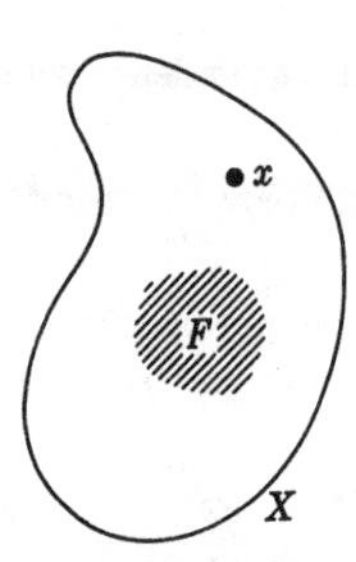

Figure 5.5

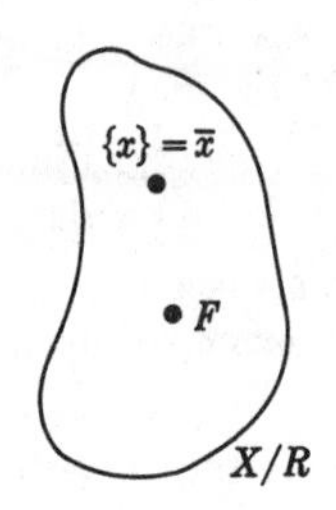

Figure 5.6

The sets U and V may be chosen in $X - F$, since $X - F$ is a T_2-subspace of X, and since a subset of $X - F$, an open set, is open in $X - F$ if and only if it is open in X. Then $\overline{U} = \{\overline{u} \mid u \in U\}$ and $\overline{V} = \{\overline{v} \mid v \in V\}$ are open disjoint subsets of X/R such that $\overline{x} \in \overline{U}$ and $\overline{y} \in \overline{V}$.

If either $x \in F$, or $y \in F$, then the other point could not be in F. For if x and y are both in F, then $\overline{x} = \overline{y} = F$, contradicting $\overline{x} \neq \overline{y}$. Suppose $\overline{x} = F$. Then there are open subsets U and V of X such that

$$\overline{x} = F \subset U, \qquad y \in V, \qquad \text{and} \qquad U \cap V = \phi.$$

It follows that $\overline{U} = \{\overline{u} \mid u \in U\}$ and $\overline{V} = \{\overline{v} \mid v \in V\}$ are disjoint open subsets of X/R such that $\overline{x} \in \overline{U}$ and $\overline{y} \in \overline{V}$. Therefore X/R is T_2.

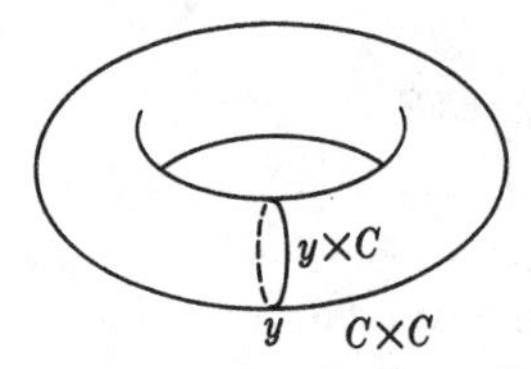

Figure 5.7

Example 11. In Example 14 of Chapter 4, a circle is obtained by identifying 0 and 1 in [0, 1]. Since {0, 1} is a closed subset of [0, 1], we know that the circle is at least a T_2-space. Since the circle is a subspace of a metric space R^2, and any metric space is regular, we have the stronger result that a circle is a regular space. The *torus* $C \times C$ (Fig. 5.7), where C is a circle, is regular since it is the product of regular spaces. The torus is also seen to be regular because it is a subspace of R^3.

EXERCISES

1. In regard to Example 10,
   a) verify that the family of $\mathfrak{N}_x$ forms an open neighborhood system for a topology on R;
   b) show that R with this topology is T_2.

2. Find an example of a regular space X, τ and an equivalence relation R on X such that the identification space X/R is not regular.
3. Let $X = \{1, 2, 3\}$. Find all topologies on X which are T_3. Find all topologies on X which are regular.
4. Suppose a space is both T_0 and T_3. Is the space necessarily regular? Note that the example given of a T_3-space which was not regular was not T_0 (Example 9).
5. Prove that the property of being regular is not preserved by continuous functions. Prove, however, that the property of being either T_3 or regular is preserved by homeomorphisms.
6. Decide which of the following spaces are T_3 or regular.
   a) any totally ordered set with the order topology (Example 6)
   b) any pseudometric space
   c) the set of real numbers with the topology described in Example 11, Chapter 3
   d) the plane R^2 with the topology described in Example 5, Chapter 3
7. Let X, τ be a topological space. For each $A \subset X$, define $w(A)$ to be

$$\{x \mid \text{every neighborhood of } x \text{ meets every neighborhood of } A\}.$$

   a) Prove that if A is open, then $w(A) = \text{Cl } A$.
   b) Prove that X, τ is T_3 if and only if $w(A) = \text{Cl } A$ for each $A \subset X$.

5.4 T_4- AND NORMAL SPACES

We now come to one of the most important, but also one of the most awkward, of the separation properties. It is awkward because it does not have many of the properties relative to the derived topological spaces that we have come to expect of separation properties (see Section 5.1, Exercise 6, and Propositions 3 and 5).

Definition 5. A topological space X, τ is said to be T_4 if given any two disjoint closed subsets F and F' of X, there are disjoint open sets U and V such that

$$F \subset U \qquad \text{and} \qquad F' \subset V$$

(Fig. 5.8). A space is said to be *normal* if it is both T_1 and T_4.

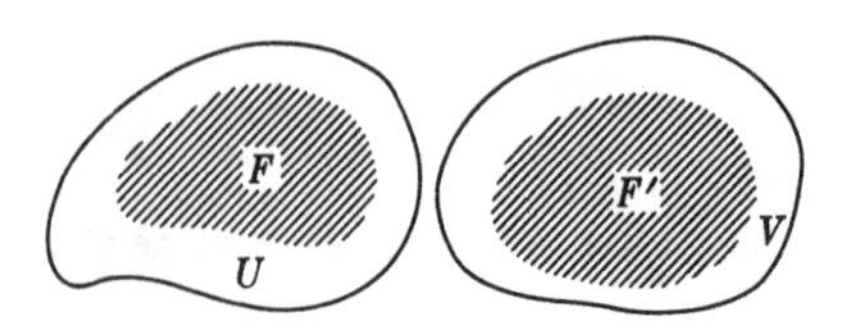

Figure 5.8

Example 12. Any space X, τ of more than one point with the trivial topology is T_4 (there are no nonempty disjoint closed subsets of X), but is not normal, since no one-point subset of X is closed. Any metric space is normal (Propositions 5 and 13 of Chapter 2).

Note that any space with the discrete topology has all of the separation properties introduced so far.

We now prove another criterion for a space to be T_4.

Proposition 7. A space X, τ is T_4 if and only if given any closed subset F of X and any open subset U of X with $F \subset U$, there is an open set V such that

$$F \subset V \subset \operatorname{Cl} V \subset U.$$

(Note the similarity between this proposition and Proposition 4 regarding T_3-spaces. Note also the difference, namely, F is a set rather than a point. This difference helps us explain the rather bad "hereditary" properties of normal spaces.)

Proof. Suppose X is T_4, F is a closed subset of X, and U is an open set which contains F. Then $X - U$ is a closed set and $(X - U) \cap F = \phi$. Hence there are open sets W and V such that

$$X - U \subset W, \qquad F \subset V, \qquad \text{and} \qquad W \cap V = \phi.$$

Then $F \subset V \subset \operatorname{Cl} V \subset U$ (since $X - U \subset W$ and $W \cap V = \phi$). Therefore V has the desired properties.

Suppose now that given any closed set F and any open set U which contains F, there is an open set V such that $F \subset V \subset \operatorname{Cl} V \subset U$. Let F and F' be any two disjoint closed subsets of X. Then $X - F$ is an open set which contains F'. By hypothesis, then there is an open set V such that

$$F' \subset V \subset \operatorname{Cl} V \subset X - F.$$

Hence $X - \operatorname{Cl} V$ is an open set which contains F and is disjoint from V. Therefore X is T_4.

It would be very nice to have an analog of Propositions 3 and 5 for normal spaces. Unfortunately, not only is it false that the product of normal spaces is normal; it is even false that every subspace of a normal space is normal. Examples of normal spaces for which some subspace is not normal are somewhat sophisticated for this text. The following example, however, gives a regular space which is not normal, but which is the product of normal spaces.

Example 13. Let R be the set of real numbers with the topology τ as described in Example 11 of Chapter 3. Let R^2 be given the product topol-

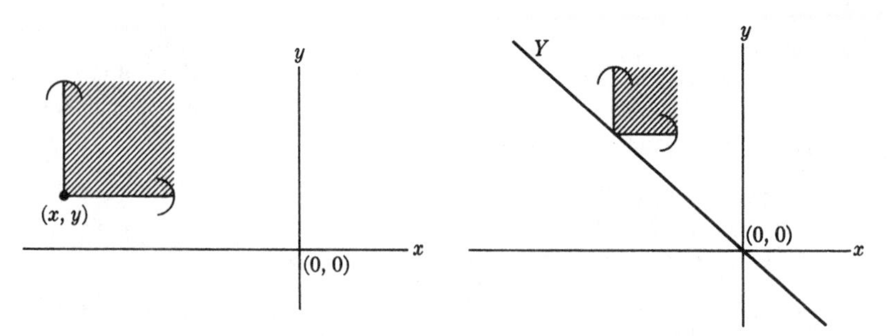

Figure 5.9

Figure 5.10

ogy. Then a typical basic neighborhood of $(x, y) \in R^2$ is as shown in Fig. 5.9. We know that R, τ is normal (Exercise 1). Each basic neighborhood U of (x, y) is not only open, but also closed. For if (x', y') is any point of $R^2 - U$, then it is readily seen that there is a basic neighborhood of (x', y') contained entirely in $R^2 - U$; hence $R^2 - U$ is open, and thus U is closed. Therefore $\text{Cl}\, U = U$. Hence, by Proposition 4, R^2 is T_3. R^2 is T_1 since each one-point subset of R^2 is closed. Alternately, R, τ is regular; thus R^2 with the product topology is regular by Proposition 5.

Let $Y = \{(x, y) \mid y + x = 0\}$. Then the subspace topology of Y is discrete, since if $(x, y) \in Y$, there is a basic neighborhood U of (x, y) such that

$$U \cap Y = \{(x, y)\}$$

(Fig. 5.10). Since Y is a closed subset of X, each subset of Y is a closed subset of X (since a subset of Y is closed in Y if and only if it is closed in X, but every subset of Y is closed in Y). Let

$$F = \{(x, y) \mid x + y = 0 \text{ and } x \text{ is rational}\}$$

and

$$F' = \{(x, y) \mid x + y = 0 \text{ and } x \text{ is irrational}\}.$$

Since F and F' are subsets of Y, they are closed. Also, $F \cap F' = \phi$. If R^2 is T_4, there must then be open sets U and V such that

$$F \subset U, \qquad F' \subset V, \qquad \text{and} \qquad U \cap V = \phi.$$

Although a rigorous argument of the impossibility of such sets will not be given here, it can be made fairly clear why there cannot be such sets. For if $(x, y) \in F$, then there would be a basic neighborhood of (x, y) contained in U. There are, however, points of F' "arbitrarily close" to (x, y). Hence some basic neighborhood contained in V of a point in F' would be bound to overlap with the basic neighborhood of (x, y) in U. Therefore $U \cap V$ could not be ϕ.

Although we do not have that every subset of a normal space is normal, we do have the weaker statement which follows.

Proposition 8. If Y is a closed subset of a normal space X, τ then the subspace Y is normal.

Proof. Since X is T_1, Y is T_1 because every subspace of a T_1-space is T_1. Since Y is closed, a subset F of Y is closed in Y if and only if F is closed in X. Therefore if F and F' are disjoint closed subsets of Y, they are also disjoint closed subsets of X. There are thus open sets U and V such that

$$F \subset U, \qquad F' \subset V, \qquad \text{and} \qquad U \cap V = \phi.$$

But then

$$F \subset Y \cap U, \qquad F' \subset Y \cap V,$$

and $Y \cap U$ and $Y \cap V$ are disjoint subsets of Y which are open in Y. Therefore Y is T_4; hence Y is normal.

Proposition 9. If the product space $\times_I X_i$ of the family of nonempty spaces $\{X_i, \tau_i\}$, $i \in I$, is normal, then X_i, τ_i is normal for each $i \in I$.

Proof. If $\times_I X_i$ is normal, then $\times_I X_i$ is T_1. But then each X_i, τ_i is homeomorphic to a closed subspace of $\times_I X_i$ (Exercise 6). Such a subspace then is normal by Proposition 8. Hence X_i, τ_i is normal (Exercise 5).

EXERCISES

1. Prove that the set R of real numbers with the topology described in Example 11 of Chapter 3 is normal.
2. Prove that a space X, τ is T_4 if and only if given any two disjoint closed subsets F and F' of X, there are open sets U and V such that
$$F \subset U, \qquad F' \subset V, \qquad \text{and} \qquad \text{Cl}\, U \cap \text{Cl}\, V = \phi.$$
3. Suppose X, τ is a normal space and F is a closed subset of X. Let X/F be the identification space formed by identifying all the points of F with one another; more figuratively, X/F is the identification space obtained by squashing F to a point (as in Proposition 6). Prove that X/F is normal.
4. Prove that every subspace of a metric space is normal. Thus if we were to find a normal space which had a subspace which was not normal, we would know such a space was not a metric space.
5. Prove that the property of being normal is preserved by homeomorphisms but not by continuous functions.
6. We saw in Proposition 20, Chapter 4, that if $\times_I X_i$ is the product space of the family of nonempty spaces $\{X_i, \tau_i\}$, $i \in I$, then each X_i, τ_i is homeo-

morphic to a subspace of $\times_I X_i$. Prove that if $\times_I X_i$ is T_1, each X_i, τ_i is homeomorphic to a closed subspace of $\times_I X_i$.

7. Are any of the spaces given in Exercise 6, Section 5.3, normal besides that given in (c)?
8. Review the proof of Proposition 5. Discuss why an analogous proof will not hold for normal spaces. That is, try to find out what goes wrong in attempting to apply to normality the techniques which gave us the "hereditary" properties of the other separation axioms.
9. Suppose X, τ and Y, τ' are normal and f is a continuous function from a subspace A of X into Y. Let $Z = X \cup Y$ have the topology τ'' defined by using $\tau \cup \tau'$ as a basis. Prove that the identification space formed from Z using f, that is, by setting x equivalent to $f(x)$, is normal.

5.5 NORMALITY AND THE EXTENSION OF FUNCTIONS

One of the central and most difficult questions in all of topology is that of function extensions. Specifically, suppose that Y is a subspace of a space X, τ and that f is a continuous function from Y into some space Z, τ'. Does there exist a function F from X into Z such that F is continuous and $F(y) = f(y)$ for each $y \in Y$? That is, is there a continuous function

$$F: X, \tau \to Z, \tau'$$

such that $F \mid Y = f$? The answer is sometimes yes and sometimes no. For most instances, the answer is not known.

Example 14. If Y is a subspace of X, τ and i is the identity function on Y, then i is a continuous function from Y into X. Of course i can be extended to the identity function I for all of X. In this case, I is an *extension* of i since $I \mid Y = i$. This is a rather trivial and therefore uninteresting type of extension.

A function may have several extensions. For if we let $X = \{1, 2\}$ with the discrete topology and $Y = \{1\}$, then I' defined by $I'(1) = 1$, $I'(2) = 1$, is an extension of i which is different from I.

Example 15. Let X be the closed interval $[0, 1]$ with the usual absolute value topology, and let $Y = Z = \{0, 1\}$ with the subspace topology (which is the discrete topology in this case). Let i be the identity function on Y. Then i is continuous as a function from Y to Z. Although we cannot prove it at this time (we will be able to do so later in the book), i cannot be extended to a continuous function from $[0, 1]$ into Z. We may see this informally as follows: If there were a continuous function F from X onto Z (as there would have to be if i could be extended), then, since $\{0\}$ and $\{1\}$ are both open subsets of Z, $F^{-1}(\{0\})$ and $F^{-1}(\{1\})$ would both be open subsets of X; moreover,

$$X = F^{-1}(\{0\}) \cup F^{-1}(\{1\}) \quad \text{and} \quad F^{-1}(\{0\}) \cap F^{-1}(\{1\}) = \phi.$$

Thus $X = [0, 1]$ would be expressible as the union of two disjoint, nonempty, open subsets. The reader should try to express [0, 1] as the union of two such subsets in order to see the intuitive difficulties of such a decomposition.

Topological spaces which are T_4 are, however, bound up essentially with some very important extension properties. In fact, T_4-spaces can be characterized by certain of their extension properties. This is proved in the following proposition, one of the most important propositions in topology.

Proposition 10 (*Urysohn's lemma*). A topological space X, τ is T_4 if and only if given any disjoint nonempty closed subsets A and B of X, there is a continuous function f from X into $Z = [0, 1]$ (with the absolute value topology) such that

$$f(a) = 0 \quad \text{for any } a \in A \qquad \text{and} \qquad f(b) = 1 \quad \text{for any } b \in B.$$

Before proving this proposition, we note that it is indeed a proposition dealing with function extensions. Explicitly, if X, τ is a T_4-space and Y is a subspace of X which can be expressed as the union of two disjoint nonempty closed subsets of X, then the function $g: Y \to [0, 1]$ such that

$$g(a) = 0 \quad \text{for all } a \in A \qquad \text{and} \qquad g(b) = 1 \quad \text{for all } b \in B$$

can be extended to a continuous function $f: X \to [0, 1]$. Although this may appear as a rather modest result about function extensions because of the restrictions that have been placed upon Y and g, very general and important results flow from Urysohn's lemma. We shall mention a few of these after the proof.

Proof (Proposition 10). Suppose X has the property described and A and B are any two disjoint nonempty subsets of X. Then there is a continuous function f from X into $Z = [0, 1]$ such that

$$f(a) = 0 \quad \text{for all } a \in A \qquad \text{and} \qquad f(b) = 1 \quad \text{for all } b \in B.$$

Now $U' = \{x \mid 0 \leq x < 1/2\}$ and $V' = \{x \mid 1/2 < x \leq 1\}$ are disjoint open subsets of Z; therefore, since f is continuous, $U = f^{-1}(U')$ and $V = f^{-1}(V')$ are disjoint open subsets of X. But $A \subset U$ and $B \subset V$, and hence X is T_4.

Suppose now that X is T_4. Recall that a space X, τ is T_4 if and only if given any closed subset F of X and any open set U which contains F, there is an open set V which contains F such that

$$F \subset V \subset \operatorname{Cl} V \subset U$$

(Proposition 7). Suppose A and B are disjoint nonempty closed subsets of X (Fig. 5.11). Consider the set of rational numbers q such that $0 \leq q \leq 1$ and q is of the form $q = n/2^k$, where n and k are positive integers. For example, $1/2$, $3/2^2 = 3/4$, and $5/2^3 = 5/8$ are such rational numbers. With each such rational q we will associate an open subset $U(q)$ of X such that

1) $A \subset U(q)$;
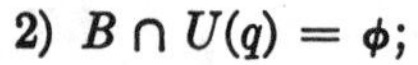
2) $B \cap U(q) = \phi$;
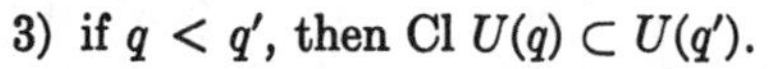
3) if $q < q'$, then $\operatorname{Cl} U(q) \subset U(q')$.

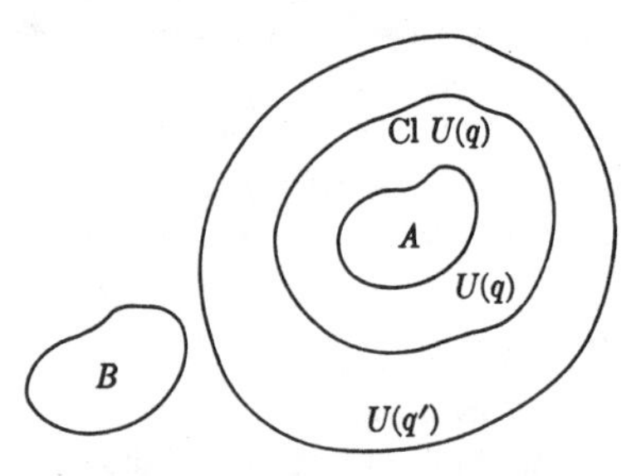

Figure 5.11

Since X is T_4 and A and B are disjoint closed subsets of X, there are disjoint open sets U and V such that $A \subset U$ and $B \subset V$. We let $U = U(0)$ and $X - B = U(1)$. Using Proposition 7, we can find an open set, which we let be $U(\frac{1}{2})$, such that

$$\operatorname{Cl} U(0) \subset U(\tfrac{1}{2}) \subset \operatorname{Cl} U(\tfrac{1}{2}) \subset U(1).$$

Similarly, we can find $U(\frac{1}{4})$ and $U(\frac{3}{4})$ such that

$$\operatorname{Cl} U(0) \subset U(\tfrac{1}{4}) \subset \operatorname{Cl} U(\tfrac{1}{4}) \subset U(\tfrac{1}{2})$$

and

$$\operatorname{Cl} U(\tfrac{1}{2}) \subset U(\tfrac{3}{4}) \subset \operatorname{Cl} U(\tfrac{3}{4}) \subset U(1).$$

We will continue finding the $U(q)$ by induction on k, the exponent of 2 in $q = n/2^k$. Note that we have already defined $U(q)$ for $k = 1$ and $k = 2$.

Assume we have defined $U(q)$ for k. We now define $U(q)$ for $k + 1$ (and thus for $n = 1, 3, \ldots, 2^{k+1} - 1$). Note that the definition of $U(q)$ needs to be given only for odd n; for if n were even, the numerator and denominator of q could be divided by 2. Because the $U(q)$ have already been constructed for $q = n/2^k$, n odd, we have

$$\operatorname{Cl} U\left(\frac{n-1}{2^{k+1}}\right) \subset U\left(\frac{n+1}{2^{k+1}}\right)$$

[since n is odd, $\operatorname{Cl} U((n-1)/(2^{k+1})) = \operatorname{Cl} U(((n-1)/2)/2^k)$, which has already been defined]. We therefore can find an open set, which we let be

$U(n/2^{k+1})$ such that

$$\text{Cl}\, U((n-1)/2^{k+1}) \subset U(n/2^{k+1}) \subset \text{Cl}\, U(n/2^{k+1}) \subset U((n+1)/2^{k+1})$$

(Fig. 5.12). We thus have an inductive definition of $U(q)$ for each q as described. By construction, the collection of $U(q)$ have properties (1) through (3) given above.

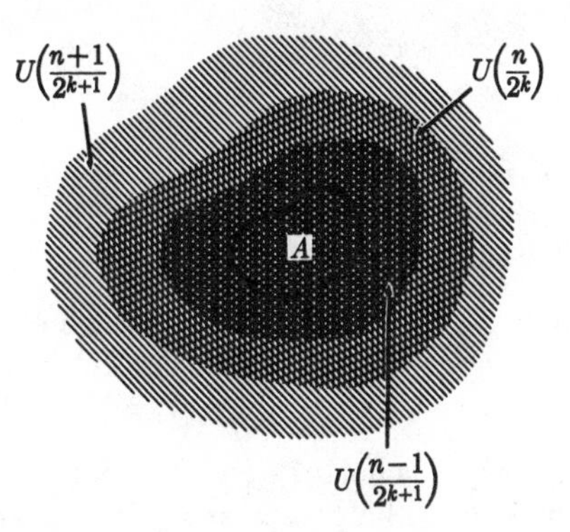

Figure 5.12

We now define a function f from X into $Z = [0, 1]$ such that $f(a) = 0$ for all $a \in A$ and $f(b) = 1$ for all $b \in B$. If $x \in X$, define $f(x) = 1$, if $x \in B$. If x is not in B, then $x \in U(1)$. For each x not in B, define

$$f(x) = \text{greatest lower bound } \{q \mid q = n/2^k \text{ and } x \in U(q)\}$$

(this set of real numbers has a lower bound, 0, and hence has a greatest lower bound). Certainly $0 \leq f(x) \leq 1$. If $x \in A$, then $x \in U(0)$; therefore $f(0) = 0$. We now prove that f is continuous.

Suppose $f(x_0) = y_0$. First, assume that y_0 is neither 0 nor 1. Then, given any positive number p, there are rationals q and q' of the form $n/2^k$ such that

$$y_0 \in (q, q') \subset (y_0 - p, y_0 + p)$$

(that is, the set of "binary" rationals is dense in [0, 1]. Alternately, any real number can be approximated to an arbitrary degree of accuracy by a rational of the form $n/2^k$). Then (Fig. 5.13) $V = U(q') - \text{Cl}\, U(q)$ is a neighborhood of x_0, and

$$f(V) \subset (y_0 - p, y_0 + p).$$

If y_0 is either 0 or 1, then the corresponding neighborhoods of 0 and 1, respectively, are $[0, q')$ and $(q, 1]$; but the argument is the same. What we have shown is that, given any neighborhood H of y_0 (any neighborhood

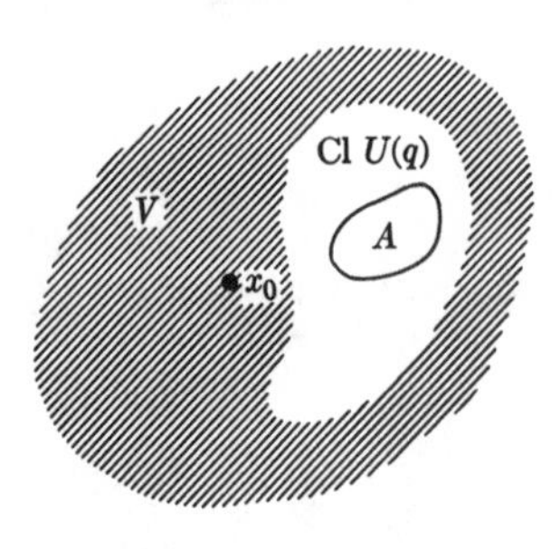

Figure 5.13

of y_0 contains one of the interval neighborhoods of y_0 we have used), there is a neighborhood V of x_0 such that $f(V) \subset H$. By Proposition 6 of Chapter 4, then, f is continuous.

Corollary. Suppose X, τ is a normal space which contains more than one point. Then there is always a nonconstant continuous function f from X into $[0, 1]$.

Proof. Since X is normal, X is both T_1 and T_4. Since X is T_1, each one-point subset of X is closed. Suppose x and y are distinct points of X. Then $\{x\}$ and $\{y\}$ are disjoint closed nonempty subsets of X. Therefore, by Proposition 10, there is a continuous function f from X into $[0, 1]$ such that $f(x) = 0$ and $f(y) = 1$. Thus f satisfies the corollary.

The following proposition stated without proof appears stronger than Proposition 10, but would require Proposition 10 for its proof. It gives the most important extension property of a normal space.

Proposition 11 (*Tietze's Extension Theorem*). Let X be any normal space and Y be any closed subset of X. Suppose f is *any* continuous function from Y into R, the space of real numbers with the absolute value topology. Then there is a continuous extension F of f from X into R.

Note that Proposition 11 includes Proposition 10 as a special case. For if A and B are disjoint closed nonempty subsets of X, then setting $Y = A \cup B$ and defining $f: Y \to R$ by

$$f(a) = 0 \quad \text{for } a \in A \qquad \text{and} \qquad f(b) = 1 \quad \text{for } b \in B,$$

we see that f is continuous as a function from Y into R and can be extended by Proposition 11.

Corollary. Let X, τ be any normal space and A be any closed subset of X. Suppose f is a continuous function from A into R^n, the product of the space R of real numbers with itself n times. Then there is a continuous extension F of f from X into R^n.

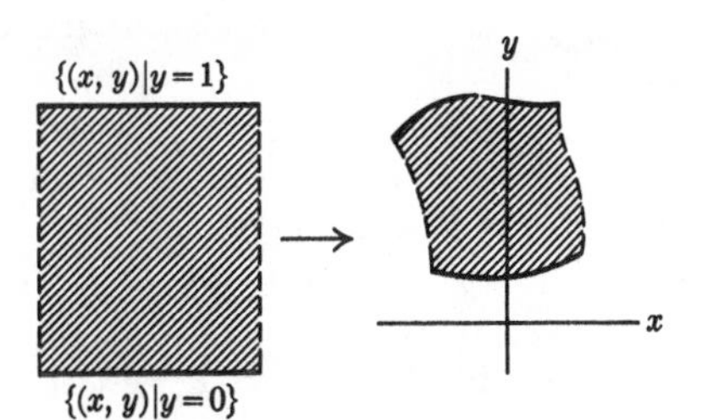

Figure 5.14

Proof. Let p_i be the projection into the ith coordinate from R^n into R [defined by $p_i(x_1, \ldots, x_i, \ldots, x_n) = x_i$]. Then, setting $f_i = p_i \circ f$, f_i is a continuous function from A into R. Each f_i therefore has a continuous extension F_i to all of X. Define

$$F(x) = (F_1(x), \ldots, F_n(x))$$

for each $x \in X$. Then F is an extension of f; moreover, F is continuous, by Proposition 21 of Chapter 4.

Example 16. Let R be the space of real numbers with the absolute value topology and R^2 be the plane with the product topology. Both of these spaces are normal. Let $Z \subset R$ be the set of integers, and let $Z^2 \subset R^2$ be the set of points of R^2 of the form (m, n), where m and n are integers. The subspace topology on both Z and Z^2 is the discrete topology, and hence any function f from Z into Z^2 is continuous. Both Z and Z^2 are of the same cardinality; thus we have a one-one function f from Z onto Z^2, and f is continuous. By the corollary to Proposition 11, then f has a continuous extension F from R into R^2. The reader might find from a little experimentation that this is a case where the proof that F exists is much simpler than trying to construct a specific F.

Example 17. A continuous function from the interval $[0, 1]$ into any space X, τ is called a *path* in X. The space $[0, 1] \times [0, 1]$ (with the product topology) is a normal space, and

$$A = \{(x, y) \mid y = 1\} \cup \{(x, y) \mid y = 0\}$$

is a closed subset of $[0, 1] \times [0, 1]$. If f is any continuous function from A into R^2, then f has a continuous extension F (Fig. 5.14). Technically, this means that any two paths in R^2 are *homotopic* (Chapter 11).

EXERCISES

1. Using only Proposition 10, prove the following: Suppose A and B are disjoint closed subsets of a normal space X, τ and f is a continuous function from $A \cup B$

into R^n such that $f \mid A$ and $f \mid B$ are each constant functions. Then f has a continuous extension F to all of X.

2. Any space which can be substituted for R in Proposition 11 is called an *absolute retract.* Which of the following are definitely absolute retracts? Which could not possibly be absolute retracts?

   a) R^n, with the product topology from the space R of real numbers with the absolute value topology
   b) I^n, with the product topology, where $I = [0, 1]$
   c) any finite set with the discrete topology
   d) $(0, 1)$ with the usual topology
   e) $R^2 - \{(0, 0)\}$ with the Pythagorean topology

3. Which of the following statements about absolute retracts are true?

   a) If X, τ is an absolute retract and x and y are any points of X, then there is a continuous function f from $[0, 1]$ into X such that $\{x, y\} \subset f([0, 1])$.
   b) The product space of a countable family of nonempty spaces is an absolute retract if and only if each component space is an absolute retract.

4. Prove that the set of binary rationals as described in the proof of Proposition 10 is dense in the space of real numbers.

5. A space X is said to be *completely normal* (sometimes called T_5) if every subspace of X is normal. Prove that X is completely normal if and only if X is T_1 and given any two subsets A and B of X such that $\text{Cl } A \cap B = A \cap \text{Cl } B = \phi$, there exist disjoint open sets U and V such that $A \subset U$ and $B \subset V$.

6 CONVERGENCE

6.1 THE NEED FOR A GENERALIZED NOTION OF CONVERGENCE

The reader will recall that we have already discussed convergence of sequences in metric spaces in Chapter 2. He may therefore suspect that extending the theory of convergence to general topological spaces will merely consist of rewording the definition of a convergent sequence in terms of a general space. For example, we might say: A sequence $\{s_n\}$, $n \in N$, in a space X, τ converges to a limit y in X if every neighborhood of y contains all but finitely many of the s_n.

Actually, however, not only will we find it necessary to generalize the notion of convergence of a sequence, but we will also have to generalize the very notion of a sequence as well. The purpose of this section is to illustrate this point. We begin by proving a proposition concerning sequences in metric spaces.

Proposition 1. Let A be a subset of a metric space X, D. Then $y \in \text{Cl } A$ if and only if there is a sequence $\{s_n\}$, $n \in N$, such that

$$s_n \to y \qquad \text{and} \qquad s_n \in A$$

for each $n \in N$.

Proof. Suppose $y \in \text{Cl } A$. Then if $y \in A$, let $\{s_n\}$, $n \in N$, be the sequence defined by $s_n = y$ for all $n \in N$. Then

$$s_n \to y \qquad \text{and} \qquad s_n \in A$$

for all $n \in N$. Suppose $y \in \text{Cl } A - A$. By Proposition 12, Chapter 3, $\text{Cl } A = A \cup A'$; therefore $y \in A'$. Then each neighborhood of y contains at least one element of A. Let U_n be the D-$1/n$-neighborhood of y for each positive integer n. For each n, select $s_n \in U_n \cap A$. The sequence $\{s_n\}$, $n \in N$, thus obtained converges to y, and each s_n is in A.

On the other hand, suppose y is such that there is a sequence $\{s_n\}$, $n \in N$, such that $s_n \to y$, and $s_n \in A$ for each $n \in N$. Since $s_n \to y$, each neighborhood of y contains all but finitely many of the s_n. Therefore each neighborhood of y contains some point of A. By Proposition 13, Chapter 3, then $y \in \text{Cl } A$.

If the notion of a sequence were sufficient for the study of general topological spaces, we would expect that this proposition relating closures and sequences generalizes to arbitrary topological spaces. That is, if $A \subset X$, where X, τ is a topological space, then $y \in \text{Cl } A$ if and only if, etc. There are a number of reasons why we would like this proposition to generalize. The fact is, though, that it does not generalize using sequences. This is shown by the following example.

Example 1. Let X be the set of all functions from the set R of real numbers into R. We make no assumption about the continuity of these functions. We will define a topology on X by specifying an open neighborhood system. Suppose f is any element of X. Let F be any finite subset of R and p be any positive real number. Define

$$U(f, F, p) = \{g \in X \mid |g(x) - f(x)| < p \text{ for all } x \in F\}.$$

Let $\mathfrak{N}_f$ be the set of all $U(f, F, p)$ for all finite subsets F of R and all positive numbers p. Note that for a given f, $U(f, F, p)$ depends on *both* F and p; hence $U(f, F, p)$ is not a p-neighborhood in the metric sense.

We will now show that this definition of $\mathfrak{N}_f$ for each $f \in X$ gives us an open neighborhood system for a topology on X. In accordance with Proposition 7 of Chapter 3, we must show that (i) through (iv) of Definition 5, Chapter 3 are satisfied.

Statements (i) and (ii) are clearly satisfied.

iii) Suppose $U(f, F_1, p_1)$ and $U(f, F_2, p_2)$ are any two members of $\mathfrak{N}_f$. Then

$$U\big(f, F_1 \cup F_2, \min(p_1, p_2)\big)$$

is an element of $\mathfrak{N}_f$ which is contained in $U(f, F_1, p_1) \cap U(f, F_2, p_2)$. For suppose

$$g \in U\big(f, F_1 \cup F_2, \min(p_1, p_2)\big);$$

we may assume $p_1 \leq p_2$. It follows that

$$|g(x) - f(x)| < p_1 \leq p_2$$

for each $x \in F_1$ and each $x \in F_2$. Therefore

$$g \in U(f, F_1, p_1) \cap U(f, F_2, p_2).$$

iv) Suppose $U(f, F, p) \in \mathfrak{N}_f$ and $g \in U(f, F, p)$. Let

$$F = \{x_1, \ldots, x_n\}$$

and

$$q_i = p - |f(x_i) - g(x_i)|, \; i = 1, \ldots, n.$$

Set $p' = \min(q_1, \ldots, q_n)$. Then

$$U(g, F, p') \subset U(f, F, p)$$

(Exercise 1). Therefore the $\mathfrak{N}_f$ do form an open neighborhood system for a topology on X.

Let A be the set of all elements f in X such that $f(x) = 0$ or 1 for any $x \in R$, and $f(x) = 0$ for at most countably many $x \in R$. Suppose g is the function defined by $g(x) = 0$ for all $x \in R$. We will show that $g \in \text{Cl } A$. For let $U(g, F, p)$ be any basic neighborhood of g. Let h be the function defined by $h(x) = 0$ for $x \in F$ and $h(x) = 1$ if $x \notin F$. Then

$$h \in A \cap U(g, F, p).$$

Therefore any neighborhood of g meets A; hence $g \in \text{Cl } A$.

Suppose there is a sequence $\{f_n\}$, $n \in N$, such that $f_n \in A$ for each $n \in N$, and $f_n \to g$. Let $B_n = \{x \mid f_n(x) = 0\}$. Each B_n is a countable subset of R; hence $\bigcup_N B_n$ is also a countable subset of R. Since R is uncountable, we can find $z \in R - \bigcup_N B_n$. Let $F = \{z\}$ and $p = 1/2$. Consider $U(g, F, p)$. Then no matter what n is, $z \notin B_n$, and hence $f_n(z) = 1$. Therefore

$$|f_n(z) - g(z)| = 1.$$

There is no positive integer n, then, for which $f_n \in U(g, F, p)$. But $U(g, F, p)$ is a neighborhood of g; thus if $f_n \to g$, $U(g, F, p)$ would have to contain all but finitely many of the f_n. It is impossible then that $f_n \to g$.

What we have here, then, is a topological space for which Proposition 1 of this chapter does not hold. What are we to do? There are two alternatives, either (1) we can restrict ourselves merely to sequences and say that Proposition 1 has no generalization, or (2) we can try to generalize the notion of a sequence in such a way that Proposition 1 can be generalized to any topological space. It is this latter alternative that we choose.

EXERCISES

1. In Example 1, prove (iv) in the proof that the $\mathfrak{N}_f$ form an open neighborhood system.
2. A topological space X, τ is said to be *first countable* if there is an open neighborhood system for τ such that $\mathfrak{N}_x$ is a countable collection for each $x \in X$.
   a) Prove that every metric space is first countable.
   b) Prove that Proposition 1 holds for any topological space which is first countable. Thus the space in Example 1 is not first countable.
   c) Prove that any subspace of a first countable space is first countable.

d) Prove that the product space of a countable family of nonempty spaces is first countable if and only if each component space is first countable.

e) Which of the spaces mentioned in Section 5.3, Exercise 6 are first countable?

3. Do you think that the following might be an appropriate generalization of sequences? Let I be any set. Then an *I-sequence* in a space X, τ will be a function s from I into X. We will denote the I-sequence by $\{s_i\}$, $i \in I$, where s_i denotes $s(i)$. We will say that $\{s_i\}$, $i \in I$, converges to y if any neighborhood of y contains all but finitely many of the s_i. For example, suppose $I = (0, 1)$ and s is the identity function on I considered as a function from I into the space of real numbers. Does $\{s_i\}$, $i \in I$, converge to 1 according to our definition of convergence of a I-sequence? Does it seem as though it should if this is a suitable generalization of the notion of a sequence?

4. Let X be any uncountable set. For each $x \in X$, define

$$\mathfrak{N}_x = \{N \subset X \mid x \in N \text{ and } N \text{ excludes at most countably many points of } X\}.$$

Prove that the collection of $\mathfrak{N}_x$ forms an open neighborhood system for a topology on X. Does Proposition 1 apply to X with this topology?

5. Find two distinct topologies on the set R of real numbers such that the only sequences which converge relative to each of the topologies are those sequences which are constant from some term on, and these sequences converge only to their constant value. (Use Exercise 4 and the notion of convergence introduced in the first paragraph of this chapter.) Since the two topologies have the same convergent sequences converging to the same limits, sequences alone are inadequate to characterize either topology.

6.2 NETS

Note that the positive integers form a partially ordered set (in this case, totally ordered) such that if n and n' are any integers, there is an integer m with $n \leq m$ and $n' \leq m$. Since all partial orderings share the properties of "less than or equal to," denoted by $\leq$, we will use $\leq$ to denote any partial ordering. The set of positive integers do have other properties which are not shared by every partially ordered set; for example, the positive integers are totally ordered, well-ordered, and countable. But in any generalization, experience and the problem to be solved indicates which properties must be generalized and which are incidental to the question at hand. The experience and labor of many mathematicians over many years leads us to the following definition.

Definition 1. Let I be any partially ordered set. (Recall that $\leq$ is used to designate any partial ordering.) I is said to be a *directed set* (more accurately, an *upward directed set*) if given any i and any j in I, there is $k \in I$ such that $i \leq k$ and $j \leq k$. (Note that any totally ordered set is directed, since if I is totally ordered, given any i and j in I, either $i \leq j$ or $j \leq i$; thus if, say, $i \leq j$, we have $i \leq j$ and $j \leq j$. If I is not totally ordered, then the more general condition as given here is needed.)

Let X be any set, and I be a directed set. A function s from I into X is said to be a *net* in X. The expression $s(i)$ is usually denoted by s_i, and the net itself is usually denoted by $\{s_i\}$, $i \in I$.

A net $\{s_i\}$, $i \in I$, is said to have a property P *residually* if there is some $i_0 \in I$ such that if $i_0 \leq i$, then s_i has property P. The net $\{s_i\}$, $i \in I$, is said to have a property P *cofinally* if, for any i, $i' \in I$, there is some $k \in I$ with $i \leq k$ and $i' \leq k$ such that s_k has property P.

Actually, a net $\{s_i\}$, $i \in I$, has a property cofinally if and only if given any $i \in I$, there is $k \in I$ such that $i \leq k$ and s_k has property P (Exercise 5).

Example 2. The set N of positive integers, since it is totally ordered, is a directed set. Any sequence is therefore a net; that is, a net is a valid generalization of a sequence. A sequence $\{s_n\}$, $n \in N$, has a property residually if and only if there is an integer M such that for any integer n greater than M, s_n has the property. The sequence $\{s_n\}$, $n \in N$, has a property cofinally if and only if given any positive integer n, there is an integer $m \geq n$ such that s_m has the property.

In particular, if X, τ is a topological space and $\{s_n\}$, $n \in N$, is a sequence in X, then $s_n \to y$ if and only if $\{s_n\}$, $n \in N$, has the property of being in every neighborhood of y residually. If s is the identity function on N, that is, considering N as a sequence in itself, $\{s_n\}$, $n \in N$, has the property of being odd cofinally; of course, $\{s_n\}$, $n \in N$, also has the property then of being even cofinally.

Example 3. Let X be any set, finite or infinite. Let $P(X)$ be the family of all subsets of X. Then we may partially order $P(X)$ by set inclusion $\subset$. In order that set inclusion may direct $P(X)$ as we will want to have it later, we will say that if W and Y are subsets of X, then $W \leq Y$ if $Y \subset W$. If W and Y are any two elements of $P(X)$, that is, subsets of X, then

$$W \leq W \cap Y$$

and

$$Y \leq W \cap Y;$$

therefore $\leq$ makes $P(X)$ into a directed set. Note that even though $P(X)$ is directed, it is not totally ordered; that is, it is not true that given any two subsets W and Y of X, we must have either $W \leq Y (Y \subset W)$, or

$Y \leq W (W \subset Y)$. If X is finite, then $P(X)$ is finite as well, and hence it is quite possible to have a finite directed set.

If $x \in X$, set

$$P(X, x) = \{W \in P(X) \mid x \in W\}.$$

Then $P(X, x)$ is also a directed set (directed by $\leq$ just as $P(X)$ is}. One possible function s from $P(X, x)$ into X would be a selection function where, if $W \in P(X, x)$, then $s(W) \in W$. Such a selection function would then give a net in X, $\{s_W\}$, $W \in P(X, x)$.

If X, τ is a topological space, then we might set

$$T(X, x) = \{U \mid U \text{ is a neighborhood of } x\}.$$

The set $T(X, x)$ is also a directed set [in fact, a directed subset of $P(X, x)$]. Using a selection function $s: T(X, x) \to X$, where $s(U) \in U$ for each $U \in T(X, x)$, we get a net $\{s_U\}$, $U \in T(X, x)$, in X. We might suspect that this net converges to x, since it has the property of being in every neighborhood of x residually (Exercise 2).

Example 4. Let $\{s_n\}$, $n \in N$, be the sequence in the set R of real numbers defined by $s_n = (-1)^n$. If R is given any topology whatsoever, then $\{s_n\}$, $n \in N$, has the property of being in every neighborhood of 1 cofinally. This sequence is also in every neighborhood of -1 cofinally, but it is residually in every neighborhood of both 1 and -1 if and only if every neighborhood of 1 is a neighborhood of -1 and every neighborhood of -1 is a neighborhood of 1.

EXERCISES

1. Which of the following sets with the orderings as given are directed sets?
   a) the set of positive integers partially ordered by "divides"
   b) the interval $[0, 1]$ ordered by $\leq$
   c) the interval $(0, 1)$ ordered by $\leq$
   d) the set $\{1, 2, 3, 4\}$, where the order is defined by the relation

$$R = \{(1, 1), (2, 2), (3, 3), (4, 4), (2, 3)\}.$$

2. Prove the assertion in Example 3 that $\{s_U\}$ has the property of being in any neighborhood of x residually. Using this example, make a tentative definition of what we mean by saying that a net converges to an element y of a space X, τ.

3. Suppose I and I' are sets which are directed by $\leq$ and $\leq'$, respectively. Suppose (i, i') and (j, j') are elements of $I \times I'$. Define

$$(i, i') \leq (j, j')$$

if $i \leq j$ and $i' \leq' j'$. Prove that $I \times I'$ with the relation $\leq$ as defined is a directed set. You must prove that $I \times I'$ is both partially ordered and directed.

4. Let $\{s_n\}$, $n \in N$, be a sequence in the set of integers with the property that if $m \leq n$, $s_m \leq s_n$. Which of the following properties must such a sequence have residually? cofinally? neither cofinally nor residually?

   a) The property of being odd; b) The property of being positive;
   c) The property that $s_n \leq n$; d) The property that s_n is prime.

   Suppose $\{s_n\}$, $n \in N$, is the sequence defined by $s_n = 4n + 1$. Which of the properties (a) through (d) does this sequence have residually? cofinally?

5. Prove that a net $\{s_i\}$, $i \in I$, has a property P cofinally if and only if for any $i \in I$, there is an element k of I such that $i \leq k$ and s_k has property P.

6. The following define functions from $[0, 1]$ into R^2 and hence define nets in R^2. If R^2 has its usual topology, indicate any points to which you feel the nets should converge. Explain informally the reasons for your answers in each case.

   a) $f(x) = (x, 2)$ b) $f(x) = \left(1/(x+1), x^2\right)$ c) $f(x) = (\cos x, \sin x)$

6.3 SUBSEQUENCES AND SUBNETS

Fundamental in any study of either sequences or nets is the concept of a *subsequence*, or its generalization, a *subnet*.

Definition 2. Suppose $\{s_i\}$, $i \in I$, is a net in a set X. Let J be a directed set and k a function from J to I such that

i) if $j \leq j'$, then $k(j) \leq k(j')$;

ii) if $i, i' \in I$, then there is $j \in J$ such that $i \leq k(j)$ and $i' \leq k(j)$.

That is, k is order-preserving, and considered as a net in I, k is cofinal in I. Then the composition $s \circ k$ from J into X is said to be a subnet of the net $\{s_i\}$, $i \in I$. The subnet $s \circ k$ is usually written as $\{s_{k_j}\}$, $j \in J$.

Note that each s_{k_j} is also an s_i for some i (specifically, for $i = k_j$), and that the s_{k_j} have the property of being cofinal (but not necessarily residual) in the set of s_i.

Example 5. Let $\{s_n\}$, $n \in N$, be any sequence. Let $2N$ be the set of positive even integers, and let k be the identity mapping from $2N$ into N. We first verify that k has properties (i) and (ii) of Definition 2.

i) If m and n are positive even integers and $m \leq n$, then $k(m) = m$ and $k(n) = n$.

ii) Given any integer n, there is a positive even integer greater than n; therefore there is $m \in 2N$ such that

$$n \leq k(m) = m.$$

The function $s \circ k$ is therefore a subnet of $\{s_n\}$, $n \in N$. In the case where a subnet is a sequence, it is customary to call such a subnet a *subsequence*; thus $s \circ k$ is a subsequence of $\{s_n\}$, $n \in N$. Explicitly, $\{s_{k_m}\}$, $m \in 2N$, is the same as $\{s_{2n}\}$, $n \in N$. A subsequence of a sequence is a sequence in its own right, and a subnet of a net is itself a net.

Example 6. Let X, τ be any topological space, and let $x \in X$. Suppose $P(X, x)$ is as defined in Example 3. Let s be any selection function from $P(X, x)$ into X; that is, $s(W) \in W$ for each $W \in P(X, x)$. Then $\{s_W\}$, $W \in P(X, x)$, is a net in X. Let $T(X, x)$ be (as in Example 3 also) the set of neighborhoods of x. Then

$$T(X, x) \subset P(X, x).$$

Let k be the identity mapping from $T(X, x)$ into $P(X, x)$. Then $s \circ k$ does *not* define a subnet of $\{s_W\}$, $W \in P(X, x)$, unless $\{x\}$ is itself a neighborhood of x. For if $\{x\}$ is not open, then $\{x\} \notin T(X, x)$; hence there is no $U \in T(X, x)$ such that $\{x\} \leq k(U)$ as is required by (ii) of Definition 2.

Example 7. Let $\{s_n\}$, $n \in N$, be any sequence. Let k be a function from N into N defined by

$$k(n) = n \quad \text{for } n < 10, \qquad k(n) = 10 \quad \text{for all } n \geq 10.$$

Then $s \circ k$ satisfies (i) of Definition 2, but not (ii). On the other hand, if we define $k'\colon N \to N$ by

$$k'(n) = n \quad \text{if } n \text{ is even}, \qquad k'(n) = 2 \quad \text{if } n \text{ is odd},$$

then $s \circ k'$ satisfies (ii), but not (i), of Definition 2.

Example 8. Let $\{s_i\}$, $i \in I$, be a net where I is the set of real numbers greater than or equal to 1. Let k be the function from the directed set $I \times I$ (directed as in Section 6.2, Exercise 3) defined by $k(i, i') = ii'$. We verify that k satisfies (i) and (ii) of Definition 2.

i) If $(i_1, i_2) \leq (i_3, i_4)$, then $i_1 \leq i_3$ and $i_2 \leq i_4$. Since all of the numbers concerned are greater than or equal to 1,

$$k(i_1, i_2) = i_1 i_2 \leq k(i_3, i_4) = i_3 i_4.$$

ii) If i and i' are any real numbers greater than or equal to 1, $i' \leq i$, then $i \leq i^2$ and $i' \leq i^2$; hence $i \leq k(i, i)$ and $i' \leq k(i, i)$. Therefore $s \circ k$ is a subnet of $\{s_i\}$, $i \in I$. Note that here the set which "indexes" the subnet is actually "richer" than the original index set.

We now prove some of the fundamental properties of subnets.

Proposition 2. Suppose a net $\{s_i\}$, $i \in I$, has a property P cofinally. Then there is a subnet $\{s_{k_j}\}, j \in J$, of $\{s_i\}$, $i \in I$, which has the property P residually.

Proof. Let J be the set of all $j \in I$ such that s_j has property P, and let k be the identity map from J into I. If J has the order induced from I, then J is at least a partially ordered set. Since the property P is cofinal, given any j, $j' \in J \subset I$, there is $j'' \in I$ such that $j \leq j''$, $j' \leq j''$, and $s_{j''}$ has property P. Therefore

$$j'' \in J \qquad j \leq j'', \qquad \text{and} \qquad j' \leq j'';$$

hence J is a directed set. Since $k: J \to I$ is the identity mapping, k is certainly order-preserving. Since P is cofinal, k also satisfies (ii) of Definition 2. For if i and i' are any elements of I, there is $j \in I$ such that $i \leq j$, $i' \leq j$, and s_j has P, and hence

$$j \in J, \qquad i \leq k(j), \qquad \text{and} \qquad i' \leq k(j).$$

Therefore $s \circ k$ is a subnet of $\{s_i\}$, $i \in I$. By definition of J, each s_{k_j} has the property P; thus $\{s_{k_j}\}, j \in J$, has the property P residually.

Proposition 3. Suppose a net $\{s_i\}$, $i \in I$, has a property P residually. Then every subnet of $\{s_i\}$, $i \in I$, also has the property P residually.

Proof. Since $\{s_i\}$, $i \in I$, has P residually, there is $i_0 \in I$ such that if $i_0 \leq i$, then s_i has P. Suppose $\{s_{k_j}\}$, $j \in J$, is a subnet of $\{s_i\}$, $i \in I$. Applying (i) and (ii) of Definition 2, we can find $j_0 \in J$ such that if $j_0 \leq j$, then $i_0 \leq k(j)$. Hence if $j_0 \leq j$, s_{k_j} has the property P. The net $\{s_{k_j}\}$, $j \in J$, therefore has the property P residually.

Corollary. A net $\{s_i\}$, $i \in I$, has a property P residually if and only if every subnet of $\{s_i\}$, $i \in I$, has the property P residually.

Proof. Each net is a subnet of itself (Exercise 1); hence if each subnet of $\{s_i\}$, $i \in I$, has P residually, then $\{s_i\}$, $i \in I$, does also. The converse is Proposition 3.

EXERCISES

1. Prove that every net is a subnet of itself. [*Hint:* Use $J = I$ and let k be the identity mapping.]
2. Prove the converse of Proposition 2. That is, if a subnet $\{s_{k_j}\}, j \in J$, of the net $\{s_i\}$, $i \in I$, has a property residually, then $\{s_i\}$, $i \in I$, has the property cofinally.

3. Let I and J be directed sets and $\{s_i\}$ and $\{t_j\}$ be nets indexed by I and J, respectively, in some set X. Let $I \times J$ be directed as in Section 6.2, Exercise 3. Define a function $s \times t$ from $I \times J$ into $X \times X$ by $(s \times t)(i, j) = (s_i, t_j)$.
   a) Prove that $s \times t$ defines a net in $X \times X$.
   b) Prove that if $\{s_i\}$, $i \in I$, and $\{t_j\}$, $j \in J$, both have a property P residually or cofinally, then $\{(s_i, t_j)\}$, $(i, j) \in I \times J$, has the property P in the same way that both nets have it.
   c) Suppose M and M' are directed sets and k and k' are functions from M and M' into I and J, respectively, such that $s \circ k$ and $t \circ k'$ are subnets of $\{s_i\}$, $i \in I$ and $\{t_j\}$, $j \in J$. Define the obvious function

$$k \times k' : M \times M' \to I \times J.$$

   Prove that $(s \times t) \circ (k \times k')$ is a subnet of $\{(s_i, t_j)\}$, $(i, j) \in I \times J$.

4. Find an example of a net which has a property P cofinally, but such that no *subsequence* of the net has the property residually. This in turn will be accomplished if we find a net no subnet of which is a subsequence. To find such a net, consider Example 1 of this chapter. Let $g \in X$ be the function which is identically 0, and let $T(X, g)$ be the set of all neighborhoods of g. Let s be a selection function from $T(X, g)$ into X. Prove that the net $\{s_V\}$, $V \in T(X, g)$ has the property of being in every neighborhood of g residually, but that no subsequence of this net has the property; in fact, prove that there are no subsequences of $\{s_V\}$, $V \in T(X, g)$, at all.

5. Find a net in the set N of positive integers which has the property cofinally of being equal to every positive integer. Describe explicitly the subnet of this net which is residually equal to 3.

6.4 CONVERGENCE OF NETS

Thus far we have primarily studied the idea of a net in an arbitrary set without regard to any topological structure that might be on the set. But just as, in Chapter 2, we were interested in the convergence of sequences in metric spaces, so now we are interested in finding a notion of convergence for nets in general topological spaces which generalizes the notion of convergence of sequences. We note that the criterion for convergence of a sequence in a metric space can be restated: A sequence $\{s_n\}$, $n \in N$, converges to y if and only if given any neighborhood U of y, then $\{s_n\}$, $n \in N$, is residually in U. We therefore make the following definition.

Definition 3. Let X, τ be any topological space, and suppose $\{s_i\}$, $i \in I$, is a net in X. Then $\{s_i\}$, $i \in I$, is said to *converge* to a point y of X if $\{s_i\}$, $i \in I$, is residually in every neighborhood of y. If $\{s_i\}$, $i \in I$, converges to y, we write $s_i \to y$. In other words, $s_i \to y$ if given any neighborhood U of y, there is $i_0 \in I$ such that if $i_0 \leq i$, $s_i \in U$. The point y is called the *limit* of $\{s_i\}$, $i \in I$.

Example 9. If X is any topological space, $x \in X$, and $T(X, x)$ is the directed set of neighborhoods of x, then for any selection function $s: T(X, x) \to X$, the net $\{s_U\}$, $U \in T(X, x)$, converges to x. For let V be any neighborhood of x. Then since $s_U \in U$ for each $U \in T(X, x)$ and $V \leq U$ means $U \subset V$, if $V \leq U$, $s_U \in V$.

Example 10. Again consider Example 1. We have seen that no sequence of elements of A converges to g. We now show that there is a net $\{s_i\}$, $i \in I$, such that $s_i \to g$ and $s_i \in A$ for each $i \in I$. Let $T(X, g)$ be as defined previously. We wish to prove the existence of a selection function s from $T(X, g)$ into X such that

$$s(W) \in W \cap A \qquad \text{for each} \quad W \in T(X, g).$$

This will be accomplished if we show that for any finite subset F of R, the set of real numbers, and for any positive number p,

$$U(g, F, p) \cap A \neq \phi$$

(for the family of $U(g, F, p)$ is $\mathfrak{N}_g$, and hence any neighborhood of g contains a neighborhood of this form). This was already done, however, in proving that $g \in \text{Cl } A$. Therefore the net $\{s_W\}$, $W \in T(X, g)$, where $s_W \in W \cap A$, converges to g. We thus see that even though no sequence of elements of A converges to $g \in \text{Cl } A$, there is a net of elements of A which converges to g. We seem therefore to be well on our way to generalizing Proposition 1.

The reader may feel that at least sequences are sufficient for doing whatever has to be done pertaining to the ordinary space R of real numbers (that is, R with the absolute value metric), and that nets are only of use in dealing with "screwball" topological spaces such as that given in Example 1. This is not at all the case, but to help convince the reader that nets are of great value even in real analysis, we give the following example.*

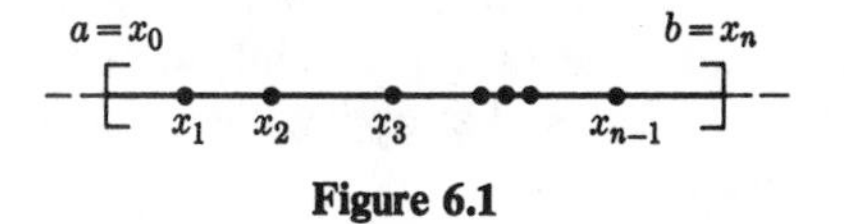

Figure 6.1

Example 11. Let $[a, b]$ be a closed interval in R, the space of real numbers with the absolute value metric. A *partition* P of $[a, b]$ is a finite collection of points

$$x_0 = a < x_1 < x_2 < \cdots < x_{n-1} < x_n = b$$

---

* Actually, Riemann integrals can be adequately handled entirely in terms of sequences, but the use of nets is more elegant. Many important notions, however, depending on the concept of convergence cannot be handled adequately without appeal to something more general than sequences.

(Fig. 6.1). The *mesh* of P, denoted by $m(P)$, is defined to be

$$\max(|x_i - x_{i+1}|,\ i = 0, \ldots, n-1, \text{ where } n \text{ is the number of points in the partition});$$

thus the maximum mesh of any partition of $[a, b]$ would be $b - a$. Suppose P_1 and P_2 are two partitions of $[a, b]$. Then P_1 is said to be *finer* than P_2, if $P_2 \subset P_1$; if $P_2 \subset P_1$, then $m(P_1) \leq m(P_2)$, since P_1 has at least as many points as P_2. Set $P_1 \leq P_2$ if P_1 is finer than P_2. If we let $\mathcal{P}$ denote the family of all partitions of $[a, b]$, the reader can show that $\leq$ makes $\mathcal{P}$ into a directed set.

Let f be any function from $[a, b]$ into R. For any partition P, say P consists of

$$x_0 = a < x_1 < \cdots < x_{n-1} < x_n = b,$$

define

$$s(f, P) = \sum_{i=0}^{n-1} f(x_i)(x_{i+1} - x_i)$$

and

$$S(f, P) = \sum_{i=0}^{n-1} f(x_{i+1})(x_{i+1} - x_i).$$

Then $S(f, -)$ and $s(f, -)$ define nets in R, that is,

$$\{S(f, P)\},\quad P \in \mathcal{P}, \qquad \text{and} \qquad \{s(f, P)\},\quad P \in \mathcal{P}.$$

If the former net converges, its limit is called the *upper Riemann integral* of f over $[a, b]$; if the latter net converges, its limit is called the *lower Riemann integral* of f over $[a, b]$. If both nets converge to a common limit, this limit is called the *Riemann integral* of f over $[a, b]$, commonly denoted by $\int_a^b f(x)\, dx$.

Admittedly, this example has been somewhat sketchy. The interested reader, however, can find this topic developed at length in most books in real analysis. It should indicate, though, that nets do furnish a powerful and effective tool in defining and studying a concept known to the reader from elementary calculus.

We now prove some of the more fundamental properties of the convergence of nets.

Proposition 4. Suppose $\{s_i\}$, $i \in I$, is a net in X such that $\{s_i\}$, $i \in I$, is residually constant; that is, there is $y \in X$ and $i_0 \in I$ such that if $i_0 \leq i$, $s_i = y$. Then $s_i \to y$.

Proof. Since any neighborhood of y contains y, $\{s_i\}$, $i \in I$, is residually in any neighborhood of y; hence $\{s_i\}$, $i \in I$, converges to y.

Proposition 5. If $s_i \to y$, then every subnet of $\{s_i\}$, $i \in I$, also converges to y.

Proof. Since $s_i \to y$, $\{s_i\}$, $i \in I$, has the property of being in every neighborhood of y residually. Then, by Proposition 3, every subnet of $\{s_i\}$, $i \in I$, also has the property of being in every neighborhood of y residually. Therefore every subnet of $\{s_i\}$, $i \in I$, converges to y.

Proposition 6. If every subnet of a net $\{s_i\}$, $i \in I$, has a subsubnet which converges to y, then $s_i \to y$. This is to say that if $\{s_i\}$, $i \in I$, does not converge to y, then there is a subnet of $\{s_i\}$, $i \in I$, *no* subnet of which converges to y.

Proof. Since $\{s_i\}$, $i \in I$, does not converge to y, there is a neighborhood U of y such that there does not exist any $i_0 \in I$ such that $i_0 \leq i$ implies $s_i \in U$. Let $J = \{j \in I \mid s_j \notin U\}$, and let k be the identity mapping from J into I. Then $s \circ k$ is a subnet of $\{s_i\}$, $i \in I$ (Exercise 1). But each s_{k_j} is not an element of U. Therefore there cannot be a subnet of $\{s_{k_j}\}$, $j \in J$, which converges to y.

We now prove the long-awaited generalization of Proposition 1.

Proposition 7. If A is any subset of a topological space X, τ, then

$$x \in \operatorname{Cl} A$$

if and only if there is a net $\{s_i\}$, $i \in I$, such that

$$s_i \to x \quad \text{and} \quad s_i \in A \qquad \text{for each} \quad i \in I.$$

Proof. Suppose first there is a net $\{s_i\}$, $i \in I$, such that

$$s_i \to x \quad \text{and} \quad s_i \in A \qquad \text{for each} \quad i \in I.$$

Then each neighborhood of x contains at least one point of A. Therefore $x \in \operatorname{Cl} A$.

Suppose $x \in \operatorname{Cl} A$. Let $T(X, x)$ be the directed set of neighborhoods of x, and let s be a selection function from $T(X, x)$ into X such that

$$s(W) \in W \cap A \qquad \text{for each} \quad W \in T(X, x).$$

We know that such a selection function exists because $x \in \operatorname{Cl} A$; hence every neighborhood of x contains some point of A. Then the net $\{s_W\}$, $W \in T(X, x)$, converges to x (Example 9).

Proposition 7 strengthens our opinion that we have not only generalized sequences properly, but have also generalized the notion of convergence properly.

It was shown that any sequence which converges in a metric space converges to a unique limit (Proposition 6, Chapter 3). We might then wonder, In what types of spaces do convergent nets have unique limits? The next proposition answers this question.

Proposition 8. A space X, τ is T_2 if and only if given any convergent net $\{s_i\}$, $i \in I$, in X, the limit of $\{s_i\}$, $i \in I$, is unique.

Proof. Suppose X, τ is T_2, but that there is some net $\{s_i\}$, $i \in I$, in X such that $\{s_i\}$, $i \in I$, converges to distinct points x and y. Since X is T_2, there are neighborhoods U and V of x and y, respectively, such that $U \cap V = \phi$. Since $s_i \to x$ and $s_i \to y$, $\{s_i\}$, $i \in I$, is residually in both U and V. Therefore there are i_0 and i_0' such that $i_0 \leq i$ implies $s_i \in U$ and $i_0' \leq i$ implies $s_i \in V$. Since I is a directed set, there is $j \in I$ such that $i_0 \leq j$ and $i_0' \leq j$. Therefore $s_j \in U \cap V$, a contradiction.

Suppose X, τ is not T_2. Then there are distinct points x and y of X such that every neighborhood of x meets every neighborhood of y. Let $T(X, x)$ and $T(X, y)$ be the directed sets of neighborhoods of x and y. Then $T(X, x) \times T(X, y)$ is a directed set [directed by defining $(U, V) \leq (U', V')$ if $U' \subset U$ and $V' \subset V$ as in Section 6.2, Exercise 3]. Since $U \cap V \neq \phi$ for each $(U, V) \in T(X, x) \times T(X, y)$, there is a selection function

$$s: T(X, x) \times T(X, y) \to X$$

such that $s(U, V) \in U \cap V$. Then

$$\{s_{(U,V)}\},\ (U, V) \in T(X, x) \times T(X, y),$$

is a net in X which converges to both x and y (Fig. 6.2).

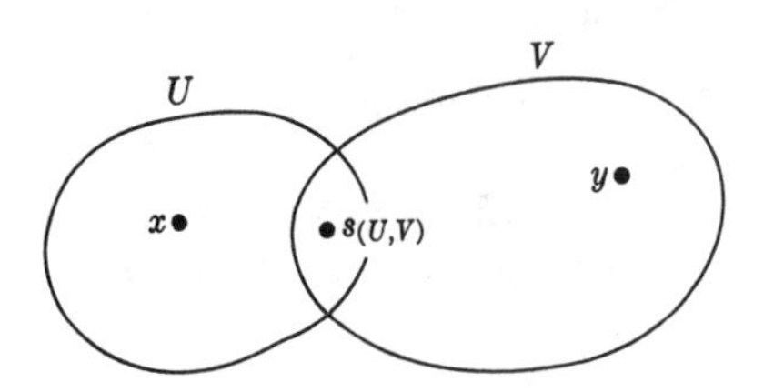

Figure 6.2

Example 12. Lest it seem peculiar to the reader that a net should be able to converge to several points, let him remember that if X is a set with the trivial topology, then any sequence $\{s_n\}$, $n \in N$, in X converges to *every* point of X. For if $x \in X$, then the only neighborhood of x is X, and every sequence in X is residually in X. Admittedly, however, the nicest spaces are those where convergent nets have unique limits. This is why many topologists restrict their attention only to spaces which are at least T_2.

EXERCISES

1. In Proposition 6, prove that $s \circ k$ is a subnet of $\{s_i\}$, $i \in I$.
2. Suppose X, D is a metric space and $\{s_i\}$, $i \in I$, is a net in X.
   a) Suppose $s_i \to x$. Prove that a subsequence of $\{s_i\}$, $i \in I$, converges to x.
   b) Prove that if every subsequence of $\{s_i\}$ converges to x, then $s_i \to x$.
   c) Prove (a) and (b) when it is merely assumed that X, τ is a first countable space (Section 6.1, Exercise 2).
3. Let X be a set. Suppose a "rule" of convergence is given which satisfies
   i) $s_i \to x$ implies every subnet of $\{s_i\}$, $i \in I$, converges to x, and
   ii) if a net $\{s_i\}$, $i \in I$, is residually constantly equal to y, then $s_i \to y$.

   Define a subset A of X to be closed if and only if for each net $\{s_i\}$, $i \in I$, such that $s_i \in A$ for each $i \in I$ and $s_i \to y$, $y \in A$.
   a) Show that the set of closed subsets of X defines a topology τ on X.
   b) Show that each net which converges according to the original rule of convergence also converges with respect to τ.
   c) Show that some net which did not converge with respect to the original rule might still converge with respect to τ.
4. Prove that the only nets which converge in a space with the discrete topology are nets which are residually constant.
5. Let N be the set of positive integers. Discuss the convergence of nets in N when N is given each of the following topologies.
   a) $\tau = \{N, \phi, \{0\}\}$
   b) $\tau = \{U \subset N \mid U \text{ contains all but finitely many elements of } N\}$
   c) the subspace topology from R with the absolute value topology
6. Let N, the set of positive integers, have the topology in Problem 5(b). Show that every net in N has a convergent subnet.
7. Suppose X is any set and τ and τ' are two possible topologies for X. Prove that $\tau' \subset \tau$ if and only if every net in X which converges with respect to τ also converges with respect to τ'.
8. Find a criterion in terms of nets for a space to be T_1. Do likewise for T_0.

6.5 LIMIT POINTS

Definition 4. Let $\{s_i\}$, $i \in I$, be any net in a space X, τ. A point y of X is said to be a *limit point* of $\{s_i\}$, $i \in I$, (not to be confused with limit) if $\{s_i\}$, $i \in I$, is *cofinally* in every neighborhood of y. That is, y is a limit point of $\{s_i\}$, $i \in I$, if given any neighborhood U of y and any elements i and i' of I, there is $i'' \in I$ such that $i \leq i''$, $i' \leq i''$, and $s_{i''} \in U$.

Note that if $s_i \to y$, then y is a limit point of $\{s_i\}$, $i \in I$. On the other hand, a net need not converge to a limit point, as the following example demonstrates.

Example 13. Let R be the space of real numbers with the absolute value topology. Let $\{s_n\}$, $n \in N$, be the sequence in R defined by $s_n = (-1)^n$. If n is odd, then $s_n = -1$, and if n is even, $s_n = 1$. Then $\{s_n\}$, $n \in N$, has both -1 and 1 as limit points. For if U is any neighborhood of 1 and m and m' are any two elements of N, then there is an even integer m'' greater than both m and m', and $s_{m''} \in U$. Thus 1 is a limit point of $\{s_n\}$, $n \in N$; similarly, -1 is also a limit point. We note that even though the space involved is T_2, a net, here a sequence, may have a number of different limit points.

However, $\{s_n\}$, $n \in N$, does not converge to either 1 or -1. For suppose $s_n \to 1$. Since R is T_2, we may find neighborhoods U and V of 1 and -1, respectively, such that $U \cap V = \phi$. Then $\{s_n\}$, $n \in N$, is residually in U, a contradiction to the fact that $\{s_n\}$, $n \in N$, is cofinally constantly -1 and $-1 \notin U$. Similarly, $s_n \not\to -1$.

We note, however, that if we let N' represent the set of positive even integers, N'' represent the set of positive odd integers, and k' and k'' be the identity mappings from N' into N and N'' into N, respectively, then $s \circ k'$ and $s \circ k''$ are subsequences of $\{s_n\}$, $n \in N$, which converge to 1 and -1, respectively. We might conjecture then that even though a net need not converge to one of its limit points, some subnet of that net might. We prove this conjecture in the next proposition.

> **Proposition 9.** Let X, τ be a topological space and $\{s_i\}$, $i \in I$, be a net in X. Then y is a limit point of $\{s_i\}$, $i \in I$, if and only if $\{s_i\}$, $i \in I$, has a subnet which converges to y.

Proof. Suppose $\{s_i\}$, $i \in I$, has a subnet which converges to y. Then there is a directed set J and a function k from J into I as in Definition 2 such that $s \circ k$ is a subnet of $\{s_i\}$, $i \in I$, and $s_{k_j} \to y$. Suppose U is any neighborhood of y. Then $\{s_{k_j}\}$, $j \in J$, is residually in U, that is, there is $j_0 \in J$ such that $j_0 \leq j$ implies $s_{k_j} \in U$. Suppose i and i' are any two elements of I. Then since I is directed, there is $i'' \in I$ such that $i \leq i''$ and $i' \leq i''$. But $k(j_0)$ is an element of I, and $\{s_{k_j}\}$, $j \in J$, is cofinal in $\{s_i\}$, $i \in I$. We can therefore find $j' \in J$ such that

$$k(j_0) \leq k(j') \qquad \text{and} \qquad i'' \leq k(j').$$

Since k is order-preserving, $j_0 \leq j'$; hence $s_{k_{j'}} \in U$. Since $\leq$ is transitive, $i \leq k(j')$ and $i' \leq k(j')$. In sum then, given i and i' in I, we have found $k(j') \in I$ such that

$$i \leq k(j'), \qquad i' \leq k(j'), \qquad \text{and} \qquad s_{k_{j'}} \in U.$$

Therefore $\{s_i\}$, $i \in I$, is cofinally in U, and hence y is a limit point of $\{s_i\}$, $i \in I$.

Suppose now that $\{s_i\}$, $i \in I$, has y as a limit point. Let J be the set of ordered pairs (i, U) such that $s_i \in U$, where U is a neighborhood of y. The set J may be ordered by letting

$$(i, U) \leq (i', U') \qquad \text{if} \quad i \leq i' \quad \text{and} \quad U' \subset U$$

[note that this is the ordering induced from $I \times T(X, y)$]. It is easily seen that J is partially ordered (Exercise 1). We must, however, show that J is directed. Let (i, U) and (i', U') be elements of J. Then $U \cap U'$ is a neighborhood of y with

$$U \cap U' \subset U \qquad \text{and} \qquad U \cap U' \subset U'.$$

Since I is directed, and since by assumption y is a limit point of $\{s_i\}$, $i \in I$, there is $i'' \in I$ such that $i \leq i''$, $i' \leq i''$, and $s_{i''} \in U \cap U'$. Therefore

$$(i'', U \cap U') \in J, \qquad (i, U) \leq (i'', U \cap U'),$$

and

$$(i', U') \leq (i'', U \cap U').$$

Hence J is a directed set.

Define the function k from J into I by $k(i, U) = i$. It is trivial that k satisfies condition (i) of Definition 2. Let i and i' be elements of I and let U be any neighborhood of y. Since y is a limit point of $\{s_i\}$, $i \in I$, there is $i'' \in I$ such that $i \leq i''$, $i' \leq i''$, and $s_{i''} \in U$. Then

$$(i'', U) \in J, \qquad i \leq k(i'', U),$$

and

$$i' \leq k(i'', U).$$

Therefore condition (ii) of Definition 2 is satisfied, and $s \circ k$ is a subnet of $\{s_i\}$, $i \in I$. We now show that $s_{k(i,U)} \to y$.

Let U be any neighborhood of y. Take any i_0 such that $s_{i_0} \in U$. Then $(i_0, U) \in J$. If $(i_0, U) \leq (i, V)$, then $V \subset U$ and $s_i \in V$; therefore

$$s_{k(i,V)} \in V \subset U.$$

Thus $s \circ k$ is residually in U, and hence $s \circ k$ converges to y.

Corollary. If a subnet of $\{s_i\}$, $i \in I$, has y as a limit point, then so does $\{s_i\}$, $i \in I$.

The proof is left as an exercise.

EXERCISES

1. Prove that J in Proposition 9 is partially ordered.

2. Prove the corollary to Proposition 9.
3. Let $\{s_i\}$, $i \in I$ be a net in a space X, τ and let A be the set of limit points of $\{s_i\}$, $i \in I$. Need $\{s_i \mid i \in I\} \cup A$ be closed?
4. Find all the limit points of each of the following sequences.
   a) $s_n = 1/n$ in the set of real numbers with the order topology
   b) $s_n = (-1)^n$ in the set of real numbers with the trivial topology
   c) $s_n = (-1)^n$ in the set of real numbers with the discrete topology
   d) $s_n = n$ in the set of real numbers with the absolute value topology
5. Let X, τ be a space with the property that any net in X which has a limit point converges to that limit point. Discuss the various possibilities for the topology on X.
6. Let R be the set of real numbers with the absolute value topology. A sequence $\{s_n\}$, $n \in N$, is said to be *bounded* if there are real numbers m and M such that $m \leq s_n \leq M$ for all $n \in N$. Prove that any bounded sequence in R has a limit point. Prove that every convergent sequence in R is bounded, but that not every bounded sequence is convergent.
7. Prove or disprove: There exists a sequence of real numbers which has every real number as a limit point. Prove or disprove: There exists a net of real numbers which has R as its set of limit points.

6.6 CONTINUITY AND CONVERGENCE

The following proposition relates the convergence of nets and the continuity of functions. It is a generalization of Proposition 10 of Chapter 2, thus strengthening the assertion that nets are a good generalization of sequences.

> **Proposition 10.** Let f be a function from a space X, τ to a space Y, τ'. Then f is continuous if and only if for every net $\{s_i\}$, $i \in I$, in X such that $s_i \to x$, the net $\{f(s_i)\}$, $i \in I$, in Y converges to $f(x)$.

Proof. Suppose f is continuous, but also suppose there is a net $\{s_i\}$, $i \in I$, in X such that $s_i \to x$, but $\{f(s_i)\}$, $i \in I$, does not converge to $f(x)$. Then there is a subnet of $\{f(s_i)\}$, $i \in I$, no subnet of which converges to $f(x)$ (Proposition 6). Let

$$\{f(s_{k_j})\}, \quad j \in J,$$

be such a subnet. Then there is a neighborhood V of $f(x)$ for which $\{f(s_{k_j})\}$, $j \in J$, is residually not in V. But since f is continuous, $f^{-1}(V)$ is a neighborhood of x. Since $s_i \to x$, $\{s_i\}$, $i \in I$, is residually in $f^{-1}(V)$. But $\{s_{k_j}\}$, $j \in J$, is a subnet of $\{s_i\}$, $i \in I$, which is not residually in $f^{-1}(V)$. Therefore $\{s_i\}$, $i \in I$, has a subnet which does not converge to x, a contradiction to Proposition 5.

Suppose for each net $\{s_i\}$, $i \in I$, in X such that $s_i \to x$, $f(s_i) \to f(x)$, but f is not continuous. Then there is a neighborhood V of $f(x)$ such that for no neighborhood U of x do we have $f(U) \subset V$. Let $T(X, x)$ be the directed set of all neighborhoods of x. Let s be a selection function from $T(X, x)$ into X such that $f(s(U)) \notin V$ for all $U \in T(X, x)$. Then $s_U \to x$, but $f(s_U) \not\to f(x)$, a contradiction.

We now use Proposition 10 to prove several results about nets in derived topological spaces.

Proposition 11. Let X, τ be any space and R be an equivalence relation on X. For each $x \in X$, let $\bar{x}$ denote the equivalence class of x. Then if $s_i \to x$ in X, $\bar{s}_i \to \bar{x}$ in the identification space X/R.

Proof. The function defined by $x \to \bar{x}$ is continuous. We then apply Proposition 10.

Proposition 12. Suppose $\{s_i\}$, $i \in I$, is a net in the product space $\times_J X_j$. We will denote the jth coordinate of s_i by s_i^j; thus $\{s_i^j\}$, $i \in I$, will be a net in X_j. Then

$$s_i \to y = (y_1, y_2, \ldots, y_j, \ldots)$$

(remember we have restricted our attention in this text to the product of countably many spaces) if and only if

$$s_i^j \to y_j.$$

Proof. Suppose $s_i \to y$. Then the projection mapping from $\times_J X_j$ into X_j, τ_j is continuous. Therefore, by Proposition 10,

$$p_j(s_i) = s_i^j \to p_j(y) = y_j.$$

Suppose $s_i^j \to y_j$ for each $j \in J$. Let U be a typical basic neighborhood for y in the product topology. Then $U = \times_J W_j$, where each W_j is an open subset of X_j; in particular, each W_j is a neighborhood of y_j. Also $W_j = X_j$ for each $j \in J$, except finitely many, say $j_1, \ldots, j_n$. For each $j \in J$, except $j_1, \ldots, j_n$, $s_i^j \in W_j$. For $j_1, \ldots, j_n$ we can find $i_1, \ldots, i_n$ in I such that if $i_q \leq i$, $s_i^{j_q} \in W_{j_q}$, $q = 1, \ldots, n$. Since I is directed, we can find, using induction if necessary, $i_0 \in I$ such that if $i_0 \leq i$, $s_i^{j_q} \in W_{j_q}$, $q = 1, \ldots, n$. Thus if $i_0 \leq i$, $s_i^j \in W_j$ for each $j \in J$, and hence if $i_0 \leq i$,

$$s_i \in \mathop{\times}_J W_j.$$

Therefore $\{s_i\}$, $i \in I$, is residually in $\times_J W_j$. Since $\times_J W_j$ was a typical basic neighborhood and every neighborhood of y contains such a basic neighborhood, $\{s_i\}$, $i \in I$, is residually in every neighborhood of y; therefore $s_i \to y$.

It is not true that if Y is a subspace of X, $\{s_i\}$, $i \in I$, converges in X, and $s_i \in Y$ for each $i \in I$, then $\{s_i\}$, $i \in I$, *considered as a net in* Y also converges. This is not even true for sequences, as we see from the following example.

Example 14. Let R be the set of real numbers with the absolute value topology. Then the sequence defined by $s_n = 1/n$ converges to 0 in R, but does not converge at all in the subspace $(0, 1)$, since $(0, 1)$ does not contain the limit 0. The following is true, however.

Proposition 13

a) If $s_i \to y$ in X and Y is a subspace of X such that $s_i \in Y$ for each i and $y \in Y$, then $s_i \to y$ in Y also.

b) If $s_i \to y$ in X and Y is a closed subspace of X, then if each $s_i \in Y$, then $y \in Y$ as well, and $s_i \to y$ in Y.

Proof. The proof of (a) is left as an exercise. Statement (b) follows immediately from (a) and Proposition 7.

EXERCISES

1. Prove Proposition 13.
2. Let $\{X_j, \tau_j\}$, $j \in J$, be a countable family of nonempty spaces, and consider the product $\times_J X_j$ of the sets $\{X_j\}$, $j \in J$. How much of Proposition 12 is true if $\times_J X_j$ is given a topology coarser than the product topology? How much of Proposition 12 is true if $\times_J X_j$ is given a topology finer than the product topology?
3. Prove or disprove that a function f from a space X, τ onto a space Y, τ^1 is a homeomorphism if and only if a net $\{s_i\}$, $i \in I$, converges to $x \in X$ if and only if $\{f(s_i)\}$, $i \in I$, converges to $f(x)$ in Y.
4. Using the results of this chapter, prove that if f is a continuous function from X, τ into Y, τ', then $f(\text{Cl } A) \subset \text{Cl } f(A)$ for any $A \subset X$.
5. Also using methods from this chapter, prove Proposition 21, Chapter 4.
6. Suppose f is a continuous function from X, τ into Y, τ'. Let $\{s_i\}$, $i \in I$, be a net in X, and suppose A is the set of limit points of this net. Prove that $f(A)$ is a set of limit points for $\{f(s_i)\}$, $i \in I$. Is it necessarily a complete set of limit points for $\{f(s_i)\}$, $i \in I$, or might there be others as well? [*Hint:* Consider the sequence defined by $s_n = 1/n$ in $(0, 1)$, and the identity function from $(0, 1)$ into the space of real numbers.]
7. Let X/R be an identification space of a space X. Discuss the conditions R must satisfy for the following to hold: For any net $\{s_i\}$, $i \in I$, in X which converges to a point x, $\{\bar{s}_i\}$, $i \in I$, the identification net in X/R converges *only* to $\bar{x}$.

6.7 FILTERS

There is an alternative approach to the concept of convergence in a general topological space through the notion of a *filter*. While the study of filters is of great importance in point set topology, we will accomplish much of what filters might be useful for by using nets. Nevertheless, we will introduce the notion of a filter now and study some of its basic properties for two reasons: (1) the notion of a filter is sufficiently important that anyone studying even introductory point set topology should at least know what a filter is; and (2) the reader should come to realize that, even in mathematics, there may be many means to the same end. A proposition in mathematics may have many proofs, and different machinery can be developed to accomplish the same task. We will in this section try to stress the relations between nets and filters and the analogies in their use. We would expect that since nets and filters are both designed for the study of convergence, there will have to be many theorems about filters completely analogous to theorems stated in terms of nets.

Definition 5. Let X be any set. A collection $\mathcal{A}$ of nonempty subsets of X is said to be a *filter on* X if

i) $\mathcal{A} \neq \phi$;

ii) if A and B are in $\mathcal{A}$, then $A \cap B$ is also in $\mathcal{A}$;

iii) if $A \in \mathcal{A}$ and $A \subset B$, then $B \in \mathcal{A}$.

If X, τ is a topological space and $\mathcal{A}$ is a filter on X, then $\mathcal{A}$ is said to *converge* to x, denoted by $\mathcal{A} \to x$, if every neighborhood of x is a member of $\mathcal{A}$. $\mathcal{A}$ is said to have x as a *limit point* if every neighborhood of x meets every member of $\mathcal{A}$. That is, $\mathcal{A} \to x$ if given any neighborhood U of x, $U \in \mathcal{A}$. x is a limit point of $\mathcal{A}$ if given any neighborhood U of x, and any $A \in \mathcal{A}$, then $U \cap A \neq \phi$.

Example 15. If X is any set and Y is any nonempty subset of X, then the family $\mathcal{A}$ of all subsets of X which contain Y is a filter on X. We verify that $\mathcal{A}$ satisfies Definition 5. Since $Y \neq \phi$, each set which contains Y is nonempty.

i) Since $Y \subset Y$, $Y \in \mathcal{A}$; hence $\mathcal{A} \neq \phi$.

ii) If A and B are in $\mathcal{A}$, then $Y \subset A$ and $Y \subset B$; thus $Y \subset A \cap B$, and therefore $A \cap B \in \mathcal{A}$.

iii) If $A \in \mathcal{A}$, then $Y \subset A$. Therefore if $A \subset B$, then $Y \subset A \subset B$, and thus $B \in \mathcal{A}$. Hence $\mathcal{A}$ is a filter on X.

If X, τ is a space and $x \in X$, then $T(X, x)$, the family of all neighborhoods of x, is not a filter on X, since given any neighborhood U of x, it is not necessarily true that any subset of X which contains U is also a neighborhood of x. If we let $T^*(X, x)$ be the family of all subsets A of X such that A contains a neighborhood of x, then $T^*(X, x)$ is a filter on x. Moreover, since $T(X, x) \subset T^*(X, x)$,

$$T^*(X, x) \to x.$$

(Compare this to Examples 3, 6, and 9.)

Example 16. Let X be any set and let $\mathfrak{D}$ be a nonempty collection of nonempty subsets of X with the property that if B and B' are in $\mathfrak{D}$, then there exists $B'' \in \mathfrak{D}$ such that $B'' \subset B \cap B'$. Let

$$\mathfrak{A} = \{A \mid B \subset A,\, B \in \mathfrak{D}\}.$$

Then $\mathfrak{A}$ is a filter on X (Exercise 1). $\mathfrak{A}$ is said to be the filter *generated* by $\mathfrak{D}$ and $\mathfrak{D}$ is said to be a *basis* for the filter $\mathfrak{A}$.

Example 17. Suppose $\{s_i\}$, $i \in I$, is a net in a set X. Let $\mathfrak{D}$ be the family of all subsets of the form $B_j = \{s_i \mid j \leq i\}$, for all $j \in I$. Then the family $\mathfrak{D}$ is a nonempty collection of nonempty subsets of X having the property that if B_j and $B_{j'}$ are members of $\mathfrak{D}$, then there is $B_{j''} \in \mathfrak{D}$ such that

$$B_{j''} \subset B_j \cap B_{j'}$$

(Exercise 2). $\mathfrak{D}$ thus forms the basis for a (unique) filter $\mathfrak{A}$ as in Example 16; specifically,

$$\mathfrak{A} = \{A \mid B_j \subset A \text{ for some } j \in I\}.$$

$\mathfrak{A}$ is said to be the filter *generated* by the net $\{s_i\}$, $i \in I$.

Proposition 14. Let $\{s_i\}$, $i \in I$, be a net in a space X, τ and let $\mathfrak{A}$ be the filter generated by $\{s_i\}$, $i \in I$. Then

a) $\mathfrak{A} \to x$ if and only if $s_i \to x$;
b) $\mathfrak{A}$ has x as a limit point if and only if x is a limit point of $\{s_i\}$, $i \in I$.

Proof

a) Suppose $s_i \to x$. Then given any neighborhood U of x, there is $j \in I$ such that B_j (using the notation of Example 17) is a subset of U. Since $B_j \subset U$, $U \in \mathfrak{A}$. Therefore every neighborhood of x is in $\mathfrak{A}$, and hence $\mathfrak{A} \to x$.

Suppose $\mathfrak{A} \to x$. Then given any neighborhood U of x, there is $j \in I$ such that $B_j \subset U$. Hence if $j \leq i$, $s_i \in U$. Thus $\{s_i\}$, $i \in I$, is residually in every neighborhood of x, and therefore $s_i \to x$.

The proof of (b) is left as an exercise.

In Example 17 we associated a filter with any net. In order for this association to be at all meaningful, Proposition 14 was a necessity. For if Proposition 14 were not true, then a net might converge without the corresponding filter converging, or we might have net and filter converging to different points. But if nets and filters are merely to be different approaches to the same concept, this would be intolerable. We now show that starting with a filter $\mathcal{A}$, we can associate a net with $\mathcal{A}$ which has the same convergence properties as $\mathcal{A}$.

Let $\mathcal{A}$ be a filter on a set X. We will construct a net *based* on $\mathcal{A}$. $\mathcal{A}$ is a collection of nonempty subsets of X. If A and B are in $\mathcal{A}$, let

$$A \leq B \qquad \text{if} \quad B \subset A.$$

It is easy to verify that this makes $\mathcal{A}$ into a directed set. Since each $A \in \mathcal{A}$ is a nonempty set, we can find a selection function s from the directed set $\mathcal{A}$ into X such that $s(A) \in A$. Then $\{s_A\}$, $A \in \mathcal{A}$, is a net in X called a *net based on* $\mathcal{A}$.

Proposition 15. $\mathcal{A} \to y$ if and only if every net $\{s_A\}$, $A \in \mathcal{A}$, based on $\mathcal{A}$ also converges to y.

Proof. Suppose $\mathcal{A} \to y$ and $\{s_A\}$, $A \in \mathcal{A}$, is a net based on $\mathcal{A}$. Let U be any neighborhood of y; since $\mathcal{A} \to y$, $U \in \mathcal{A}$. Then if $U \leq A$, $A \subset U$, for any $A \in \mathcal{A}$. Hence if $U \leq A$, $s_A \in A \subset U$; therefore $\{s_A\}$, $A \in \mathcal{A}$, is residually in U. Thus $s_A \to y$.

Suppose $\mathcal{A} \not\to y$. Then there is a neighborhood U of y which is not a member of $\mathcal{A}$. If $A \in \mathcal{A}$, select $s_A \in A - U$; such a selection is always possible, for if $A - U = \phi$, then $A \subset U$, which would make U a member of $\mathcal{A}$. Then $\{s_A\}$, $A \in \mathcal{A}$, is a net based on $\mathcal{A}$ which does not converge to y.

Proposition 16. Suppose $\mathcal{A}$ is a filter on a space X, τ. Then x is a limit point of $\mathcal{A}$ if and only if there is a filter $\mathcal{A}'$ such that $\mathcal{A} \subset \mathcal{A}'$ and $\mathcal{A}' \to x$.

Proof. Suppose x is a limit point of $\mathcal{A}$. Let

$$\mathfrak{D}' = \{A \cap U \mid A \in \mathcal{A} \text{ and } U \text{ is a neighborhood of } x\}.$$

Then $\mathfrak{D}'$ is the basis for a filter $\mathcal{A}'$ on X (Exercise 3). Since $A \cap U \subset U$ for each $A \in \mathcal{A}$ and any neighborhood U of x, $U \in \mathcal{A}'$. Therefore every neighborhood of x is a member of $\mathcal{A}'$, and thus $\mathcal{A}' \to x$. It remains to be shown that $\mathcal{A} \subset \mathcal{A}'$. This follows at once from the fact that X is a neighborhood of x; hence if $A \in \mathcal{A}$, $A \cap X = A$ is a member of $\mathfrak{D}'$, and hence of $\mathcal{A}'$.

Suppose on the other hand that there is a filter $\mathcal{A}'$ such that $\mathcal{A} \subset \mathcal{A}'$ and $\mathcal{A}' \to x$. Let U be any neighborhood of x and A be an element of $\mathcal{A}$. Then U and A are both members of $\mathcal{A}'$; hence

$$A \cap U \in \mathcal{A}'$$

by (ii) of Definition 5. Since $A \cap U \in \mathcal{A}'$, then $A \cap U \neq \phi$. Therefore, given any neighborhood U of x, and any member A of $\mathcal{A}$, $A \cap U \neq \phi$. Hence x is a limit point of $\mathcal{A}$.

Comparing Proposition 16 with Proposition 9, we find that the filter analog of a subnet is a *finer filter*, where a filter $\mathcal{A}'$ is *finer* than a filter $\mathcal{A}$ if $\mathcal{A} \subset \mathcal{A}'$.

EXERCISES

1. Prove that $\mathcal{A}$ in Example 16 is a filter.
2. Prove in Example 17 that there is $B_{j''} \in \mathfrak{D}$ such that $B_{j''} \subset B_j \cap B_{j'}$.
3. In Proposition 16, prove $\mathfrak{D}'$ is a filter basis.
4. Prove (b) in Proposition 14.
5. Suppose f is a function from a space X, τ into a space Y, τ' and $\mathcal{A}$ is a filter on X.
   a) Set $f(\mathcal{A}) = \{f(A) \mid A \in \mathcal{A}\}$. Prove that $f(\mathcal{A})$ is the basis for a filter $\mathcal{A}'$ on Y.
   b) Prove that f is continuous if and only if given any filter $\mathcal{A}$ on X such that $\mathcal{A} \to x$, $\mathcal{A}'$ [the filter for which $f(\mathcal{A})$ is a basis] $\to f(x)$.
6. Suppose A is a subset of X, τ. Prove $x \in \text{Cl } A$ if and only if there is a filter $\mathcal{A}$ on X such that $\mathcal{A} \to x$ and $A \in \mathcal{A}$.
7. Prove that a space X, τ is T_2 if and only if any convergent filter on X converges to a unique limit.
8. Which of the following are filters? Which are filter bases? For those which are neither filters nor filter bases, indicate which properties are lacking.
   a) the family of subsets of a set X which contain $Y \subset X$
   b) the set of all closed half-planes in R^2 which contain $(0, 0)$
   c) the set of all open half-planes of R^2
   d) the union of two filters on a set X
9. State and prove the filter analogs of Propositions 4 and 5 of this chapter.
10. Find a criterion in terms of filters for a space to be T_1.

6.8 ULTRANETS AND ULTRAFILTERS

There is another concept involving nets which will prove useful in the discussion of compact spaces; this concept is that of an *ultranet*. Actually, the filter analog of an ultranet is more natural, since it is a bit difficult to motivate the notion of an ultranet, except to say that it works. We therefore introduce the concept of an ultrafilter first and then pass to the net analog.

Definition 6. An *ultrafilter* on a set X is a maximal filter on X. That is, a filter $\mathcal{A}$ on a set X is an ultrafilter on X if given any filter $\mathcal{A}'$ finer than $\mathcal{A}$ (i.e., $\mathcal{A} \subset \mathcal{A}'$), $\mathcal{A} = \mathcal{A}'$.

Example 18. Let X be any set and $x \in X$. Then the family of all subsets of X which contain x forms an ultrafilter $\mathcal{A}$ on X. For if $\mathcal{A}'$ is any filter finer than $\mathcal{A}$ and $A' \in \mathcal{A}'$, then either $x \in A'$ or $x \notin A'$. Now if $x \in A'$, then $A' \in \mathcal{A}$. If $x \notin A'$, then $x \in X - A'$; hence $X - A' \in \mathcal{A} \subset \mathcal{A}'$. But then

$$A' \cap (X - A') = \phi \in \mathcal{A}',$$

a contradiction to the fact that each member of $\mathcal{A}'$ is nonempty. Therefore x is an element of each member of $\mathcal{A}'$; hence $\mathcal{A}' \subset \mathcal{A}$, and consequently

$$\mathcal{A} = \mathcal{A}'.$$

Proposition 17. A necessary and sufficient condition that a filter $\mathcal{A}$ on a set X be an ultrafilter is that given any subset A of X, either

$$A \in \mathcal{A} \qquad \text{or} \qquad X - A \in \mathcal{A}.$$

Proof. Suppose $\mathcal{A}$ is a filter on X with the property that either A or $X - A$ is a member of $\mathcal{A}$ for any $A \subset X$. Suppose $\mathcal{A}'$ is a filter on X which is finer than $\mathcal{A}$. To prove that $\mathcal{A} = \mathcal{A}'$, it will suffice to prove that each element of $\mathcal{A}'$ is also an element of $\mathcal{A}$. Let $A' \in \mathcal{A}'$. If $A' \in \mathcal{A}$, we are done. If $A' \notin \mathcal{A}$, then

$$X - A' \in \mathcal{A} \subset \mathcal{A}'.$$

But then A' and $X - A'$ are both members of $\mathcal{A}'$; hence

$$A' \cap (X - A') = \phi$$

is a member of $\mathcal{A}'$, a contradiction.

Suppose now that $\mathcal{A}$ is an ultrafilter on X, but there is a subset A of X such that neither A nor $X - A$ is a member of $\mathcal{A}$. We will find a filter $\mathcal{A}'$ on X which is strictly finer than $\mathcal{A}$. If $A \cap B \neq \phi$ for each $B \in \mathcal{A}$, then we can take

$$\mathcal{D} = \{A \cap B \mid B \in \mathcal{A}\}$$

as a filter basis for a filter $\mathcal{A}'$ which is strictly finer than $\mathcal{A}$ (since $A \in \mathcal{A}'$, but $A \notin \mathcal{A}$).

Suppose, however, that $A \cap B_1 = \phi$ for some $B_1 \in \mathcal{A}$. We will show that

$$(X - A) \cap B \neq \phi$$

for every $B \in \mathcal{A}$. Suppose $(X - A) \cap B_2 = \phi$ for some $B_2 \in \mathcal{A}$. Then

$$\begin{aligned} B_1 \cap B_2 = {} & ((B_1 \cap B_2) \cap A) \cup ((B_1 \cap B_2) \cap (X - A)) \\ & \subset (B_1 \cap A) \cup (B_2 \cap (X - A)) = \phi, \end{aligned}$$

a contradiction since $B_1 \cap B_2$ is a member of $\mathcal{A}$ and hence must be nonempty. Therefore

$$\mathcal{D} = \{(X - A) \cap B \mid B \in \mathcal{A}\}$$

is a basis for a filter $\mathcal{A}'$ on X which is strictly finer than $\mathcal{A}$ (since $\mathcal{A}'$ contains $X - A$).

Proposition 18. If X is any set, then every filter $\mathcal{A}$ on X is contained in an ultrafilter.

Proof. Consider the family $\mathfrak{A}$ of all filters $\mathcal{A}'$ on X such that $\mathcal{A} \subset \mathcal{A}'$. $\mathfrak{A}$ can be partially ordered by "is finer than." Suppose $\mathcal{C} = \{\mathcal{A}'_k\}$, $k \in K$, is a chain in $\mathfrak{A}$. We now show that

$$\mathcal{D} = \{B \mid B \in \mathcal{A}'_k \text{ for some } k \in K\}$$

is a basis for a filter $\mathcal{A}''$ on X. First, $\mathcal{D}$ is certainly a nonempty collection of nonempty sets, since each $\mathcal{A}_k$ is a nonempty collection of nonempty sets. Suppose B and B' are members of $\mathcal{D}$. Then $B \in \mathcal{A}'_k$ and $B' \in \mathcal{A}'_{k'}$ for some k and k' in K. But $\mathcal{C}$ is a chain, and hence either $\mathcal{A}'_k \subset \mathcal{A}'_{k'}$ or $\mathcal{A}'_{k'} \subset \mathcal{A}'_k$; assume the latter. Then B and B' are both in $\mathcal{A}'_k$, and thus $B \cap B' \in \mathcal{A}'_k$. Hence

$$B \cap B' \in \mathcal{D}.$$

Therefore $\mathcal{D}$ is a basis for a filter $\mathcal{A}''$ on X. Moreover, $\mathcal{A}''$ is clearly finer than any member of $\mathcal{C}$. Hence $\mathcal{A}''$ is an upper bound in $\mathcal{C}$ for $\mathfrak{A}$.

Each chain in $\mathfrak{A}$ thus has an upper bound. Applying Zorn's lemma, $\mathfrak{A}$ therefore contains a maximal element $\overline{\mathcal{A}}$. Then $\overline{\mathcal{A}}$ is an ultrafilter which contains $\mathcal{A}$.

Proposition 19. Suppose f is any function from a set X onto a set Y and $\mathcal{A}$ is an ultrafilter on X. Let $f(\mathcal{A})$ be the filter basis as described in Section 6.7, Exercise 5, and let $\mathcal{A}'$ be the filter on Y that it determines. Then $\mathcal{A}'$ is an ultrafilter on Y.

Proof. Suppose $A \subset Y$. In order to show that $\mathcal{A}'$ is an ultrafilter, we must show that either A or $Y - A$ is a member of $\mathcal{A}'$ (Proposition 17). Since $\mathcal{A}$ is an ultrafilter on X, either $f^{-1}(A)$ or $f^{-1}(Y - A) = X - f^{-1}(A)$ is a member of $\mathcal{A}$. If $f^{-1}(A) \in \mathcal{A}$, then

$$f(f^{-1}(A)) = A \in f(\mathcal{A}) \subset \mathcal{A}'.$$

If $X - f^{-1}(A)$, then $Y - A \in \mathcal{A}'$. Therefore $\mathcal{A}'$ is an ultrafilter on Y.

Proposition 20. If $\mathcal{A}$ is an ultrafilter on a space X, τ and y is a limit point of $\mathcal{A}$, then $\mathcal{A} \to y$.

Proof. Since y is a limit point of $\mathcal{A}$, by Proposition 16 there is a filter $\mathcal{A}'$ finer than $\mathcal{A}$ which converges to y. But since $\mathcal{A}$ is an ultrafilter, $\mathcal{A} = \mathcal{A}'$.

Recall that the net analog of a finer filter is a subnet. The net analog of a member of a filter is, of course, an element of the net. Combining these observations with Proposition 17, we make the following definition.

Definition 7. A net $\{s_i\}$, $i \in I$, in a set X is said to be an *ultranet* if, given any subset A of X, $\{s_i\}$, $i \in I$, is residually in either A or $X - A$.

Example 19. If X is any set and $x \in X$, then if I is any directed set, the net $\{s_i\}$, $i \in I$, defined by $s_i = x$ for all $i \in I$, is an ultranet in X. For if $A \subset X$, either $x \in A$, or $x \in X - A$; hence $\{s_i\}$, $i \in I$, is residually in either A or $X - A$. Note that the filter generated by this ultranet is the ultrafilter of Example 18.

Proposition 21

a) A net is an ultranet if and only if the filter it generates is an ultrafilter.
b) A filter $\mathcal{A}$ is an ultrafilter only if every net based on $\mathcal{A}$ is an ultranet.

Proof

a) Suppose $\{s_i\}$, $i \in I$, is an ultranet in the set X. Let $A \subset X$ and let $\mathfrak{D}$ be as in Example 17. Since $\{s_i\}$, $i \in I$, is an ultranet, there is $j \in I$ such that $B_j \subset A$ or $B_j \subset X - A$ (since $\{s_i\}$, $i \in I$, is residually in either A or $X - A$). Therefore either $A \in \mathcal{A}$, the filter for which $\mathfrak{D}$ is a basis, or $X - A \in \mathcal{A}$; hence $\mathcal{A}$ is an ultrafilter.

   Suppose the filter $\mathcal{A}$ generated by $\{s_i\}$, $i \in I$ is an ultrafilter. Let $A \subset X$. Then since $\mathcal{A}$ is an ultrafilter, either $A \in \mathcal{A}$, or $X - A \in \mathcal{A}$. By definition of $\mathcal{A}$, there must be $j \in I$ such that $B_j \subset A$ or $B_j \subset X - A$. In any case, $\{s_i\}$, $i \in I$, is residually in either A or $X - A$, and is therefore an ultranet on X.

The proof of (b) is left as an exercise.

We now give the ultranet analogs of Propositions 18, 19, and 20.

Proposition 22

a) Every net contains a subnet which is an ultranet.
b) If f is a function from a set X onto a set Y and $\{s_i\}$, $i \in I$, is an ultranet in X, then $\{f(s_i)\}$, $i \in I$, is an ultranet in Y.
c) If $\{s_i\}$, $i \in I$, is an ultranet in a space X, τ and y is a limit point of $\{s_i\}$, $i \in I$, then $s_i \to y$.

Proof

b) Suppose $A \subset Y$. Then $\{s_i\}$, $i \in I$, is residually in either $f^{-1}(A)$ or $X - f^{-1}(A)$. Therefore $\{f(s_i)\}$, $i \in I$, is residually in either A or $Y - A$, and is hence an ultranet in Y.

c) Let U be any neighborhood of y. Since $\{s_i\}$, $i \in I$, is an ultranet, it is residually in either U or $X - U$. Since y is a limit point of $\{s_i\}$, $i \in I$, the net could not be residually in $X - U$. Therefore $\{s_i\}$, $i \in I$, is residually in U; hence $s_i \to y$.

a) Let $\{s_i\}$, $i \in I$, be any net in a set X and let $\mathcal{A}$ be the filter generated by $\{s_i\}$, $i \in I$. By Proposition 18, $\mathcal{A} \subset \mathcal{A}'$, where $\mathcal{A}'$ is an ultrafilter. We first show that $\{s_i\}$, $i \in I$, is cofinally in A for each $A \in \mathcal{A}'$. Let $A \in \mathcal{A}'$. If $\{s_i\}$, $i \in I$, is not cofinally in A, then $\{s_i\}$, $i \in I$, is residually in $X - A$. But then

$$X - A \in \mathcal{A} \subset \mathcal{A}'.$$

Therefore A and $X - A$ are both elements of $\mathcal{A}'$, an impossibility; hence $\{s_i\}$, $i \in I$, is cofinally in A. Let

$$J = \{(i, A) \mid i \in I,\ A \in \mathcal{A}', \text{ and } s_i \in A\}.$$

Since $\mathcal{A}'$ and I are both directed sets ($\mathcal{A}'$ is directed by letting $A \leq A'$ if $A' \subset A$), J is directed. Define $k: J \to I$ by $k(i, A) = i$. Then $s \circ k$ is easily verified to be a subnet of $\{s_i\}$, $i \in I$. By definition of $s \circ k$, $s \circ k$ is residually in each $A \in \mathcal{A}'$. But $\mathcal{A}'$ is an ultrafilter and hence contains any subset of X or its complement. Thus $s \circ k$ is residually in any subset of X or its complement, and is therefore an ultranet in X.

EXERCISES

1. In Proposition 22, verify in the proof of (a) that $s \circ k$ is a subnet of $\{s_i\}$, $i \in I$.
2. Prove (b) of Proposition 21. Show that the converse of (b) is false.
3. a) Let $\{s_i\}$, $i \in I$, be a net in X, and suppose $\mathcal{A}$ is the filter generated by $\{s_i\}$, $i \in I$. Is $\{s_i\}$, $i \in I$, then a net based on $\mathcal{A}$? Is it the only net based on $\mathcal{A}$?
   b) Suppose $\mathcal{A}$ is a filter on a set X and $\{s_A\}$, $A \in \mathcal{A}$, is a net based on $\mathcal{A}$. Is $\mathcal{A}$ necessarily the filter which $\{s_A\}$, $A \in \mathcal{A}$, generates?
4. Which of the following sequences in R, the set of real numbers, are ultranets? Note that the property of being an ultranet is independent of the topology on R.

   a) $s_n = 1/n$ b) $s_n = n$ c) $s_n = 1/n^2$ d) $s_n = (-1)^n$

5. a) Prove that a function f from a space X, τ to a space Y, τ' is continuous if and only if given any ultranet $\{s_i\}$, $i \in I$, such that $s_i \to y$, $f(s_i) \to f(y)$.
   b) Prove that $f\colon X, \tau \to Y, \tau'$ is continuous if and only if given any ultrafilter $\mathfrak{A}$ in X such that $\mathfrak{A} \to y$, the filter generated by $f(\mathfrak{A})$ converges to $f(y)$.
6. Prove or disprove: A space X, τ is T_2 if and only if every convergent ultranet in X converges to a unique limit.
7. Let N be the set of positive integers and $\mathfrak{F}$ be the set of subsets of N containing all but finitely many elements of N. Prove that $\mathfrak{F}$ is a filter. Prove that any ultrafilter containing $\mathfrak{F}$ is nontrivial, and hence there exists a nontrivial ultrafilter on N.

7 COVERING PROPERTIES

7.1 OPEN COVERS AND REFINEMENTS

Some of the most important aspects of certain types of topological spaces can be expressed as *covering properties*. The nice definitions given in this chapter were not always used in the study of topology. As is usually the case with a new discipline, those who pioneered in topology thought certain properties were important for a space to have. The best means of expressing those properties, best from the point of view of most elegant and most workable, were only developed from years of experience. The student should be sophisticated enough to realize that areas of mathematical study are not born full-grown but, as with human infants, require a period of growth of many years before reaching maturity.

Compactness, the most important covering property, was once defined as follows: A space X, τ is *compact* if for every infinite subset $A \subset X$, there is at least one $y \in X$ such that given any two neighborhoods U and U' of y, $U \cap A$ and $U' \cap A$ have the same cardinality. Even the novice in topology will realize that this is a rather cumbersome definition. As more became known about the property that this definition was intended to convey, equivalent expressions of it became known. Compactness is now defined as a covering property.* Certain other concepts valuable in the study of topological spaces can also be best expressed as covering properties.

A *cover* of a space X, τ is exactly what its name implies, a collection of subsets of X which cover X, that is, whose union is X. Usually, however, we wish the members of the cover to be sets of a particular form, generally, open sets. We therefore state the following.

Definition 1. Let X, τ be a topological space. An *open cover* of X is a collection $\{U_i\}$, $i \in I$, of open subsets of X such that

$$\bigcup_I U_i = X.$$

Let $\{U_i\}$, $i \in I$ be an open cover of the space X, τ. A collection

* The old definition of compactness, however, is actually not equivalent to the definition of compactness as a covering property.

$\{V_j\}, j \in J$, is said to be an *open subcover* of $\{U_i\}, i \in I$, if

$$\{V_j \mid j \in J\} \subset \{U_i \mid i \in I\}$$

(that is, each V_j is a U_i) and $\{V_j\}, j \in J$, is itself an open cover of X. The collection $\{V_j\}, j \in J$, is said to be a *refinement* of $\{U_i\}, i \in I$, if $\{V_j\}, j \in J$, is an open cover, and for each V_j, there is U_i such that $V_j \subset U_i$.

Note that an open subcover is a refinement, but a refinement is not necessarily an open subcover.

Example 1. Let R be the set of real numbers with the topology induced by the absolute value metric. Then

$$\{N(x, 4) \mid x \in R\},$$

that is, the set of all 4-neighborhoods in R, is an open cover of R. The set

$$\{N(n, 4) \mid n \text{ is an integer}\}$$

is an open subcover of $\{N(x, 4) \mid x \in R\}$. The set

$$\{N(x, 1) \mid x \in R\}$$

is a refinement of $\{N(x, 4) \mid x \in R\}$, since every 1-neighborhood is contained in some 4-neighborhood. In fact, every 1-neighborhood in R is contained in a 4-neighborhood of an integer; hence $\{N(x, 1) \mid x \in R\}$ is a refinement of $\{N(n, 4) \mid n \text{ is an integer}\}$, even though the cardinality of $\{N(x, 1) \mid x \in R\}$ is greater than that of $\{N(n, 4) \mid n \text{ an integer}\}$.

Example 2. Let X be a set with the discrete topology. Then $\{\{x\} \mid x \in X\}$ is an open cover of X. Moreover, this open cover has no proper subcover, nor any proper refinement. If X has the trivial topology, then the only open covers of X are $\{X, \phi\}$ and $\{X\}$. (See Exercise 1.)

Example 3. Let N be the set of positive integers with the topology determined by calling a subset U of N open if U contains all but at most finitely many elements of N. Let $\{U_i\}, i \in I$, be any open cover of N. Pick any U_i. Then U_i contains all but at most finitely many of the positive integers; say U_i excludes $n_1, \ldots, n_p$. Since $\{U_i\}, i \in I$, is an open cover, every element of N is in at least one of the U_i, and hence there are at most p other members of $\{U_i\}, i \in I$, say $U_{i_1}, \ldots, U_{i_p}$ such that

$$N = U_i \cup U_{i_1} \cup \cdots \cup U_{i_p}.$$

Thus $\{U_i, U_{i_1}, \ldots, U_{i_p}\}$ is a finite open subcover of $\{U_i\}, i \in I$. We therefore see that every open cover of N (with the prescribed topology) has a finite open subcover.

EXERCISES

1. Prove the assertions made in Example 2.
2. Let R^2 be the coordinate plane with the topology induced by the Pythagorean metric. Which of the following are open subcovers of

$$\{N((x, y), 1) \mid (x, y) \in R^2\}?$$

Which are refinements of

$$\{N((x, y), 3) \mid (x, y) \in R^2\}?$$

In the event a collection is not a subcover, or not a refinement, explain which properties are lacking.

a) $\{N((m, n), \frac{1}{2}) \mid m$ and n are integers$\}$
b) $\{N((0, 0), p) \mid p$ a positive real number$\}$
c) $\{N((x, y), 1) \mid x$ and y are rational$\}$
d) $\{N((x, y), \frac{1}{9}) \mid x$ and y are rational$\}$
e) the family of all sets of the form $\{(x, y) \mid |x - a| + |y - b| < 1\}$, where (a, b) is any point of R^2
f) the family of all subsets of R^2

3. Prove that R^2 with the Pythagorean topology has a countable cover consisting of p-neighborhoods. Prove that the set of real numbers with the order topology has a countable cover consisting of intervals of the form $(-p, p)$, where $p > 0$.
4. Suppose the open interval $(0, 1)$ is given the absolute value topology. Form $\{U_n\}, n = 1, 2, 3, \ldots$, where $U_n = (1/(n + 1), 1)$. Prove that $\{U_n\}, n \in N$, is an open cover of $(0, 1)$. Show that no finite number of the U_n cover $(0, 1)$, even though any finite number of the U_n may be omitted and what remains still give an open cover of $(0, 1)$.
5. Suppose τ is a topology on the set N of positive integers with the property that any open cover of N has an open subcover which contains at most two elements. Describe all possibilities for τ.

7.2 COUNTABILITY PROPERTIES

One would rightly suspect that covering properties take the following general form: If $\{U_i\}, i \in I$, is any open cover of a space X, τ, then there is an open subcover (or refinement) of $\{U_i\}, i \in I$, satisfying some special condition. One of the most natural conditions the subcover might satisfy is a cardinality condition. Such a cardinality condition is given in the following definition.

Definition 2. A space X, τ is said to be a *Lindelöf space* if every open cover of X has a countable open subcover.

Example 4. The space presented in Example 3 is certainly a Lindelöf space, since every open cover of N not only has a countable subcover, but even has a finite subcover.

In order to discuss Lindelöf spaces more completely, more terminology is needed.

Definition 3. Let X, τ be a topological space. X is said to be *first countable* if there is an open neighborhood system for τ such that $\mathfrak{N}_x$ is countable for each $x \in X$. (See Section 6.1, Exercise 2.) X is said to be *second countable* if there is a basis for τ which consists of countably many sets. X is said to be *separable* if X contains a countable dense subset. (For the definition of a dense subset, see Section 3.6).

Example 5. Let R be the set of real numbers with the absolute value topology τ. For each $x \in R$, let

$$\mathfrak{N}_x = \{N(x, 1/n) \mid n \text{ a positive integer}\}.$$

It is easily verified that the collection of $\mathfrak{N}_x$ forms an open neighborhood system for τ. However, each $\mathfrak{N}_x$ is countable; hence R is first countable. (Actually, from Section 3.3, Exercise 6, we have the more general result that any metric space is first countable.) The set of rational numbers forms a countable dense subset of R, and hence R is also separable. Proposition 5 will tell us that R is second countable as well, and therefore is Lindelöf (Proposition 3).

Example 6. Let X be any uncountable set with the discrete topology. For each $x \in X$, set $\mathfrak{N}_x = \{\{x\}\}$. Then the collection of $\mathfrak{N}_x$ forms an open neighborhood system for the discrete topology on X. Thus X is first countable. Since $\text{Cl } A = A$ for every $A \subset X$ (since every subset of X is closed), the only dense subset of X is X itself. But X is uncountable, and hence there is no countable dense subset of X; X is therefore not separable. X is neither second countable, nor Lindelöf (Exercise 6).

Proposition 1. Any second countable space is first countable.

Proof. Let X, τ be any second countable space, and let $\mathcal{B}$ be a countable basis for τ. Then the collection of sets of the form

$$\mathfrak{N}_x = \{B \in \mathcal{B} \mid x \in B\} \qquad \text{for all} \quad x \in X$$

forms an open neighborhood system for τ (Chapter 3, Proposition 6). Since $\mathcal{B}$ is countable, each $\mathfrak{N}_x$ is also countable. Therefore X is first countable.

Proposition 2. Any second countable space is separable.

Proof. Let X, τ be a second countable space, and let $\mathcal{B}$ be a countable basis for τ. For each $B \in \mathcal{B}$, select $x_B \in B$. Then $\{x_B \mid B \in \mathcal{B}\}$ is a countable subset of X. The proof that it is also dense is left as an exercise.

Proposition 3. Any second countable space is Lindelöf.

Proof. Let X, τ be a second countable space with $\mathcal{B}$ as a countable basis. Suppose $\{U_i\}$, $i \in I$, is any open cover of X. We select a subcover of $\{U_i\}$, $i \in I$, as follows: Number the elements of $\mathcal{B}$ sequentially, that is, B_1, $B_2, \ldots, B_n, \ldots$ Select B_k from $\mathcal{B}$ if there is a member U_i of the open cover such that $B_k \subset U_i$. For each B_k selected, choose one U_i for which $B_k \subset U_i$ and call it U_{k_i}. Since the collection of B_k selected must be countable, the collection of U_{k_i} is also countable. It remains to be shown that

$$\{U_{k_i} \mid B_k \text{ was selected}\}$$

is actually a subcover of $\{U_i\}$, $i \in I$. Since $\{U_i\}$, $i \in I$, is an open cover of X and each U_i is the union of elements of $\mathcal{B}$, the collection of selected B_k actually forms a refinement of $\{U_i\}$, $i \in I$; therefore $\{U_{k_i} \mid B_k$ was selected$\}$ is an open subcover of $\{U_i\}$, $i \in I$.

For general topological spaces, no other implications hold between Lindelöf, first and second countable, and separable, other than those given in Propositions 1, 2, and 3.

The following proposition describes how these properties behave with respect to subspaces and product spaces.

Proposition 4

a) Any subspace of a first countable space is first countable.
b) Every subspace of a second countable space is second countable, and hence is also separable.
c) Every closed subspace of a Lindelöf space is Lindelöf; however, it is not true that every subspace of a Lindelöf space is necessarily Lindelöf.
d) The product space of a countable family of nonempty spaces is second countable if and only if each component space is second countable. (This is an example of a proposition which does not generalize to the product of an arbitrary family of spaces.)
e) The product of a countable family of nonempty Lindelöf spaces is not necessarily Lindelöf, but if a product space is Lindelöf and each component space is T_1, then each component space is also Lindelöf.
f) Any open subspace of a separable space is separable.
g) The product of a countable family of nonempty spaces is separable if and only if each component space is separable.

Proof

a) This statement follows from Proposition 5, Chapter 4.
b) This statement follows from Section 4.1, Exercise 3.
c) Suppose A is a closed subspace of a Lindelöf space X, τ and $\{U_i\}$, $i \in I$, is an open cover of A, each U_i being open in A. Then $U_i = A \cap V_i$ for each $i \in I$, where V_i is open in X. Since A is closed, $X - A$ is open in X. Therefore

$$\{X - A\} \cup \{V_i \mid i \in I\}$$

is an open cover of X. There is thus a countable subcover

$$\{X - A\} \cup \{V_{i_n} \mid n = 1, 2, 3, \ldots\} \quad \text{of } \{X - A\} \cup \{V_i \mid i \in I\}.$$

But then $\{V_{i_n} \mid n = 1, 2, \ldots\}$ is a countable open cover of A. Therefore $\{A \cap V_{i_n} \mid n = 1, 2, \ldots\}$ is a countable open subcover of $\{U_i\}$, $i \in I$; hence A is Lindelöf.

d) Suppose $\times_N X_n$ is the product space of the countable family of nonempty second countable spaces $\{X_n, \tau_n\}$, $n \in N$. Suppose $\mathcal{B}_n$ is a countable basis for τ_n, for each $n \in N$. Let

$$\mathcal{B} = \left\{ \times_N V_n \mid V_n = X_n \text{ for all but at most finitely many } n,\ V_n \in \mathcal{B}_n \quad \text{if} \quad V_n \neq X_n \right\}.$$

It is easily verified that $\mathcal{B}$ is a basis for the product topology on $\times_N X_n$ and $\mathcal{B}$ is countable. Therefore $\times_N X_n$ is second countable. On the other hand, if a nonempty product space is second countable, then since each component space is homeomorphic to a subspace of the product and each subspace of a second countable space is second countable, each component space is also second countable.

e) This statement follows from (c) and the fact that if each component space is T_1, then each component space is homeomorphic to a closed subspace of the product space (*cf.* Section 5.4, Exercise 6).

f) Suppose $\{x_n \mid n \in N\}$ is a countable dense subset of the separable space X, τ. Let U be any open subspace of X, and set

$$D = \{x_n \mid n \in N\} \cap U.$$

Suppose V is any subset of U which is open in U. Since U is open, V is open in X; therefore V contains some x_n, and hence a point of D (Proposition 14, Chapter 3). Therefore, by Proposition 14, Chapter 3, D is dense in U.

The proof of (g) is left as an exercise.

The next proposition shows that in metric spaces, the properties of being Lindelöf, separable, and second countable are all equivalent. We have already seen that any metric space is first countable. Example 6 together with Exercise 6 furnishes an example of a metric space which is not second countable.

Proposition 5. If X, D is a metric space, then the following statements are equivalent:

a) X is Lindelöf. b) X is separable. c) X is second countable.

Proof. Since it has already been shown in Propositions 2 and 3 that any second countable space is both separable and Lindelöf, it will suffice to show that if X is either separable or Lindelöf, then X is second countable.

Statement (b) implies statement (c). Suppose X is separable and let $\{x_n \mid n \in N\}$ be a countable dense subset of X. Let $B(n, m) = N(x_n, 1/m)$, where m and n are in N. We shall show that

$$\mathcal{B} = \{B(n, m) \mid n, m \in N\}$$

is a basis for the metric topology on X. Let U be any open subset of X and let x be any point of U. Since U is open, there is a positive number p such that $N(x, p) \subset U$. Choose any integer $m > 2/p$. Since $N(x, 1/2m)$ is open and $\{x_n \mid n \in N\}$ is dense, there is some

$$x_n \in N(x, 1/2m)$$

(Proposition 14, Chapter 3). Then $x \in N(x_n, 1/m)$. Since $m > 2/p$, $1/m < p/2$; thus $N(x_n, 1/m) \subset N(x, p)$. Therefore $N(x_n, 1/m) \subset U$ as well. But then U is the union of members of $\mathcal{B}$ [for x was an arbitrary element of U and $N(x_n, 1/m) \in \mathcal{B}$]. Since U was an arbitrary open set, $\mathcal{B}$ is a basis for the metric topology. Moreover $\mathcal{B}$ is countable, and hence X is second countable.

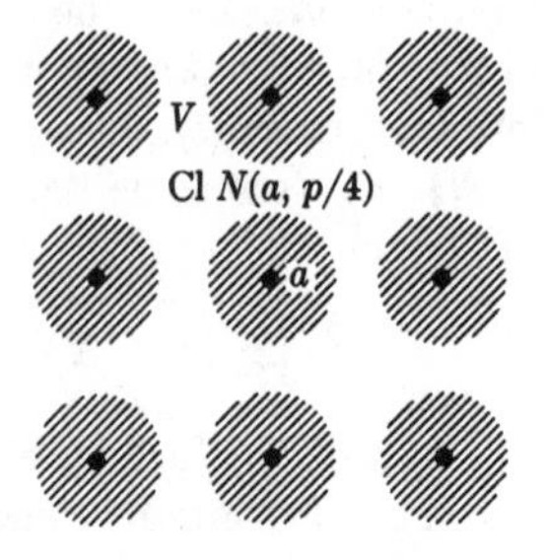

Figure 7.1

Statement (a) implies statement (b). Suppose X is Lindelöf. Choose some $p > 0$, and let E be a maximal subset of X having the property that $D(a, b) \geq p$ for all $a, b \in E$. Such a maximal subset can be shown to exist by Zorn's lemma. For each $a \in E$, consider

$$N(a, p/2) \qquad \text{and} \qquad V = X - \cup \{\text{Cl}\, N(a, p/4) \mid a \in E\}$$

(Fig. 7.1). V is open (Exercise 2). Then $\{V\} \cup \{N(a, p/2) \mid a \in E\}$ is a covering of X by open sets. Since X is Lindelöf, there is a countable subcovering. But if $N(a, p/2)$ were omitted from the original cover for any $a \in E$, the remaining sets would fail to cover X since none of them would contain a. Therefore $\{N(a, p/2) \mid a \in E\}$ must itself be countable; hence E is countable.

Carry out the construction described above for $p = 1/n$, $n = 1, 2, 3, \ldots$, and get $\{E_n\}$, $n \in N$, where E_n is the set corresponding to $p = 1/n$; that is, E_n is a maximal set having the property that $D(a, b) \geq 1/n$ for any $a, b \in E_n$. Let $S = \bigcup_N E_n$. Since S is the union of countably many countable sets, S is countable. We now show that S is dense in X.

Suppose $x \in X$ and $q > 0$; we will show that there is $z \in S$ such that $z \in N(x, q)$. Take $n > 1/q$. Then there is $z \in E_n$ such that $z \in N(x, q)$. For if not, then x has the property that $D(x, w) \geq 1/n$ for each $w \in E_n$, but $x \notin E_n$, and thus E_n would not be maximal. If U is any nonempty open subset of X, choose $x \in U$ and $q > 0$ such that $N(x, q) \subset U$. Then $N(x, q)$, and hence U, contains an element of S. Therefore S is dense (Proposition 14, Chapter 3).

Corollary. The space R of real numbers with the absolute value topology is second countable (Example 5) as is the product space R^n for any n (Proposition 4d).

We close this section with a proposition that will be needed in the proof of a key result in a later chapter.

Proposition 6. A T_3 Lindelöf space is T_4.

Proof. Let X, τ be a T_3 Lindelöf space and let A and B be disjoint closed subsets of X. If $x \in A$, then $X - B$ is a neighborhood of x. Since X is T_3, there is a neighborhood U_x of x such that $\text{Cl}\, U_x \subset X - B$ (Chapter 5, Proposition 4). Similarly, if $x \in B$, there is a neighborhood U_x of x such that $\text{Cl}\, U_x \subset X - A$. If x is not an element of either A or B, then $X - (A \cup B)$ is a neighborhood of x; hence we may find a neighborhood U_x of x such that $\text{Cl}\, U_x \subset X - (A \cup B)$ (and thus $\text{Cl}\, U_x \cap (A \cup B) = \phi$). The family of U_x for each $x \in X$ is an open cover for X. Since X is Lindelöf, this cover has a countable subcover $\{U_{x_n} \mid n = 1, 2, 3, \ldots\}$.

Let $U_1, U_2, \ldots$ be the U_{x_n} (relabeled for convenience) which meet A, and let $V_1, V_2, \ldots$ be the U_{x_n} which meet B. Then for each positive integer n, $\text{Cl}\, U_n \cap B = \phi$ and $\text{Cl}\, V_n \cap A = \phi$; moreover $A \subset \bigcup_N U_n$ and $B \subset \bigcup_N V_n$. Define $W_1 = U_1$ and set $Y_1 = V_1 - \text{Cl}\, W_1$. Let $W_2 = U_2 - \text{Cl}\, Y_1$ and $Y_2 = V_2 - (\text{Cl}\, W_1 \cup \text{Cl}\, W_2)$. Suppose W_n and Y_n

have been defined. Then set

$$W_{n+1} = U_{n+1} - (\mathrm{Cl}\, Y_1 \cup \mathrm{Cl}\, Y_2 \cup \cdots \cup \mathrm{Cl}\, Y_n)$$

and

$$Y_{n+1} = V_{n+1} - (\mathrm{Cl}\, W_1 \cup \mathrm{Cl}\, W_2 \cup \cdots \cup \mathrm{Cl}\, W_{n+1}).$$

W_n is always an open set since

$$\begin{aligned} W_n &= U_n \cap (X - (\mathrm{Cl}\, Y_1 \cup \cdots \cup \mathrm{Cl}\, Y_{n-1})) \\ &= U_n \cap (X - \mathrm{Cl}(Y_1 \cup \cdots \cup Y_{n-1})); \end{aligned}$$

hence W_n is the intersection of two open sets, and is therefore open. Similar reasoning shows that Y_n is open for each n.

Set $H = \bigcup_N W_n$ and $K = \bigcup_N Y_n$. Since H and K are the union of open sets, they are open. Suppose $a \in A$. Then $a \in U_n$ for some n, and

$$W_n = U_n - (\mathrm{Cl}\, Y_1 \cup \cdots \cup \mathrm{Cl}\, Y_{n-1}).$$

But for any k, $\mathrm{Cl}\, Y_k \subset \mathrm{Cl}\, V_k$ and $\mathrm{Cl}\, V_k \cap A = \phi$. Therefore $a \notin \mathrm{Cl}\, Y_k$ for any k. We have then that $a \in W_n$. Therefore $A \subset \bigcup_N W_n = H$. Similarly, $B \subset K$. In order to show that X is T_4, we now have merely to prove that $H \cap K = \phi$.

Suppose $x \in H \cap K$. Then $x \in W_n \cap Y_m$ for some m and n. Suppose $m \geq n$. Then

$$x \in Y_m = V_m - (\mathrm{Cl}\, W_1 \cup \cdots \cup \mathrm{Cl}\, W_n \cup \cdots \cup \mathrm{Cl}\, W_m);$$

hence x could not be in $\mathrm{Cl}\, W_n$, a contradiction. On the other hand, if $m < n$, then

$$x \in W_n = U_n - (\mathrm{Cl}\, Y_1 \cup \cdots \cup \mathrm{Cl}\, Y_m \cup \cdots \cup \mathrm{Cl}\, Y_{n-1}).$$

Thus $x \notin \mathrm{Cl}\, Y_m$, again a contradiction. Therefore H and K are disjoint open subsets of X, which contain A and B, respectively, and hence X is T_4.

Example 7. Let R be the set of real numbers with the topology τ as described in Example 11 of Chapter 3. The set Q of rational numbers is a dense subset of R since any basis element of τ [i.e., an interval of the form $[a, b)$] contains a rational number. Therefore R is separable. R is also first countable [for each $x \in R$, set $\mathfrak{N}_x = \{[x, x + 1/n) \mid n \in N\}$]. R cannot be second countable, however, For if R were second countable, then the product space R^2 would also be second countable and hence Lindelöf. But product space R^2 was shown in Example 13 of Chapter 5 to be T_3, but not T_4. If R^2 were Lindelöf and T_3, then by Proposition 6 it would have to be T_4.

EXERCISES

1. Prove that the set $\{x_B \mid B \in \mathcal{B}\}$ in Proposition 2 is dense in X.
2. The following refer to the proof of Proposition 5.
   a) Prove that there is a maximal subset E as claimed.
   b) Prove that the set
   $$V = X - \cup\{\mathrm{Cl}\, N(a, p/4) \mid a \in E\}$$
   is open. There are a number of possible approaches to this problem. One approach, for example, is to show that given any $w \in V$, $N(w, 1)$ intersects $\mathrm{Cl}\, N(a, p/4)$ for at most finitely many $a \in E$. This means that there is $p' > 0$ such that $N(w, p')$ does not intersect any of the $\mathrm{Cl}\, N(a, p/4)$.
3. Prove (g) of Proposition 4.
4. By Proposition 4(g), R^2 as described in Example 7 is separable. Let
   $$A = \{(x, y) \mid x + y = 0\} \subset R^2.$$
   Prove that A is a nonseparable subspace of R^2. Is A closed? Is A Lindelöf? Give another proof that R^2 is not Lindelöf without appealing to Proposition 6.
5. A point x of a space X, τ is said to be a *condensation point* of a subset A of X if each neighborhood of x meets A in uncountably many points. Let A^- denote the set of condensation points of A. Suppose X is a Lindelöf space. Prove that if A is uncountable, then $A^- \neq \phi$. [*Hint:* Try to construct a countable open cover of X each of whose members contains countably many of the elements of A, and hence arrive at the contradiction that A is countable.]
6. Let X be an uncountable set with the discrete topology. Prove that the collection of $\mathfrak{N}_x$ as described in Example 6 forms an open neighborhood system for the discrete topology. Find a metric on X which induces the discrete topology. Prove that X is not Lindelöf, and hence that X is neither separable nor second countable (Proposition 5).
7. Let X be the set of continuous functions from the space R of real numbers with the absolute value topology into itself. For each $f \in X$ and $p > 0$, define
   $$N(f, p) = \{g \in X \mid |f(x) - g(x)| < p \text{ for all } x \in R\}.$$
   The family of $N(f, p)$ for all $f \in X$ and all $p > 0$ forms the basis for a topology τ on X. Try to determine if X, τ is second countable. Let
   $$Y = \{f \in X \mid f \text{ has derivatives of all orders at each } x \in X\}.$$
   Is Y a second countable subspace of X?
8. Prove directly, that is, without using Proposition 6, that the space R of real numbers with the usual absolute value metric topology is second countable. [*Hint:* Prove that $\{N(x, q) \mid q > 0, q \text{ and } x \text{ rational}\}$ gives a countable basis.]

7.3 COMPACTNESS

The most important of all covering properties is *compactness.* As was pointed out earlier in this chapter, compactness was not originally viewed as a covering property, but it is through the use of coverings that compactness can be stated in its most workable form. Compactness is, like Lindelöf, a cardinality condition.

Definition 4. A space X, τ is said to be *compact* if given any open cover $\{U_i\}$, $i \in I$, of X, there is a finite subcover of $\{U_i\}$, $i \in I$.
Suppose X, τ is any space and $A \subset X$. An *open cover of* A is a collection $\{U_i\}$, $i \in I$, of open subsets of X whose union includes A. Equivalently, $\{U_i\}$, $i \in I$, is an open cover of A if $\{U_i \cap A\}$, $i \in I$, is an open cover of the subspace A. A is said to be *compact* if every open cover of A has a finite subcover. Equivalently, A is compact if the subspace A is compact.

Note that in order for a space to be Lindelöf, any open cover had to have a countable subcover. In order for a space to be compact, any open cover has to have a finite subcover. Certainly then, any compact space is also Lindelöf.

Example 8. The open interval (0, 1) with the absolute value topology is Lindelöf since it is a subspace of a second countable space R. The interval (0, 1) is not compact, as we see from Section 7.1, Exercise 4. Another example of a Lindelöf space which is not compact is any countably infinite set with the discrete topology.
An example of a compact space is the space presented in Example 3.

Proposition 7. The subspace [0, 1] of the space R of real numbers with the absolute value topology is compact.

Proof. Let $\{U_i\}$, $i \in I$, be an open cover of [0, 1], where each U_i is open in R. Let

$$T = \{x \in [0, 1] \mid \text{finitely many of the } U_i \text{ cover } [0, x)\}.$$

Then $T \neq \phi$ and 1 is an upper bound for T. Therefore T has a least upper bound, say u. If $u = 1$, we are done (since finitely many of the U_i cover [0, 1), and hence at most one more of the U_i will be needed to get a finite cover of [0, 1]). Suppose then $0 \leq u < 1$. Then either $u \in T$ or $u \notin T$.

Case 1. $u \in T$. Then finitely many of the U_i, say $U_{i_1}, \ldots, U_{i_n}$ cover $[0, u)$. There is, however, $U_{i'}$ such that $u \in U_{i'}$; therefore

$$\{U_{i'}, U_{i_1}, \ldots, U_{i_n}\}$$

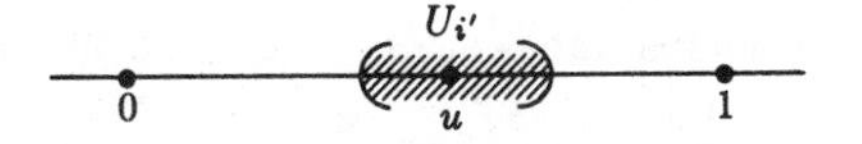

Figure 7.2

is an open cover of $[0, u)$. It is then clear (Fig. 7.2) that u could not be an upper bound for T.

Case 2. $u \notin T$. Then there is U_j such that $u \in U_j$ and finitely many of the U_i do not cover $[0, u) - U_j$. Therefore u is not the least upper bound for T.

Both cases have led to contradictions; hence it could not be that $0 \leq u < 1$. Therefore $u = 1$, and hence $[0, 1]$ is compact.

We now derive some important criteria for compactness.

Proposition 8. Let X, τ be any topological space. Then X is compact if and only if given any family $\{F_i\}$, $i \in I$, of closed subsets of X such that the intersection of any finite number of the F_i is nonempty, $\bigcap_I F_i \neq \phi$.

Proof. Suppose X is compact and let $\{F_i\}$, $i \in I$, be any family of closed subsets of X such that $\bigcap_I F_i = \phi$. Set $U_i = X - F_i$ for each $i \in I$. Then

$$X - \bigcap_I F_i = X - \phi = X = \bigcup_I (X - F_i) = \bigcup_I U_i.$$

Each U_i is the complement of a closed set and hence is open. Therefore $\{U_i\}$, $i \in I$, is an open cover of X. But X is compact; hence there are finitely many of the U_i, say $U_{i_1}, \ldots, U_{i_n}$, which cover X. Then

$$F_{i_1} \cap \cdots \cap F_{i_n} = \phi.$$

We have proved that if X is a compact space, then given any family $\{F_i\}$, $i \in I$, of closed subsets of X whose intersection is empty, the intersection of some finite family of F_i is empty.

Suppose X has the property that if the intersection of any family $\{F_i\}$, $i \in I$, of closed subsets of X is empty, the intersection of finitely many of the F_i is empty. Suppose $\{U_i\}$, $i \in I$, is any open cover of X. Then $X = \bigcup_I U_i$. Therefore setting $F_i = X - U_i$, $\{F_i\}$, $i \in I$, is a family of closed subsets of X whose intersection is empty. Hence we can find finitely many of the F_i, say $F_{i_1}, \ldots, F_{i_n}$, such that

$$F_{i_1} \cap \cdots \cap F_{i_n} = \phi.$$

Then $\{U_{i_1}, \ldots, U_{i_n}\}$ is a finite subcover of $\{U_i\}$, $i \in I$. Therefore X is compact.

Proposition 9. A space X, τ is compact if and only if every net in X has a limit point.

Proof. Suppose X is compact and let $\{x_i\}$, $i \in I$, be any net in X. Define $B_j = \{x_i \mid j \leq i\}$. Then $\{\mathrm{Cl}\, B_j\}$, $j \in J$, has the property that the intersection of any finite family of the $\mathrm{Cl}\, B_j$ is nonempty. Since X is compact, by Proposition 8, $\bigcap_I \mathrm{Cl}\, B_j \neq \phi$. Choose y in this intersection. We now show that y is a limit point of $\{x_i\}$, $i \in I$. Since $y \in \mathrm{Cl}\, B_j$ for any $j \in I$, any neighborhood U of y therefore contains at least one point of B_j. Suppose U is a neighborhood of y and j and j' are elements of I. Since I is directed, there is $j'' \in I$ such that $j \leq j''$ and $j' \leq j''$. But $y \in \mathrm{Cl}\, B_{j''}$, and hence there is $\bar{j}$ such that

$$x_{\bar{j}} \in U \cap B_{j''}.$$

Then $j \leq \bar{j}, j' \leq \bar{j}$, and $x_{\bar{j}} \in U$. Therefore $\{x_i\}$, $i \in I$, is cofinally in U; hence y is a limit point of $\{x_i\}$, $i \in I$.

Suppose, on the other hand, that X has the property that every net in X has a limit point. Let $\{F_i\}$, $i \in I$, be any family of closed subsets of X such that the intersection of finitely many of the F_i is always nonempty. Let J be the set of finite intersections of the F_i. Then J is partially ordered by $\leq$, where $A \leq B$ if $B \subset A$; moreover, J is then a directed set. Since each member of J is nonempty, we can define a selection function s from J into X such that $s(A) \in A$ for each $A \in J$. Therefore $\{s_A\}$, $A \in J$, is a net in X, and hence has a limit point y. Consider any of the F_i. If $A \in J$ and $F_i \leq A$, then $A \subset F_i$. Thus for each of the F_i, the net $\{s_A\}$, $A \in J$, is residually in F_i. Since y is a limit point of $\{s_A\}$, $A \in J$, some subnet of $\{s_A\}$, $A \in J$, converges to y (Proposition 9, Chapter 6). But since $\{s_A\}$, $A \in J$, is residually in F_i, such a subnet would be residually in F_i for each i (Proposition 3, Chapter 6). Then by Proposition 13, Chapter 6, $y \in F_i$ for each i; hence $y \in \bigcap_I F_i$. Therefore $\bigcap_I F_i \neq \phi$. By Proposition 8, then X is compact.

Corollary. A space X, τ is compact if and only if every ultranet in X converges.

Proof. If X is compact and $\{s_i\}$, $i \in I$, is an ultranet in X, then $\{s_i\}$, $i \in I$, has a limit point. But an ultranet converges to any of its limit points (Proposition 22, Chapter 6). Conversely, if every ultranet in X converges and $\{s_i\}$, $i \in I$, is any net in X, then some ultranet is a subnet of $\{s_i\}$, $i \in I$ (Proposition 22, Chapter 6). Therefore $\{s_i\}$, $i \in I$, has a subnet which converges to some point y; hence y is a limit point of $\{s_i\}$, $i \in I$ (Proposition 9, Chapter 6). Then X is compact by Proposition 9.

Example 9. We give another proof now that [0, 1] with the absolute value topology is compact. Since [0, 1] is second countable or metric, we will have shown [0, 1] is compact if we show that every sequence in [0, 1] has a limit point. Suppose $\{s_n\}$, $n \in N$, is a sequence in [0, 1].

Case 1. $\{s_n\}$, $n \in N$, is monotonically increasing, that is,

$$s_1 \leq s_2 \leq \cdots \leq s_n \leq \cdots$$

Then $\{s_n \mid n \in N\}$ has a least upper bound u, $0 \leq u \leq 1$. If U is any neighborhood of u, it is readily shown that $\{s_n\}$, $n \in N$, is residually in U, and hence $s_n \to u$. Therefore u is a limit point of $\{s_n\}$, $n \in N$.

Case 2. $\{s_n\}$, $n \in N$, is monotonically decreasing, that is,

$$s_1 \geq s_2 \geq \cdots \geq s_n \geq \cdots$$

Then $\{s_n \mid n \in N\}$ has a greatest lower bound v, $0 \leq v \leq 1$; moreover $s_n \to v$. Therefore v is a limit point of $\{s_n\}$, $n \in N$.

Case 3. If $\{s_n\}$, $n \in N$, is either monotonically increasing or decreasing from some point on, that is, for all but finitely many elements, then the exceptional elements can be discarded without penalty, and Case 1 or 2 applied.

Case 4. $\{s_n\}$, $n \in N$, is neither monotonically increasing nor monotonically decreasing from some point on. Then $\{s_n\}$, $n \in N$, is "cofinally" increasing (the quotation marks here indicate that we are applying a property informally to the sequence as a whole, rather than to individual members); hence there is a monotonically increasing subsequence of $\{s_n\}$, $n \in N$. By Case 1, this subsequence converges to a point u of $[0, 1]$. But then u is a limit point of $\{s_n\}$, $n \in N$.

Every sequence in $[0, 1]$ has a limit point, and therefore $[0, 1]$ is compact.

EXERCISES

1. Prove that the set J in Proposition 9 is a directed set.
2. Prove that the sequences in Example 9 converge as claimed. Formalize the argument in Case 4.
3. In Example 9, it is asserted that because $[0, 1]$ is second countable, we need only consider sequences. Prove: A second countable space X, τ is compact if and only if every sequence in X has a limit point.
4. Prove that a space X, τ is compact if and only if every filter on X has a limit point. Prove that X is compact if and only if every ultrafilter on X converges.
5. Let X, τ be a space and let $\mathcal{B}$ be a basis for τ. Prove that X is compact if and only if every cover of X by members of $\mathcal{B}$ has a finite subcover.
6. Decide which of the following spaces are compact.

   a) the plane R^2 with the topology which has for a subbasis

   $$\mathcal{S} = \{U \mid U = R^2 - L, \text{ where } L \text{ is any straight line}\}$$

b) the plane R^2, where an open set is any set of the form $R^2 - C$, where C contains at most countably many points of R^2, and ϕ is open
c) the subspace of rational numbers in the usual space of real numbers
d) the space in Example 1, Chapter 6

7. In Section 6.5, Exercise 6, the notion of a bounded sequence in R, the usual space of real numbers, was introduced. Let $\{s_n\}$, $n \in N$, be a bounded sequence in R, and let A be the set of limit points of $\{s_n\}$, $n \in N$. Prove that $A \cup \{s_n \mid n \in N\}$ is compact.

8. Prove that the union of finitely many compact subsets of any space is compact. Is the intersection of two compact subsets necessarily compact?

9. Prove or disprove: Let X be an infinite space with the property that the only compact subspaces of X are finite subspaces. Then X has the discrete topology.

7.4 THE DERIVED SPACES AND COMPACTNESS. THE SEPARATION AXIOMS AND COMPACTNESS

It is not necessarily true that any subspace of a compact space is compact. For example, (0, 1) is not compact (Example 8), whereas [0, 1] is compact (Proposition 7). We do though have some information about which subspaces of a compact space are compact.

Proposition 10. Any closed subset of a compact space is compact.

Proof. Let A be a closed subset of a compact space X, τ, and suppose $\{U_i\}$, $i \in I$, is any open cover of A. Then since A is closed, $X - A$ is open; hence $\{X - A\} \cup \{U_i \mid i \in I\}$ is an open cover of X. Since X is compact, $X - A$ together with finitely many of the U_i, say $U_{i_1}, \ldots, U_{i_n}$ form a cover of X. Therefore $\{U_{i_1}, \ldots, U_{i_n}\}$ is a finite subcover of A, and hence A is compact.

A partial converse to Proposition 10 is given by

Proposition 11. Any compact subset of a T_2 space is closed.

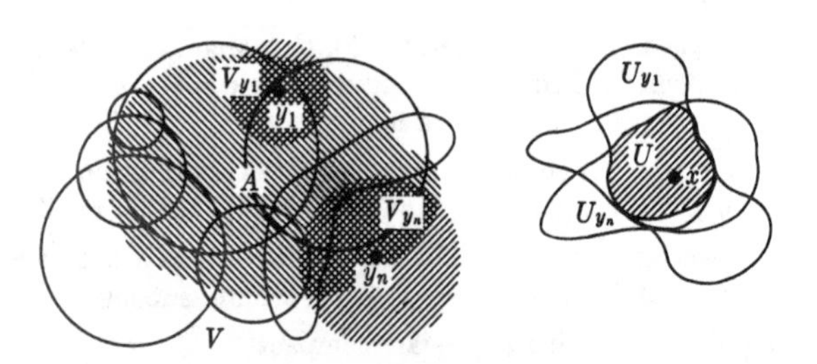

Figure 7.3

Proof. Let A be a compact subset of a T_2-space X, τ (Fig. 7.3) and suppose $x \in X - A$. We must find a neighborhood of x which does not meet A

(thus showing that $X - A$ is open). Since X is T_2, given any $y \in A$, there are neighborhoods U_y and V_y of x and y, respectively, such that

$$U_y \cap V_y = \phi.$$

Then $\{V_y\}$, $y \in A$, is an open cover of A; hence there are finitely many y, say $y_1, \ldots, y_n$, such that $\{V_{y_1}, \ldots, V_{y_n}\}$ is an open cover of A. Set

$$U = U_{y_1} \cap \cdots \cap U_{y_n} \quad \text{and} \quad V = V_{y_1} \cup \cdots \cup V_{y_n}.$$

Then $x \in U$, $A \subset V$, and $U \cap V = \phi$, and hence U is a neighborhood of x such that $U \cap A = \phi$. A is therefore closed.

Combining Propositions 10 and 11, we obtain the following.

Corollary 1. A subset of a compact T_2-space is compact if and only if it is closed.

Note in the proof of Proposition 11 that $A \subset V$, $x \in U$, and $U \cap V = \phi$. By Corollary 1, A could be any closed subset of X; thus we have also proved the following corollary.

Corollary 2. A compact T_2-space is T_3.

On the other hand, any compact space is Lindelöf, and any T_3 Lindelöf space is T_4 (Proposition 6); hence we have the following stronger result.

Corollary 3. A compact T_2-space is T_4.

If a space X, τ is not T_2, then it is not necessarily true that every compact subset of X is closed, as is shown in the following.

Example 10. Let X be any infinite set with the topology defined by calling a subset of X open if it is either empty or contains all but (at most) finitely many elements of X. The proof used in Example 3 can be used here to show that X with this topology is compact. Let U be any open subset of X other then X or ϕ. Since U contains all but finitely many elements of X, U must be compact. For since U is open, a subset V of U is open in U if and only if it is open in X; hence V is open in U if and only if V is empty or excludes at most finitely many points of U. The same argument which proves X compact can thus be used to show U compact. But since a subset of X is closed if and only if it is all of X, or finite, U is not closed. Note that X is T_1 since every one-point subset of X is closed, but X is not T_2.

We now investigate the relationship between compactness and continuous functions.

Proposition 12. Suppose f is a continuous function from a compact space X, τ onto a space Y, τ'. Then Y, τ' is compact, that is, compactness is preserved by continuous functions.

Proof. Let $\{U_i\}$, $i \in I$, be any open cover of Y. Since f is continuous, $\{f^{-1}(U_i)\}$, $i \in I$, is an open cover of X. Since X is compact, we can find finitely many U_i, say $U_{i_1}, \ldots, U_{i_n}$, such that $\{f^{-1}(U_{i_1}), \ldots, f^{-1}(U_{i_n})\}$ is an open cover of X. But then $\{U_{i_1}, \ldots, U_{i_n}\}$ is a finite open subcover of $\{U_i\}$, $i \in I$. Therefore Y is compact.

Since any homeomorphism is continuous, we have the following.

Corollary. If X, τ is compact, then any space homeomorphic to X is compact.

Example 11. Proposition 11 enables us to find many more compact spaces. For example, if X, τ is a compact space, R is an equivalence relation on X, and X/R is the identification space, then X/R is compact, since the identification mapping from X onto X/R is continuous. Since the circle is an identification space derived from [0, 1] (Chapter 4, Example 14), the circle is compact. The next proposition will give us even more compact spaces.

Proposition 13 (*Tychonoff theorem*). Let $\times_I X_i$ be the product space of the countable family of nonempty spaces $\{X_i, \tau_i\}$, $i \in I$. Then $\times_I X_i$ is compact if and only if each component space is compact.

Proof. Suppose $\times_I X_i$ is compact. Since the projection map

$$p_i: \underset{I}{\times} X_i \to X_i$$

is continuous and onto for each $i \in I$, X_i is compact for each $i \in I$ (Proposition 12).

Suppose each X_i is compact. Let $\{s_j\}$, $j \in J$, be any ultranet in $\times_I X_i$ with the ith coordinate of s_j being denoted by $s_j(i)$. Then

$$\{p_i(s_j)\} = \{s_j(i)\}, \qquad j \in J,$$

is an ultranet in X_i by Proposition 22, Chapter 6. Therefore $\{s_j(i)\}$ converges in X_i by the corollary to Proposition 9 of this chapter. But then $\{s_i\}$, $i \in I$, converges in $\times_I X_i$ by Proposition 12 of Chapter 6. Therefore $\times_I X_i$ is compact by the corollary to Proposition 9.

Example 12. We have already seen that the closed interval [0, 1] and the circle C are compact. Using Proposition 13, we can now say that $([0, 1])^n$ is compact for any n. Then cylinder $C \times [0, 1]$, the torus $C \times C$, and the cube $([0, 1])^3$ are all examples of compact spaces (Figs. 7.4, 7.5, and 7.6).

Often one of the hardest steps in proving that some function is a homeomorphism is showing that its inverse is continuous. The next proposition affords us some relief in certain special (though important) instances.

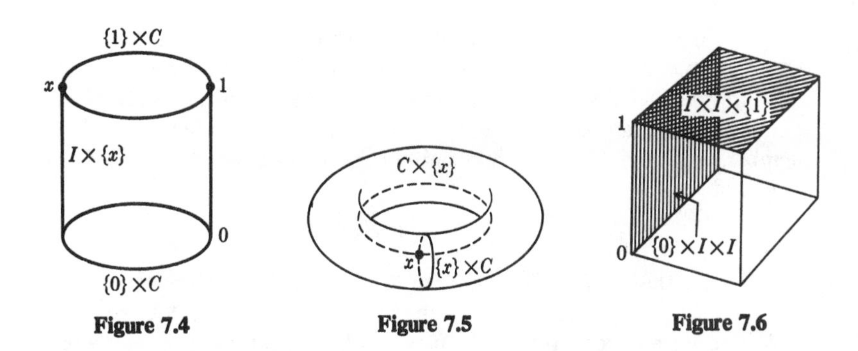

Figure 7.4

Figure 7.5

Figure 7.6

Proposition 14. Let f be a continuous one-one function from a compact space X, τ onto a T_2-space Y, τ'. Then f is a homeomorphism.

Proof. We must show that f^{-1} is continuous. We use Proposition 8, Chapter 4. Suppose F is any closed subset of X. Since F is closed, F is compact (Proposition 10); hence $f(F)$ is compact (Proposition 12). Then $f(F)$ is a compact subset of a T_2-space and is therefore closed (Proposition 11). But

$$f(F) = (f^{-1})^{-1}(F).$$

We have therefore shown that if F is any closed subset of X, $(f^{-1})^{-1}(F)$ is a closed subset of Y. Therefore f^{-1} is continuous; hence f is a homeomorphism.

Because of the great importance of the Tychonoff theorem (Proposition 13), we now present another proof which does not depend on the material of Chapter 6. We first prove another criterion for compactness.

Proposition 15. A space X, τ is compact if and only if there is a subbasis $\mathcal{S}$ such that whenever $\mathcal{C}$ is a cover of X consisting of elements of $\mathcal{S}$, then $\mathcal{C}$ contains a finite subcover of X.

Proof. If X is compact, then τ itself serves as a subbasis for τ having the required property.

Suppose now that X has a subbasis $\mathcal{S}$ having the property stated; we now prove that X is compact. Let $\mathcal{A}$ be any collection of open sets which does not contain a finite subcover of X. We will show that $\mathcal{A}$ cannot be a cover of X, and, hence, indirectly show that any open cover of X contains a finite subcover.

Let $\mathcal{H}$ be the collection of all $\mathcal{B} \subset \tau$ such that $\mathcal{A} \subset \mathcal{B}$ but no finite subset of $\mathcal{B}$ covers X. The set $\mathcal{H}$ is nonempty since $\mathcal{A} \in \mathcal{H}$; moreover, $\subset$ is a partial ordering of $\mathcal{H}$.

Assume $\mathcal{K}$ is any chain in $\mathcal{H}, \subset$. Then the union of the members of $\mathcal{K}$ is easily shown to be a member of $\mathcal{H}$ and is an upper bound for $\mathcal{K}$. Therefore, by Zorn's Lemma, $\mathcal{K}$ contains a maximal element $\mathfrak{N}$. Since $\mathcal{A} \subset \mathfrak{N}$, if we show that $\mathfrak{N}$ is not a cover of X, then $\mathcal{A}$ itself will not be a cover of X.

Suppose then that $\mathfrak{N}$ is a cover of X. Then each $x \in X$ is in some member of $\mathfrak{N}$; assume $x \in M \in \mathfrak{N}$. Now $\mathcal{S}$ is a subbasis for τ and M is a member of τ, hence we can find finitely many members $S_1, \ldots, S_n$ of $\mathcal{S}$ such that

$$x \in S_1 \cap \cdots \cap S_n \subset M.$$

Suppose that no S_i is a member of $\mathfrak{N}$, $i = 1, \ldots, n$. Then $S_i \in \tau - \mathfrak{N}$ for $i = 1, \ldots, n$. Because $\mathfrak{N}$ is a maximal element of $\mathcal{K}$, $\mathfrak{N} \cup \{S_i\}$ must contain a finite subcover of X (or it would be a member of $\mathcal{K}$ which properly contains $\mathfrak{N}$). Consequently, for $i = 1, \ldots, n$, we can find a finite subset $\mathfrak{N}_i$ of $\mathfrak{N}$ such that S_i, together with the elements of $\mathfrak{N}_i$, forms a finite open cover of X. But since $S_1 \cap \cdots \cap S_n \subset M$, it follows that

$$\{M\} \cup \left(\bigcup_{i=1}^{n} \mathfrak{N}_i \right)$$

forms a finite cover of X. This, however, is a contradiction, since $\mathfrak{N}$ contains no finite subcover of X. This contradiction stems from the assumption that no S_i is a member of $\mathfrak{N}$; therefore assume that $x \in S_i$ for some $i = 1, \ldots, n$.

The argument above shows that given any $x \in X$ for which we have some $x \in M \in \mathfrak{N}$, there is some member $S \in \mathcal{S}$ for which $x \in S \in \mathfrak{N}$. It follows then that $\mathcal{S} \cap \mathfrak{N}$ covers the same portion of X that $\mathfrak{N}$ does. Thus, if $\mathfrak{N}$ is a cover of X, then $\mathcal{S} \cap \mathfrak{N}$ is also a cover of X. But $\mathcal{S} \cap \mathfrak{N}$ is a subset of $\mathcal{S}$, and thus contains a finite subcover of X. Therefore $\mathcal{S} \cap \mathfrak{N}$, and hence $\mathfrak{N}$, cannot be a cover of X since $\mathfrak{N}$ contains no finite subcover of X.

We have shown then that any collection of open subsets of X which does not contain a finite subcover of X fails to cover X. Therefore X is compact.

Proposition 16 (*Tychonoff product theorem*, proof of which does not use nets or filters). If $\{X_i, \tau_i\}$, $i \in I$, is a nonempty family of nonempty compact spaces, then the product space $\times_I X_i, \tau$ is also compact.

Proof. The set

$$\mathcal{S} = \{p_i^{-1}(U) \mid U \in \tau_i, i \in I\}$$

forms a subbasis for the product topology τ (recall that p_i is the projection into the ith component). Let $\mathcal{A}$ be any collection of members of $\mathcal{S}$ which does not contain a finite subcover of $\times_I X_i$, and for each $i \in I$, set

$$\mathcal{A}_i = \{U \mid U \in \tau_i, p_i^{-1}(U) \in \mathcal{A}\}.$$

No finite subset of $\mathfrak{A}_i$ can cover X_i; for, otherwise,

$$\{p_i^{-1}(U)\},\ U \in \mathfrak{A}_i,$$

would be a subcollection of $\mathfrak{A}$ which covers X, and from which we could obtain a finite subcover. Since no finite subset of $\mathfrak{A}_i$ covers X_i, but X_i is compact, it follows that $\mathfrak{A}_i$ fails to cover X_i for each $i \in I$. Therefore for $i \in I$ we can find

$$x_i \in X_i - \bigcup \{A \mid A \in \mathfrak{A}_i\}.$$

Let x be that point of $\times_I X_i$ with x_i as found in the previous sentence as its ith coordinate. Then x is a point of X which is not in the union of members of $\mathfrak{A}$. Consequently, $\mathfrak{A}$ does not form a cover of X. It follows then that any collection of members of the subbasis $\mathcal{S}$ which covers $\times_I X_i$ contains a finite subcover of $\times_I X_i$; hence by Proposition 15, $\times_I X_i$ is compact.

EXERCISES

1. Decide which of the following spaces are compact. If practicable, sketch a picture of the space. The set R of real numbers, the plane R^2, or any subspace of these spaces will be assumed to have the usual metric topology. Products will have the product topology.

   a) $(0, 1) \times [0, 1]$
   b) $C \times R$, where $C = \{(x, y) \mid x^2 + y^2 = 1\} \subset R^2$
   c) $\{(x, y, z) \mid x^2 + y^2 + z^2 = 1\} \subset R^3$
   d) $\{(x, y) \mid x^2 + y^2 \leq 1\} \subset R^2$
   e) $N \times C$, where N is the set of positive integers
   f) $\{1, 2, 3, 4, 5\} \times C$, with C as in (*b*)

2. Prove that any subset of N in Example 3 is compact.
3. There is a continuous function from $[0, 1]$ onto $[0, 1] \times [0, 1]$. Prove that this function cannot be one-one.
4. It was shown that the product of normal spaces need not be normal. Prove that the product of compact normal spaces is normal.
5. Let f be the function from $[0, 1]$ onto $[0, 1]$ (with the absolute value topology) defined by $f(x) = \sin(1/x)$ if $x \neq 0$, $f(x) = 0$ if $x = 0$. Prove that f is not continuous. [*Hint:* Suppose f is a continuous function from a compact space X onto a compact space Y. Define $G_f = \{(x, y) \mid y = f(x)\}$. Prove that if f is continuous, then G_f is a closed subset of the product space $X \times Y$. The easiest way to effect this proof is through the use of Proposition 10, Chapter 6. Then G_f is a closed subset of a compact T_2-space if X and Y are both compact and T_2 (as in the case in this problem). Therefore what can be said about G_f?]

6. Prove that every compact metric space is separable.
7. Suppose X, D is any metric space. A subset Y of X is said to be *bounded* if $Y \subset N(x, p)$ for some $x \in X$ and $p > 0$. Prove that any compact subset of a metric space is both closed and bounded.
8. Suppose X, τ is a first countable space such that X is T_1 and every compact subset of X is closed. Prove that X is T_2. [*Hint:* Show that every convergent sequence in X has a unique limit.]
9. Let X, D be a compact metric space and let $\{s_n\}$, $n \in N$, be a sequence in X such that given any $p > 0$, there is $m \in N$ such that if $m < n$ and $m < n'$, then $D(s_n, s_{n'}) < p$. Prove that $\{s_n\}$, $n \in N$, converges in X. Prove that this is not necessarily true if the assumption that X is compact is removed.
10. Prove or disprove: Suppose X and Y are both compact T_2-spaces and f is a function from X into Y. Then f is continuous if and only if f considered as a subspace of $X \times Y$ is compact.
11. Suppose X, τ is a space having the property that whenever a subset A of X is compact, then $X - A$ is also compact. Which of the following properties must X also have.

    a) T_2 b) T_1 c) Every subset of X is compact.

8
MORE ABOUT COMPACTNESS

8.1 COMPACTNESS IN R^n

The product space R^n of the space R of real numbers with the absolute value topology with itself n times, better known as *Euclidean n-space*, is perhaps the most important topological space of all (or, more accurately, family of spaces, since there is a space for each positive integer n). Compact subsets of R^n therefore hold a special place among compact sets and warrant a special section to study them.

We have already seen that $[0, 1] \subset R$ is compact. Thus any subspace of R^n homeomorphic to $[0, 1]$ is compact. More generally, any continuous image of $[0, 1]$ in R^n is compact. Using the various propositions already proved, we can find many compact subsets of R^n. However, R^n has many properties not shared by all topological spaces. We would therefore expect there to be certain criteria for compactness which are more peculiar to R^n. Proposition 2 gives such a criterion. Preparatory to Proposition 2, we first prove the following.

> **Proposition 1.** Suppose $x = (x_1, \ldots, x_n)$ and $y = (y_1, \ldots, y_n)$ are any two points of R^n. Define $D(x, y) = \max(|x_i - y_i|, i = 1, \ldots, n)$ (*cf.* Examples 3 and 6 of Chapter 2). Then D is a metric on R^n. Moreover, the topology induced on R^n by D is the same as the product topology on R^n.

Proof. The proof that D is actually a metric is straightforward and is left as an exercise. Let τ be the product topology on R^n and τ' be the topology induced by D. In order to prove $\tau = \tau'$, we will use Corollary 1, Proposition 9, Chapter 3. Suppose $x = (x_1, \ldots, x_n) \in R^n$. Set

$$\mathfrak{N}_x = \left\{ \bigtimes_{i=1}^{n} N(x_i, p_i) \mid \text{where } p_i > 0, \; N(x_i, p_i) = (x_i - p_i, x_i + p_i) \subset R, i = 1, \ldots, n \right\},$$

and

$$\mathfrak{N}'_x = \{N'(x, p) \mid p > 0, \text{ where } N'(x, p) \text{ is the } D\text{-}p\text{-neighborhood of } x \text{ in } R^n\}.$$

Then taking the collection of all $\mathfrak{N}_x$ and the collection of $\mathfrak{N}'_x$ for all $x \in R^n$, we get open neighborhood systems for τ and τ', respectively.

Suppose $N' \in \mathfrak{N}'_x$. Then

$$N' = \bigtimes_{i=1}^{n} N(x_i, p)$$

and hence is a member of $\mathfrak{N}_x$. (For a picture of a typical N' in R^2, the reader should see Fig. 4, Chapter 2.) Suppose $N \in \mathfrak{N}_x$. Then

$$N = \bigtimes_{i=1}^{n} N(x_i, p_i),$$

where $p_i > 0$, $i = 1, \ldots, n$. Set $p = \min(p_1, \ldots, p_n)$. Then $N'(x, p) \in \mathfrak{N}'_x$ and $N'(x, p) \subset N$. Therefore by Corollary 1, Proposition 9, Chapter 3, $\tau = \tau'$.

Proposition 2. A subset A of R^n is said to be *bounded* if there is a positive number p such that $A \subset N'(\overline{O}, p)$, where $\overline{O}$ is the origin in R^n and $N'(\overline{O}, p)$ is the D-p-neighborhood of $\overline{O}$ described in Proposition 1. A subset A of R^n is compact if and only if A is closed and bounded.

Proof. Proposition 1 has shown that R^n with the product topology is a metric space (with metric D as in Proposition 1). In Section 7.4, Exercise 7, it was shown that any compact subset of any metric space is closed and bounded.

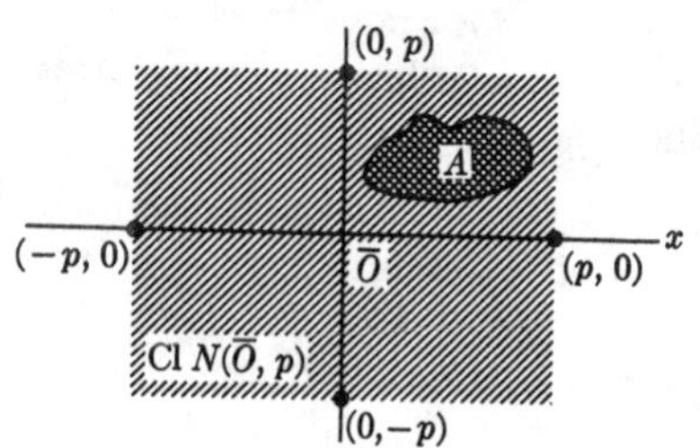

Figure 8.1

Suppose A is a closed, bounded subset of R^n (Fig. 8.1). Then, since A is bounded, $A \subset N'(\overline{O}, p)$ for some positive number p; hence

$$A \subset \operatorname{Cl} N'(\overline{O}, p).$$

Now

$$A \subset \operatorname{Cl} N'(\overline{O}, p) \subset \bigtimes_{i=1}^{n} \operatorname{Cl} N(0, p) = \bigtimes_{i=1}^{n} [-p, p].$$

But $[-p, p]$ is compact since it is homeomorphic to $[0, 1]$; hence

$$\bigtimes_{i=1}^{n} \operatorname{Cl} N(0, p)$$

is compact since it is the product of a family of compact spaces (Proposition 13, Chapter 7). Therefore A is a closed subset of a compact T_2-space (any metric space is T_2), and hence A is compact (Corollary 1, Proposition 11, Chapter 7).

Corollary. The closure of any bounded subset of R^n is compact.

Proof. Suppose A is bounded. Then

$$A \subset N'(\overline{O}, p) \subset \text{Cl}\, N'(\overline{O}, p + 1).$$

Therefore $\text{Cl}\, A \subset N'(\overline{O}, p + 1)$. $\text{Cl}\, A$ is closed, and thus $\text{Cl}\, A$ is closed and bounded, and is therefore compact.

The next proposition is true in any compact metric space, but has its application primarily in the study of real functions.

Proposition 3. Let X, D be any compact metric space and suppose $\{U_i\}$, $i \in I$, is an open cover of X. Then there is a positive number p such that $N(x, p) \subset U_i$ for some i, for any $x \in X$. That is, there is $p > 0$ such that the p-neighborhood of any point in X is a subset of at least one of the U_i. Such a number p is called a *Lebesgue number* of the cover, and is dependent on the cover for its value.

Proof. Each element x of X is contained in at least one U_i, since $\{U_i\}$, $i \in I$, is a cover of X. Since each U_i is also open, for each $x \in X$, we may select $p_x > 0$ such that $N(x, p_x) \subset U_i$ for at least one of the U_i which contain x. Since a selection has been made for each x, $\{N(x, p_x/2)\}$, $x \in X$, is itself an open cover of X (in fact, it is a refinement of the original cover). Since X is compact, we can find a finite number of the elements of X, say $x_1, \ldots, x_n$, such that

$$\{N(x_1, p_{x_1}/2), \ldots, N(x_n, p_{x_n}/2)\}$$

is an open cover of X. Let

$$p = \min(p_{x_1}/2, \ldots, p_{x_n}/2).$$

We now show that p is a Lebesgue number for $\{U_i\}$, $i \in I$. If $x \in X$, then $x \in N(x_j, p_{x_j}/2)$ for some $1 \leq j \leq n$. If $z \in N(x, p)$, then

$$D(z, x_j) \leq D(z, x) + D(x, x_j) < p + p_{x_j}/2 \leq p_{x_j}.$$

Therefore

$$N(x, p) \subset N(x_j, p_{x_j}) \subset U_i$$

for some i.

We recall that the definition of continuity of a function from one metric space to another can be expressed: A function $f\colon X, D \to Y, D'$ is continuous if given any $f(x) \in Y$ and $p > 0$, there is $q > 0$ such that

$$f(N(x, q)) \subset N(f(x), p).$$

Note that while p can be chosen arbitrarily, q is dependent on *both* x and p. Even if the same p is used for each $f(x) \in Y$, it may not be possible to find any q which will work for each $x \in X$. This point is illustrated in the following example.

Example 1. Suppose f is the function from $X = \{x \mid 0 < x\} \subset R$ into R defined by $f(x) = 1/x$ for each $x \in X$. It is not hard to verify that f is continuous. As a matter of fact, f is a homeomorphism onto its image. It is also fairly clear (and may easily be proven analytically) that for any $p > 0$ and any $q > 0$, there is $x \in X$ such that

$$f(N(x, q)) \not\subset N(f(x), p)$$

(Fig. 8.2).

We note that X in this example is not compact. If X were compact, we could indeed find a q for each p which was independent of x. To aid our discussion we make the following definition.

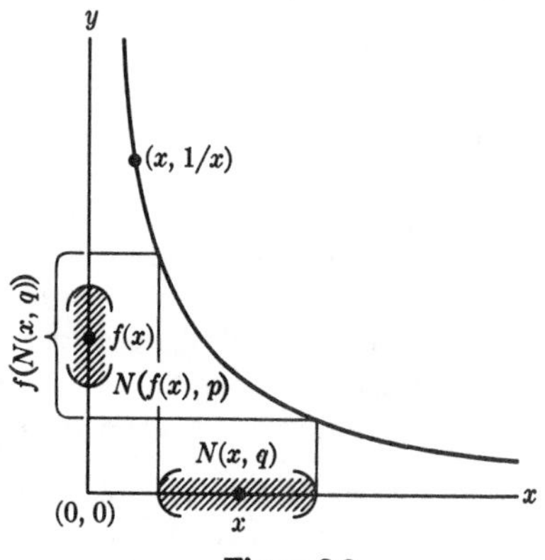

Figure 8.2

Definition 1. Let f be a function from a metric space X, D into a metric space, Y, D'. The function f is said to be *uniformly continuous* if given any $p > 0$, there is $q > 0$ (q depending *only* on p) such that for any

$$x \in X, \qquad f(N(x, q)) \subset N(f(x), p).$$

Example 2. Let f be the function from R to R defined by $f(x) = 3x$. Given any $p > 0$, if we set $q = p/3$, then

$$f(N(x, q)) \subset N(f(x), p)$$

for any $x \in R$. For if $y \in N(x, q)$, then $|x - y| < p/3$, and hence

$$|f(x) - f(y)| = |3x - 3y| = 3|x - y| < 3(p/3) = p.$$

Thus f is uniformly continuous (even though R is not compact).

Proposition 4. Suppose f is a continuous function from a compact metric space X, D into a metric space Y, D'. Then f is uniformly continuous.

Proof. Choose any $p > 0$. Then $\{N(y, p/2)\}$, $y \in Y$, is an open cover of Y. Since f is continuous, $\{f^{-1}(N(y, p/2))\}$, $y \in Y$, is an open cover of X. Let q be the Lebesgue number of this cover in accordance with Proposition 3. It is left as an exercise to prove that this q has the desired property that

$$f(N(x, q)) \subset N(f(x), p)$$

for each $x \in X$.

Corollary. Any function from a closed, bounded subset of R^n into any metric space is uniformly continuous. In particular, any function from a closed interval $[a, b] \subset R$ is uniformly continuous.

EXERCISES

1. The following refer to the proof of Proposition 1.
   a) Prove that D is a metric.
   b) Prove that the collection of all $\mathfrak{N}_x$ and the collection of all $\mathfrak{N}'_x$, are open neighborhood systems for τ and τ', respectively.
2. Complete the proof of Proposition 4.
3. A function f from a space X, τ into the space of real numbers is said to be *bounded above* if $f(x) \leq M$ for some number M and each $x \in X$. What would we mean if we said that f was *bounded below?* Prove that if X is compact and f is continuous, then f is bounded above and below. Prove that if $M =$ least upper bound $\{f(x) \mid x \in X\}$ and $m =$ greatest lower bound $\{f(x) \mid x \in X\}$, f is continuous, then there are w and y in X such that $f(w) = M$ and $f(y) = m$.
4. Which of the functions defined below from R into R are uniformly continuous?
   a) $f(x) = x + 2$, for all $x \in R$
   b) $f(x) = 4x + 7$, for all $x \in R$
   c) $f(x) = \begin{cases} x \sin(1/x), & x \neq 0 \\ 0, & \text{if } x = 0 \end{cases}$
5. Let X, D; Y, D'; and Z, D'' be metric spaces. Decide which of the following statements are true. If a statement is true, prove it; if false, find a counterexample.
   a) If the function f from X to Y and the function g from Y to Z are both uniformly continuous, then $f \circ g \colon X \to Z$ is also uniformly continuous.
   b) Suppose X and Y are both the set of real numbers and D and D' are the absolute value metric. Then if f and g are uniformly continuous functions from X to Y, then $f + g$ defined by $(f + g)(x) = f(x) + g(x)$ is also uniformly continuous.
   c) If f is a homeomorphism from X onto Y and f is uniformly continuous, then f^{-1} is also uniformly continuous.

6. A metric space X, D is said to be *totally bounded* if given any $p > 0$, the open cover $\{N(x, p)\}$, $x \in X$, has a finite subcover. Prove that any bounded subset of R^m (with the metric described earlier) is totally bounded. Prove that a compact subset of X, D is closed and totally bounded. Show that a subset of X, D may be closed and totally bounded yet not be compact.

8.2 LOCAL COMPACTNESS

There are times when a topological space possesses some property "locally" which it does not have taken as a whole. For example, a second countable space has a countable basis for its topology. A space X, τ may not be second countable, but could still have the property that there is an open neighborhood system for τ such that for any $x \in X$, $\mathfrak{N}_x$ is countable; we called such a space *first countable*. In a sense, a first countable space is a space which is locally second countable. Similarly, a space may not be compact, but still have the property that each point is contained in each member of an "appropriate" family of compact sets.

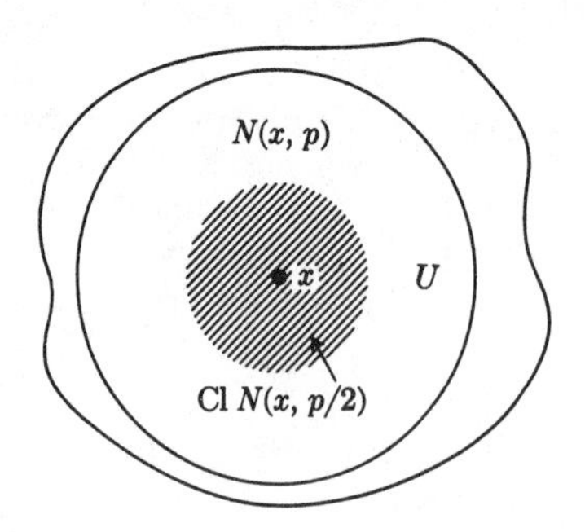

Figure 8.3

Example 3. Let R^2 be the coordinate plane with the Pythagorean metric topology. Then R^2 is not compact, since it is not bounded (Proposition 2). If $x \in R^2$ and U is any neighborhood of x, then there is $p > 0$ such that $N(x, p) \subset U$. Then

$$N(x, p/2) \subset \text{Cl}\, N(x, p/2) \subset N(x, p) \subset U$$

(Fig. 8.3). But $\text{Cl}\, N(x, p/2)$ is a closed, bounded subset of R^2, and hence is compact. We have therefore proved that if $x \in R^2$ and U is any neighborhood of x, then there is a compact set A [here $\text{Cl}\, N(x, p/2)$] such that $x \in A^\circ$ [here $N(x, p/2)$] $\subset A \subset U$. A similar property could be proved for R^n, n finite.

This example inspires the following definition of *local compactness*.

Definition 2. A space X, τ is said to be *locally compact* if given any $x \in X$ and any neighborhood U of x, there is a compact set A such that

$$x \in A^\circ \subset A \subset U.$$

Thus R^n is locally compact. The criterion for local compactness is much simpler for T_2-spaces, as we see from the next proposition.

Proposition 5. Let X, τ be a T_2-space. Then X is locally compact if and only if given any $x \in X$, there is a compact set A such that $x \in A^\circ$. (In other words, the existence of one compact subset A of X such that $x \in A^\circ$ assures us that given any neighborhood U of x, there is a compact set A' such that $x \in A'^\circ \subset A' \subset U$.)

Proof. Suppose X is locally compact and $x \in X$. Since X is a neighborhood of x, there is a compact set A such that

$$x \in A^\circ \subset A \subset X.$$

Suppose instead that given any $x \in X$, there is at least one compact set A with $x \in A^\circ$. Let U be any neighborhood of x. Then $A^\circ \cap U$ is a neighborhood of x and is a subset of U; hence we lose no generality in assuming that U is already a subset of A°. Now A is compact and T_2, and hence the subspace A is T_3 (Corollary 2, Proposition 11, Chapter 7); moreover, $U \cap A = U$ (since $U \subset A^\circ \subset A$) is a nonempty subset of A which is open in A. Therefore there is V, open in both A and in X, such that

$$x \in V \subset \text{Cl } V \text{ (in } A) \subset U \subset A^\circ$$

(Proposition 4, Chapter 5). Since A is a compact subset of a T_2-space, A is closed; thus

$$\text{Cl } V \text{ (in } X) = \text{Cl } V \text{ (in } A).$$

Then as a closed subset of a compact T_2-space, Cl V is compact. Therefore

$$x \in V = V^\circ \subset \text{Cl } V \subset U,$$

and Cl V is compact; hence X is locally compact.

Corollary. Any compact T_2-space X, τ is locally compact.

Proof. X is a compact neighborhood of any $x \in X$.

Example 4. The space X given in Example 10, Chapter 7 is only T_1, but is still locally compact. For if $x \in X$, then any neighborhood U of X is compact. Therefore $x \in U = U^\circ \subset U$ and U is compact; hence X is locally compact.

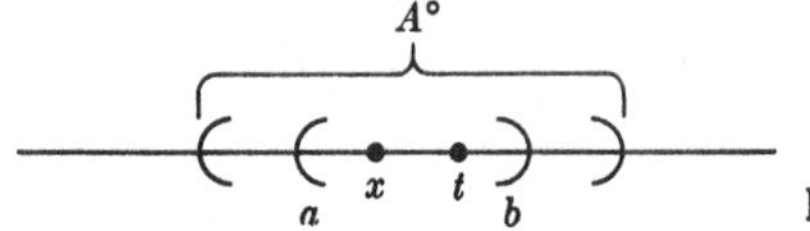

Figure 8.4

Example 5. Let Q be the subspace of rational numbers in the space R of real numbers with the absolute value topology. Then Q is not locally compact. Let $x \in Q$ and suppose A is a compact subset of Q such that $x \in A°$ (Fig. 8.4). Then A contains infinitely many elements of Q. There is $(a, b) \subset R$ such that $x \in (a, b) \cap Q \subset A°$. Choose an irrational number $t \in (a, b)$. We will now construct an open cover of A which has no finite subcover. For each $q \in A$, set

$$U(q) = \begin{cases} \{w \in R \mid q < w\}, \text{ if } t < q, \\ \{z \in R \mid z < q\}, \text{ if } q < t. \end{cases}$$

Then $\{U(q) \cap A\}$, $q \in A$, is an open cover of A which has no finite subcover. The proof of this fact is left as an exercise.

We see then that no element of Q can be contained in the interior of any compact subset of Q. Therefore Q is an example of a metric space in which not every closed bounded subset is compact. For example, $[0, 1] \cap Q$ is a closed, bounded subset of Q, but could not be compact; for if it were compact, then $\frac{1}{2}$ would be contained in the interior, $(0, 1) \cap Q$, of a compact subset of Q.

We saw in Chapter 7 that any compact T_2-space was T_4. Since local compactness is a weaker property than compactness, we should expect weaker results from local compactness than from compactness, as is the case with the following.

Proposition 6. Any locally compact T_2-space X, τ is T_3.

Proof. We apply Proposition 4 of Chapter 5. If $x \in X$ and U is any neighborhood of x, then there is a compact set A such that $x \in A° \subset A \subset U$. Since A is compact, A is closed, and hence $\text{Cl}(A°) \subset A$. Setting $V = A°$, we have $x \in V \subset \text{Cl } V \subset U$, where V is a neighborhood of x; therefore X is T_3.

We see from Example 5 that a subspace of a locally compact space need not be locally compact. We do, however, have the following proposition regarding subspaces of locally compact spaces.

Proposition 7. If a space X, τ is T_2 and locally compact, then so is every open or closed subspace.

Proof. Suppose U is an open subspace of X and $x \in U$. Then any neighborhood V of x in U is also a neighborhood of x in X. Therefore there is a compact set A such that $x \in A° \subset A \subset V \subset U$. Hence U is locally compact. Note that this part of the proof did not depend on the fact that X was T_2; thus we have shown that any open subspace of any locally compact space is locally compact.

Suppose F is a closed subspace of X and $x \in F$. Let A be any compact set such that $x \in A^\circ \subset A$. Since X is T_2, A is closed. Then $F \cap A$ is a closed subset of the compact set A and hence is compact. But $F \cap A \subset F$; hence we also have $x \in (A \cap F)^\circ$ in $F \subset A \cap F \subset F$. Since F is T_2, F is locally compact by Proposition 5.

We now prove an even stronger result.

Proposition 8. A subspace Y of a locally compact T_2-space X, τ is locally compact if and only if it is the intersection of an open set and a closed set.

Proof. Suppose Y is a locally compact subspace of X (Fig. 8.5). We will prove that Y is open in $\mathrm{Cl}\,Y$; hence $Y = U \cap \mathrm{Cl}\,Y$, where U is an open subset of X. Suppose $y \in Y$; we must find a neighborhood of y (in $\mathrm{Cl}\,Y$) which is a subset of Y. Since Y is locally compact, there is a set U' open in Y such that $y \in U'$ and $\mathrm{Cl}\,U'$ in Y is compact. Then $U' = Y \cap V$, where V is open in X. Furthermore, $\mathrm{Cl}\,U'$ in $Y = Y \cap \mathrm{Cl}(Y \cap V)$ is compact, and hence is closed. Now

$$Y \cap V \subset Y \cap \mathrm{Cl}(Y \cap V);$$

hence $\mathrm{Cl}(Y \cap V) \subset Y$. But

$$\mathrm{Cl}Y \cap V \subset \mathrm{Cl}(Y \cap V).$$

For if $z \in \mathrm{Cl}\,Y \cap V$ and W is any neighborhood of z, $V \cap W$ is a neighborhood of z. Since $z \in \mathrm{Cl}\,Y$, every neighborhood of z meets Y, and thus

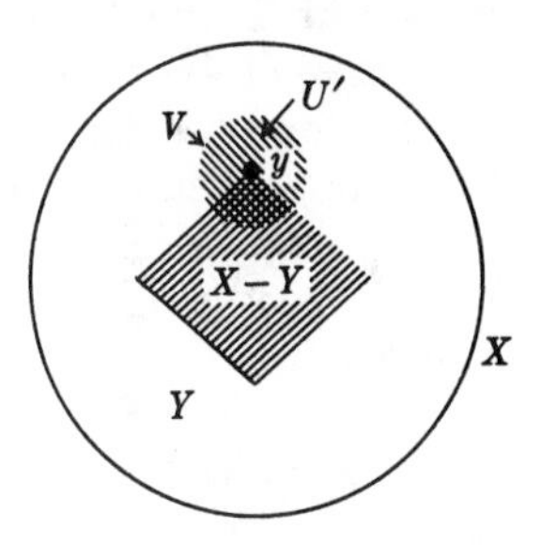

Figure 8.5

$$(V \cap W) \cap Y = W \cap (Y \cap V) \neq \phi.$$

But then every neighborhood of z meets $Y \cap V$ as well; hence

$$z \in \mathrm{Cl}(Y \cap V).$$

Therefore $\mathrm{Cl}\,Y \cap V \subset Y$. Thus $\mathrm{Cl}\,Y \cap V$ is a neighborhood of y in $\mathrm{Cl}\,Y$ such that $\mathrm{Cl}\,Y \cap V \subset Y$.

It is left as an exercise to show that the intersection of a closed subset and an open subset of X is locally compact.

We now investigate the behavior of locally compact spaces with regard to continuous functions. The following example shows that local compactness, unlike compactness, is not preserved by continuous functions.

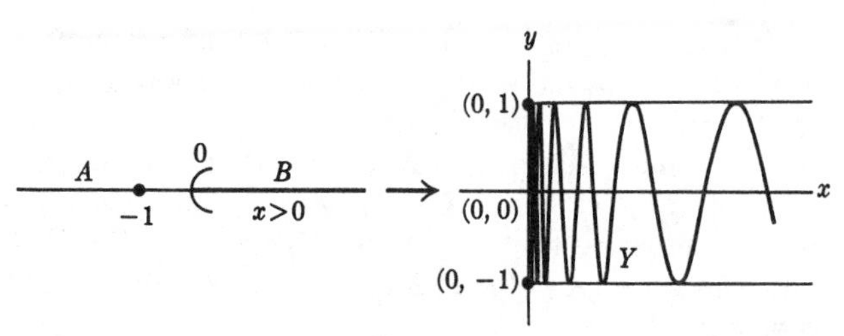

Figure 8.6

Example 6. Let $A = \{-1\}$ and $B = \{x \mid 0 < x\}$ (Fig. 8.6). Let $X = A \cup B$ be given the absolute value topology. Then X is the intersection of a closed subset of R, the usual space of real numbers, with an open subset of R [for example, $X = (\{-1\} \cup \{x \mid 0 \leq x\}) \cap (R - \{0\})$]; hence X is locally compact. Define a function f from X into R^2 (with the Pythagorean topology) by

$$f(x) = \begin{cases} (0, 0), \text{ if } x \in A, \\ (x, \sin 1/x), \text{ if } x \in B. \end{cases}$$

Then $f \mid A$ and $f \mid B$ are both continuous, and A and B are both closed subsets of X; thus f is continuous (Proposition 11, Chapter 4). Let Y be the image of f considered as a subspace of R^2. The function f is a continuous, one-one function from X onto Y. But Y is not locally compact. This can be seen from the fact that $(0, 0)$ is not contained in the interior of any compact subset of Y. This in turn follows from the fact that each neighborhood U in Y of $(0, 0)$ contains a sequence which does not have a limit point in Cl U.

A function $f: X, \tau \to Y, \tau'$ is said to be *open* if whenever U is an open subset of X, $f(U)$ is open in Y (*cf.* Section 4.6, Exercise 2). With the added assumption of openness, a continuous function will preserve local compactness.

Proposition 9. If f is a continuous, open function from a space X, τ onto a space Y, τ', then if X is locally compact, Y is also.

Proof. Suppose $y \in Y$ and U is a neighborhood of y. We must find a compact subset A of Y such that $y \in A^\circ \subset A \subset U$. Let $y = f(x)$ for some $x \in X$. By the continuity of f, there is a neighborhood V of x such that $f(V) \subset U$. Since X is locally compact, there is a compact set B such that $x \in B^\circ \subset B \subset V$. Then

$$f(x) = y \in f(B^\circ) \subset f(B) \subset U.$$

But $f(B^\circ)$ is open, since f is open; and $f(B)$ is compact, since B is compact and f is continuous. Therefore

$$y \in f(B^\circ) \subset f(B^\circ) \subset f(B) \subset U,$$

and hence Y is locally compact.

We now use Proposition 9 to study the relation between local compactness and product spaces.

Proposition 10. Suppose $\times_I X_i$ is the product space of the countable family of nonempty spaces $\{X_i, \tau_i\}$, $i \in I$. Then $\times_I X_i$ is locally compact if and only if each component space is locally compact and all of the component spaces except at most finitely many are compact.

Proof. Suppose $\times_I X_i$ is locally compact. Then the projection

$$p_i \colon \mathop{\times}_{I} X_i \to X_i$$

is continuous, onto, and open (Section 4.6, Exercise 2) for each $i \in I$. Therefore by Proposition 9, each X_i is locally compact. We must also show that all but at most finitely many of the X_i are compact. Let A be any compact subset of $\times_I X_i$ such that some point y of $\times_I X_i$ is in A°. Then there is a basic neighborhood $\times_I V_i$ of y such that $V_i = X_i$ for all but at most finitely many i and

$$\mathop{\times}_{I} V_i \subset A^\circ \subset A.$$

We therefore see that $p_i(A) = X_i$ for all but at most finitely many i. Since p_i is continuous and A is compact, X_i is compact for all but at most finitely many i.

Suppose each X_i is locally compact, and all but finitely many of the X_i are compact. Let $y \in \times_I X_i$, and let y_i be the ith coordinate of y. If U is any neighborhood of y, then U contains a basic neighborhood of y of the form $\times_I V_i$, where V_i is open in X_i for each $i \in I$ and $V_i = X_i$ for all $i \in I$, except for at most finitely many, say $i_1, \ldots, i_n$. Since each X_i is locally compact, for each $i \in I$ there is a compact subset A_i of X_i such that $y_i \in A_i^\circ \subset A_i \subset V_i$. There are at most finitely many more $i \in I$, other than $i_1, \ldots, i_n$, say $i_{n+1}, \ldots, i_m$, such that $X_{i_{n+1}}, \ldots, X_{i_m}$ are not compact. For any i not in $\{i_1, \ldots, i_n, i_{n+1}, \ldots, i_m\}$, we may let $A_i = X_i$. Then

$$y \in \mathop{\times}_{I} A_i^\circ \subset \Big(\mathop{\times}_{I} A_i\Big)^\circ \subset \mathop{\times}_{I} A_i \subset \mathop{\times}_{I} V_i.$$

But $\times_I A_i$ is the product of compact sets and is therefore compact (Proposition 13, Chapter 7). Hence $\times_I X_i$ is locally compact.

EXERCISES

1. In Example 6 show that each neighborhood U of $(0, 0)$ in Y contains a sequence which does not converge to any point of Cl U(in Y). Why does this prove that $(0, 0)$ is not contained in the interior (in Y) of any compact subset of Y?
2. Prove that any subspace which is the intersection of a closed subset and an open subset of a locally compact T_2-space is locally compact, thus completing the proof of Proposition 8. [*Hint:* Use Proposition 7.]
3. Provide the details for Example 5. In particular, show that $\{U(q) \cap A\}$, $q \in A$, is an open cover of A which has no finite subcover.
4. Which of the following subspaces of the plane R^2 with the Pythagorean topology are locally compact?
   a) $R^2 - \{(0, 0)\}$
   b) $\{(x, y) \mid x \text{ and } y \text{ are both rational}\}$
   c) $\{(x, y) \mid x \text{ and } y \text{ are both integers}\}$
   d) $R^2 - \bigcup_N \{C \mid C \text{ is a circle of radius } 1/n \text{ with center } (0,0)\}$, where N is the set of positive integers
   e) $R^2 - \{(x, y) \mid x^2 + y^2 < 1, \text{ or } x = 0 \text{ or } 1, \text{ and } y = 0 \text{ or } 1\}$
5. If X, τ is a space and R is an equivalence relation on X, is the identification mapping from X onto X/R, the identification space, necessarily open? If X is locally compact, must X/R be locally compact? Give examples to prove your points.
6. Is the union of finitely many locally compact subspaces of any space always a locally compact subspace? Is the intersection of two locally compact subspaces locally compact?
7. Suppose X, τ is a locally compact T_2-space which is second countable. Prove that X is the union of countably many compact subsets $A_1, A_2, \ldots, A_n, \ldots$ such that $A_n \subset A_{n+1}^\circ$. Example: $R^2 = \bigcup_N D_n$, where $D_n = \text{Cl } N((0, 0), n)$.
8. Find an example of a compact space which is not locally compact.

8.3 COMPACTIFICATIONS

Compact spaces are perhaps the most important of all topological spaces. It is therefore of interest to know if and how any given space can be embedded as a subspace of a compact space. If any space X, τ can be embedded as a subspace W of a compact space Y, τ', then X can be embedded as a dense subspace of some compact space. For W is a dense subspace of Cl W; this follows from Propositions 13 and 14 of Chapter 3. But Cl W is a closed subset of a compact space and hence is compact. We therefore restrict our attention to considering whether or not a given space can be embedded as a dense subspace of a compact space. Accordingly, we make the following definition.

Definition 3. Let X, τ be any topological space. By a *compactification* of X is meant a compact space Y, τ' such that X is homeomorphic to a dense subspace of Y.*

The type of compactifications that a space X, τ has will, of course, depend a lot on X. If X is not T_2, then no compactification of X could possibly be T_2, since being a subspace of a T_2-space would make X a T_2-space also. A space may have many different compactifications, perhaps infinitely many, but not all of these compactifications are of interest, and many of those which are important are beyond the ambitions of this text.

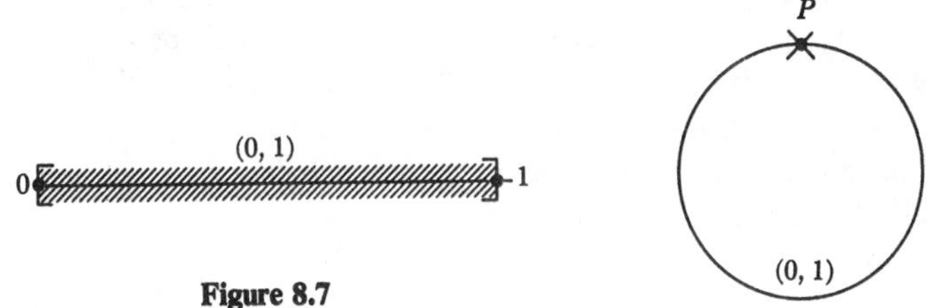

Figure 8.7 **Figure 8.8**

Example 7. Let the open interval (0, 1) have the absolute value topology. Then two compactifications of (0, 1) are the circle [note that any circle less one point is homeomorphic to (0, 1)] and the closed interval [0, 1] (Figs. 8.7, 8.8). Note that these two compactifications are not homeomorphic to one another.

The compactification which we will study in this text is the *Alexandroff*, or *one-point*, compactification of a T_2- space.

Definition 4. Let X, τ be any T_2 topological space. If X is compact, then we define the *Alexandroff compactification* of X to be X, τ itself. Suppose X is not compact. Let P be any point which is not in X, and set $Y = X \cup \{P\}$. Define a topology on Y as follows: U is open in Y if U is an open subset of X; or if $P \in U$, then $Y - U$ is a compact subset of X. Y with the topology thus defined is called the *Alexandroff*, or *one-point*, compactification of X. P is called an *ideal point.*

It must now be shown that (i) Y is a topological space; (ii) Y is compact; and (iii) X is dense in Y.

---

* Technically, a compactification of a space X is a space Y *and* a homeomorphism from X onto a dense subspace of Y. Thus, technically, if X can be embedded in Y in more than one way, Y would actually be part of several compactifications of X, i.e., one for each distinct embedding. The author feels that the definition of compactification given in the text is more appropriate for an introductory text, however.

Proposition 11. Let X, τ be any T_2- space. Then the Alexandroff compactification Y of X is a topological space and is a compactification of X in the sense of Definition 3.

Proof. If X is already compact, the proposition is trivial. Suppose X is not compact. If $y \in Y$, set

$$\mathfrak{N}_y = \{U \mid U \text{ is open in } Y \text{ and } y \in U\},$$

that is, $\mathfrak{N}_y$ is the family of all neighborhoods of y. We will show that the collection of $\mathfrak{N}_y$ forms an open neighborhood system for τ', a topology on Y in which the open sets are those described in Definition 4. Since, if $y \in X$, the neighborhoods of y in Y are the same as the neighborhoods of y in X, we need only consider the case when $y = P$, the ideal point. We now verify Definition 5 of Chapter 3 for $\mathfrak{N}_P$.

i) Since any one-point subset $\{x\}$ of X is compact, $X - \{x\}$ is a neighborhood of P; therefore $\mathfrak{N}_P \neq \phi$.

ii) By assumption, $P \in U$ for each $U \in \mathfrak{N}_P$.

iii) Suppose U and U' are neighborhoods of P. Then $Y - U = K$ and $Y - U' = K'$, where K and K' are compact subsets of X. Then

$$(Y - U) \cup (Y - U') = Y - (U \cap U') = K \cup K'.$$

But $K \cup K'$ is compact since it is the union of two compact sets (Section 7.3, Exercise 8). Therefore $U \cap U'$ is a neighborhood of P.

iv) Suppose U is any neighborhood of P and $z \in U$. If $z = P$, then $z \in U \in \mathfrak{N}_z$ and $U \subset U$. If $z \in X$, then $U - \{P\}$ is an open subset of X; for $X - U$ is compact, and X is T_2. Therefore $X - U$ is closed; hence

$$X - (X - U) = U - \{P\}$$

is open in X, and hence also in Y. Therefore

$$z \in U - \{P\} \subset U \qquad \text{and} \qquad U - \{P\} \in \mathfrak{N}_z.$$

The verification of (v) is left as a simple exercise. Therefore the collection of $\mathfrak{N}_x$ forms an open neighborhood system for a topology τ' on Y which is precisely the family of open sets defined for Y.

Since every neighborhood of any point in Y meets X, X is dense in Y. It remains to be shown that Y is compact. Let $\{U_i\}$, $i \in I$, be any open cover of Y. Then $P \in U_i$ for some i, say i'. Since $U_{i'}$ is a neighborhood of P, $Y - U_{i'}$ is a compact subset of X and $\{U_i\}$, $i \in I$, is an open cover of $Y - U_{i'}$. Then finitely many of the U_i, say $U_{i_1}, \ldots, U_{i_n}$, cover

$Y - U_{i'}$; hence

$$\{U_{i'}, U_{i_1}, \ldots, U_{i_n}\}$$

is a finite subcover of $\{U_i\}$, $i \in I$. Therefore Y is compact.

Example 8. The one-point compactification of (0, 1) as in Example 7 is the circle. Note that (0, 1) also has a two-point compactification [0, 1]. The one-point compactification of the space R of real numbers (with the usual topology) is again a circle, since R is homeomorphic to (0, 1). It is given as an exercise to prove that homeomorphic spaces have homeomorphic one-point compactifications. Usually the ideal point for the space of real numbers is taken to be ∞.

Example 9. If X is an infinite set with the discrete topology and P is an ideal point for the one-point compactification Y of X, then the neighborhoods of P will be all subsets of Y which contain all but finitely many points of X, since the finite subsets of X are the only compact subsets of X.

Note that X is always an open subset of its one-point compactification, since X is open in X. This implies that the subset containing only the ideal point of any one-point compactification is closed. We thus see that if X is T_2, then its one-point compactification is at least T_1. We now investigate conditions under which a one-point compactification is T_2.

Proposition 12. The Alexandroff compactification of any space X, τ is T_2 if and only if X is T_2 and locally compact.

Proof. Suppose X is compact. Then the Alexandroff compactification of X is X itself. Now X is T_2 if X is T_2 and locally compact; on the other hand, if X is T_2, then X is T_2 and is also locally compact by the corollary to Proposition 5. Assume that X is not compact, and let Y be the Alexandroff compactification of X. If Y is T_2, then X is T_2, since X is a subspace of Y. Now Y is T_2 and compact and is therefore locally compact. But X is an open subspace of Y; hence X is locally compact (Proposition 7).

On the other hand, suppose X is T_2 and locally compact. Let x and y be distinct points of Y. If x and y are both in X, then since X is T_2, there are neighborhoods U and V of x and y, respectively, such that $U \cap V = \phi$. Suppose $x = P$. Then $y \in X$, and hence there is a compact subset A of X such that $y \in A^\circ \subset A$. We have then that $Y - A$ is a neighborhood of $x = P$ and that A° is a neighborhood of y with

$$A^\circ \cap (Y - A) = \phi.$$

Therefore Y is T_2.

Corollary 1. If X, τ is a locally compact T_2-space, then the one-point compactification of X is normal.

Proof. Any compact T_2-space is both T_1 and T_4 (Corollary 3, Proposition 11, Chapter 7).

Corollary 2. Any locally compact T_2-space X, τ is regular.

Proof. The one-point compactification of X is normal, and hence is also regular. Since X is a subspace of a regular space, X is regular.

Note that we had already proved Corollary 2 previously (Proposition 6), but that the use of compactifications gives a simple, elegant proof for the result. Note too that since we have found examples of T_2-spaces which are not locally compact (e.g., as in Example 5), we therefore have the one-point compactifications of such spaces as examples of compactifications of T_2-spaces which are not T_2.

EXERCISES

1. Verify that the circle is the one-point compactification of $(0, 1)$ (Example 8).
2. Prove that if a T_2- space X, τ is homeomorphic to a space X', τ', then the one-point compactifications of these spaces are homeomorphic. Show that two spaces might have homeomorphic one-point compactifications even though the spaces are not homeomorphic to one another.
3. Describe the one-point compactifications of each of the following subspaces of R^2 with the usual Pythagorean topology. Where practicable, sketch the compactification.

   a) $\{(x, y) \mid x \in (0, 1], y = 0\}$ b) $\{(1/n, 1/n) \mid n = 1, 2, 3, \ldots\}$
   c) $\{(x, y) \mid x^2 + y^2 < 1\}$ d) $\{(x, y) \mid x^2 + y^2 < 1\} \cup \{(0, 1)\}$
   e) $\{(x, y) \mid -1 \leq x \leq 1\}$
4. Which of the following are compactifications of $\{(x, y) \mid x^2 + y^2 < 1\}$ with the Pythagorean topology? Each of the following spaces is to be considered as a subspace of Euclidean n-space R^n for an appropriate n.

   a) $\{(x, y, z) \mid x^2 + y^2 + z^2 = 1\}$ b) $\{(x, y, z) \mid x^2 + y^2 = 1, 0 \leq z \leq 1\}$
   c) $\{(x, y) \mid x^2 + y^2 \leq 1\}$ d) $\{(x, y, z) \mid x^2 + y^2 + z^2 \leq 1\}$
   e) $\{(x, y) \mid |x| \leq 1, |y| \leq 3\}$
6. Suppose f is a continuous function from a T_2-space X, τ into a T_2-space Y, τ'. Can f necessarily be extended to a continuous function from the one-point compactification of X into Y? Suppose f can be extended to a continuous function F from Z, the one-point compactification of X, into Y. Is $f(Z)$ necessarily a compactification of $f(X)$? Is $f(Z)$ necessarily the one-point compactification of $f(X)$ if it is a compactification?
7. Suppose f is a continuous function from X, τ onto Y, τ', and let X' and Y' be the one-point compactifications of X and Y, respectively. Define $F: X' \to Y'$ by $F(x) = f(x)$ if $x \in X$, and $f(P) = P'$, where P and P' are the ideal points of X and Y, respectively. Is F necessarily continuous?

8. a) Describe the *two-point compactification* of the open interval (0, 1) with its usual topology. How can this two-point compactification be defined rigorously, that is, constructed from (0, 1).
   b) Sketch the one-, two-, three-, and four-point compactifications of $(0, 1) \cup (2, 3) \cup (4, 5)$.

8.4 SEQUENTIAL AND COUNTABLE COMPACTNESS

There are certain properties which a space can have which are related to, but do not have the full force of, compactness. We have already seen one such property, local compactness. The purpose of this section is to introduce two other such properties, *sequential compactness* and *countable compactness.*

Definition 5. A space X, τ is said to be *countably compact* if any *countable* open cover of X has a finite subcover. X is said to be *sequentially compact* if every sequence in X has a convergent subsequence.

Any space which is compact is clearly countably compact. Also, since any open cover of a Lindelöf space has a countable subcover, a countably compact Lindelöf space is compact.

Proposition 13 will give another criterion for countable compactness.

Proposition 13. Let X, τ be any space. Suppose $B \subset X$; then a point $x \in X$ is said to be an *accumulation point* of B if every neighborhood of x contains infinitely many points of B. Then X, τ is countably compact if and only if every countably infinite subset of X has at least one accumulation point.

Proof. Suppose that B is a countably infinite subset of X, but that B does not have an accumulation point, and assume X countably compact. Select $x_1, x_2, \ldots, x_n, \ldots$, a sequence of distinct points of B. Set

$$A_n = \{x_n, x_{n+1}, \ldots\} \qquad \text{and} \qquad C_n = X - A_n.$$

Given any point $y \in X$, y is not an accumulation point of B; hence there is a neighborhood U_y of y such that $U_y \cap B$ contains at most finitely many elements. Therefore $x_n \in U_y$ for at most finitely many n. If n is sufficiently large, then $U_y \cap A_n = \phi$, and therefore $U_y \subset C_n$. Set $V_n = C_n^\circ$. Then $\{V_n\}$, $n \in N$, is a countable open cover of X. But since X is countably compact, there is a finite subcover, say $\{V_1, \ldots, V_m\}$, of $\{V_n\}$, $n \in N$. Now x_{m+1} is not an element of $C_1 \cup \cdots \cup C_m$; hence x_{m+1} cannot be an element of $V_1 \cup \cdots \cup V_m$, a contradiction since $V_1 \cup \cdots \cup V_m = X$. Consequently, if X is countably compact, B must have an accumulation point.

Suppose now that every countably infinite subset of X has an accumulation point, but suppose that we can find a countable open cover $\{U_n\}$, $n \in N$, for which there is no finite subcover. Then for any $n \in N$, $X - \bigcup_{j=1}^{n} U_j \neq \phi$. Pick $x_1 \in X - U_1$; suppose $x \in U_{n_1}$. Then pick

$$x_2 \in X - (U_1 \cup \cdots \cup U_{n_1}).$$

Suppose x_k has been chosen and $x_k \in U_{n_k}$. Choose

$$x_{k+1} \in X - (U_1 \cup \cdots \cup U_{n_k}).$$

By the manner in which they were chosen, these points must all be distinct; thus the set

$$B = \{x_k \mid k = 1, 2, \ldots\}$$

is a countably infinite subset of X. Then B has an accumulation point y, and $y \in U_n$ for some n. But if k' is large enough, $n < n_{k'}$; hence $x_k \notin U_n$ for $k > k'$. Hence U_n is a neighborhood of y which meets B in only finitely many elements, contradicting the fact that y is an accumulation point of B. There must therefore be a finite subcover of $\{U_n\}$, $n \in N$, and hence X is countably compact.

Proposition 14. If a space X, τ is sequentially compact, it is countably compact.

Proof. Suppose a space X, τ is sequentially compact and B is any countable infinite subset of X. Then we can find a sequence $\{s_n\}$, $n \in N$, where $s_n \in B$ for each n, and no two s_n are equal. Then $\{s_n\}$, $n \in N$, has a convergent subsequence; consequently, $\{s_n\}$, $n \in N$, has a limit point y (Proposition 9, Chapter 6). But then y is accumulation point of B. Hence X is countably compact by Proposition 13.

The reader may have already noted that the famous Bolzano-Weierstrass theorem from real analysis is really a statement that any compact subset of R or R^2 (that is, a closed, bounded subset) is sequentially compact.

The reader should also note that the distinction between countably compact and sequentially compact is hairline thin. For in any countably compact space, any sequence either takes some value infinitely often, or else the set of points in the sequence is an infinite set and hence has an accumulation point. This does not imply, however, that a sequence has a subsequence which converges, but only that it has a subnet which converges, and the distinction in this case is fine indeed (although we are justified in saying that any first countable space is sequentially compact if and only if it is countably compact). In "nice" topological spaces, sequential and countable compactness are equivalent; examples showing that countable compactness does not imply sequential compactness are rather esoteric.

One of the nicest types of topological space is the metric space. We have already seen that in a metric space the properties of being second countable, Lindelöf, or separable are all equivalent. We now will show that in a metric space the properties of being countably compact, sequentially compact, or compact are equivalent. We do this in two propositions.

Proposition 15. Any countably compact metric space X, D is separable.

Proof. For any positive number p, there is a maximal subset E_p of X such that for any $a, b \in E_p$, $D(a, b) \geq p$. The proof will closely follow the lines of the proof that (a) implies (b) in Proposition 5, Chapter 7. If E_p were infinite for any $p > 0$, then it would have an accumulation point y. But then $N(y, p/2)$ would contain infinitely many points of E_p; hence any two of these points would be closer together than p, a contradiction. Therefore E_p is finite for each p. However, given any $x \in X$,

$$N(x, p) \cap E_p \neq \phi,$$

or we would have a contradiction to the maximality of E_p.

Take $E_{1/n}$ for each positive integer n. Then $\bigcup_N E_{1/n}$ is a countable dense subset of X, and hence X is separable.

Corollary 1. Any countably compact metric space is compact.

Proof. By Proposition 5, Chapter 7, any separable metric space is Lindelöf. But any countably compact Lindelöf space is compact.

Since any sequentially compact metric space is countably compact (Proposition 14), we have the following.

Corollary 2. Any sequentially compact metric space is compact.

Proposition 16. If X, D is any metric space, then the following statements are equivalent:

a) X is compact.
b) X is countably compact.
c) X is sequentially compact.

Proof. It remains to be shown that if X is countably compact, then X is sequentially compact. Suppose that X is countably compact and $\{s_n\}$, $n \in N$, is a sequence in X. If $s_n = y$ for infinitely many n, then a subsequence of $\{s_n\}$, $n \in N$, converges to y (Exercise 1). Suppose then that no value is assumed by the s_n more than a finite number of times; in fact, we lose no generality in assuming that the s_n are all distinct. Since $\{s_n \mid n \in N\}$ is an infinite set, it has an accumulation point y. Set $U_n = N(y, 1/n)$ for each $n \in N$. Then $U_n \cap \{s_n\}$ is infinite for each $n \in N$. Choose

$$s_{n_1} \in \{s_n\} \cap U_1.$$

Choose $s_{n_2} \in \{s_n\} \cap U_2, n_1 < n_2$. In general, choose

$$s_{n_k} \in \{s_n\} \cap U_k, n_{k-1} < n_k.$$

Then $\{s_{n_1}, \ldots, s_{n_k}, \ldots\}$ is a subsequence of $\{s_n\}$, $n \in N$, which converges to y; hence X is sequentially compact.

There are a number of other properties which are in some way related to compactness, for example, *metacompactness*, *pseudocompactness*, and the very important property of *paracompactness*. We shall examine this latter concept briefly in Section 10.5.

EXERCISES

1. Prove that if a sequence assumes some value infinitely many times, then a subsequence of the sequence converges to that value.
2. Prove that in first countable spaces, countable compactness and sequential compactness are equivalent.
3. Suppose f is a continuous function from a space X, τ onto a space Y, τ'. Prove that if X is countably compact, then Y is also. Prove that if X is sequentially compact, then Y is also.
4. Prove that a closed subset of a countably compact or sequentially compact space is also countably or sequentially compact.
5. Suppose f is a continuous function from a countably compact space X, τ into the space R of real numbers with the absolute value topology. Prove that there are numbers m and M such that $m \leq f(x) \leq M$ for all $x \in X$, that is, prove that f is bounded.
6. Let X be any set and suppose τ and τ' are two possible topologies on X. Suppose X, τ is compact and τ' is coarser than τ. Is X, τ' compact? Suppose X, τ is countably compact or sequentially compact and τ' is coarser than τ. Is X, τ' necessarily countably compact or sequentially compact?
7. Find an example of a separable metric space which is not countably compact. Is it possible to have a nonseparable metric space which is countably compact?
8. Need the set of accumulation points of any set be closed? Need a set together with its accumulation points be closed? Prove that any set together with its accumulation points is closed provided the space is T_2.

9
CONNECTEDNESS

9.1 THE NOTION OF CONNECTEDNESS

If we tried to intuitively guess what might be meant by saying that a topological space is "connected," we would probably include the space R of real numbers with the absolute value topology among the connected spaces, and the set $\{0, 1\}$ with the discrete topology among the spaces which are not connected. This is because intuitively we would imagine a disconnected space as a space which could be broken up into disjoint parts. Of course, any space can be broken up into "parts." The real numbers, for example, can be expressed as the union of the set of irrational numbers with the set of rational numbers. Such a decomposition into disjoint parts of any kind is not sufficient to make the space disconnected. Obviously, too, if connectedness is to be a topological notion, then whether or not a space is connected should depend on its topology. Note that $\{0, 1\}$ with the discrete topology can be expressed as the union of the disjoint open sets $\{0\}$ and $\{1\}$, but it does not seem that R can be expressed as the union of two disjoint nonempty open sets. With these considerations in mind, we make the following definition.

Definition 1. A space X, τ is said to be *disconnected* if X can be expressed as the union of two disjoint, open, nonempty subsets of X. Otherwise, X is said to be *connected.* That is, X is connected if there do not exist disjoint, nonempty, open subsets U and V of X such that $X = U \cup V$. A subset A of X is said to be *connected* if the subspace A is connected.

Example 1. If X is any set consisting of two or more points and X is given the discrete topology, then X is disconnected. For choosing any $x \in X$, $\{x\}$ and $X - \{x\}$ are then disjoint, open, nonempty subsets of X whose union is X.

On the other hand, if X is given the trivial topology, then X is connected, since then the only nonempty open subset of X is X itself.

Proposition 1. The closed interval $[0, 1]$ with the absolute value topology is connected.

Proof. Suppose $[0, 1]$ is disconnected. Then $[0, 1] = U \cup V$, where U and V are disjoint, nonempty, open subsets of $[0, 1]$. Suppose $u \in U$ and $v \in V$. We may assume $u < v$ (relabeling U and V, if necessary). Let S be the set of numbers s such that $s < u$ or $[u, s] \subset U$. Then S has 1 as an upper bound, and hence S has a least upper bound a with $0 < a < 1$. Since $[0, 1] = U \cup V$, either $a \in U$ or $a \in V$. Suppose $a \in U$. Since U is open, there is $p > 0$ such that $(a - p, a + p) \subset U$. Then

$$[a - p/2, a + p/2] \subset U$$

(Fig. 9.1); hence $a + p/2 \in S$. This contradicts the assumption that a is an upper bound for S. Suppose $a \in V$. Then there is $p > 0$ such that $(a - p, a + p) \subset V$, and thus $a - p/2$ is an upper bound for S, contradicting the assumption that a is the least upper bound. Therefore $[0, 1]$ is not disconnected; hence $[0, 1]$ is connected.

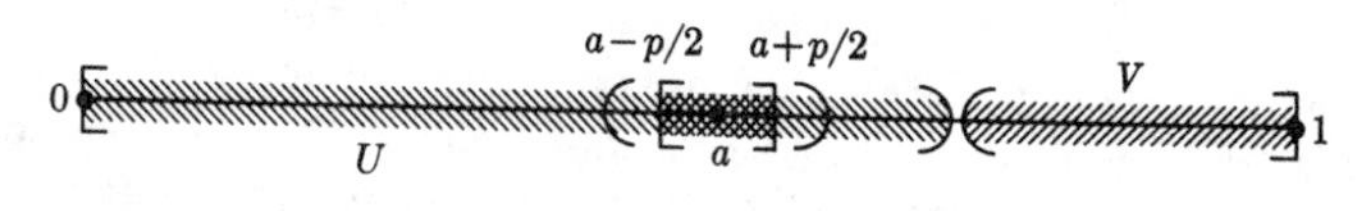

Figure 9.1

Note that since connectedness has been defined in a negative way, that is, a space is connected if it is not disconnected, most proofs that a space is connected are by contradiction: a space is assumed to be disconnected and a contradiction is proved.

Connectedness, like compactness, is a property which is preserved by continuous functions.

Proposition 2. Suppose f is a continuous function from a space X, τ onto a space Y, τ'. If X is connected, then so is Y.

Proof. Assume Y is not connected. Then $Y = U \cup V$, where U and V are nonempty, disjoint, open subsets of Y. Then $f^{-1}(U)$ and $f^{-1}(V)$ are disjoint, nonempty (since f is onto) subsets of X whose union is X. But since f is continuous, $f^{-1}(U)$ and $f^{-1}(V)$ are also open subsets of X; therefore X is disconnected, a contradiction. Hence Y must be connected.

Corollary 1. Any closed interval in R, any closed line segment in R^2, and, in general, any image of $[0, 1]$ under a continuous function is connected.

Corollary 2. If X, τ is any connected space and R is an equivalence relation on X, then the identification space X/R is connected.

Proof. The identification mapping from X onto X/R is continuous.

Example 2. Since a circle in R^2 (with the Pythagorean topology) can be thought of as an identification space formed from $[0, 1]$, the circle is also connected.

We now give some more criteria for connectedness.

Proposition 3. Let X, τ be any topological space. Then the following statements are equivalent.

a) X is connected.
b) X cannot be expressed as the union of two disjoint, nonempty, closed subsets.
c) The only subsets of X which are open and closed are X and ϕ.
d) If A is any subset of X other than X or ϕ, then $\text{Fr } A \neq \phi$.
e) Let $Y = \{0, 1\}$ have the discrete topology. Then there is no continuous function from X onto Y.

Proof. Statement (a) implies statement (b). Suppose $X = A \cup B$, where A and B are disjoint, nonempty, closed subsets of X. Then $X - A = B$ and $X - B = A$ are both the complements of closed sets, and hence are open. Thus $X = A \cup B$ is also the expression of X as the union of two disjoint, nonempty, open subsets of X. Hence X is not connected.

Statement (b) implies statement (c). Suppose A is a subset of X which is both open and closed, but that A is neither X nor ϕ. Then $X - A$ is also open and closed, and nonempty. Thus

$$X = (X - A) \cup A$$

is the expression of X as the union of two disjoint, nonempty, closed subsets, contradicting (b).

Statement (c) implies statement (d). If A is a subset of X other than X or ϕ, and $\text{Fr } A = \phi$, then since $\text{Cl } A = A^\circ \cup \text{Fr } A$, we have $\text{Cl } A = A^\circ$. On the other hand, $A^\circ \subset A$ and $A \subset \text{Cl } A$, and hence $A = A^\circ = \text{Cl } A$; thus A is both open and closed in X. Therefore if (c) holds, there can be no subset A of X, other than X or ϕ, such that $\text{Fr } A = \phi$.

Statement (d) implies statement (a). Suppose $X = U \cup V$, where U and V are disjoint, open, nonempty subsets of X. Then U and V are also closed. Therefore

$$U = U^\circ = \text{Cl } U.$$

But $\text{Fr } U = \text{Cl } U - U^\circ$ (Proposition 12, Chapter 3); hence

$$\text{Fr } U = U - U = \phi,$$

a contradiction of (d).

Statement (a) implies statement (e). Suppose there is a continuous function from X onto $Y = \{0, 1\}$. Then since X is connected, Y must be also, which is not the case.

Statement (e) implies statement (a). Suppose X is disconnected. Then $X = U \cup V$, where U and V are nonempty, disjoint, open subsets of X. Define $g: X \to Y$ by $g(x) = 0$, if $x \in U$, and by $g(x) = 1$, if $x \in V$. Then $g^{-1}(\{0\}) = U$ and $g^{-1}(\{1\}) = V$; hence g is continuous, a contradiction of (*e*).

We now see why the extension F in Example 15 of Chapter 5 cannot be continuous.

Example 3. There are a number of ways that we can show that the space N in Example 3 of Chapter 7 is connected. For example, let A be any subset of N. Suppose A is infinite, but not all of N. Since a subset of N is closed if and only if it is finite, the only closed set which contains A is N; hence $\text{Cl } A = N$. Then

$$\text{Fr } A = N - A \neq \phi$$

since $A \neq N$. Suppose A is finite but nonempty. Then A is closed and thus $\text{Cl } A = A$. But since A excludes infinitely many elements of N, $A^\circ = \phi$; hence $\text{Fr } A = A \neq \phi$. By (d) of Proposition 3, N is connected.

Proposition 4. Suppose X, τ is a space such that $X = U \cup V$, where U and V are disjoint, open, nonempty subsets of X. Let A be any connected subspace of X. Then either $A \subset U$ or $A \subset V$.

Proof. If $A \cap U \neq \phi$ and $A \cap V \neq \phi$, then $A \cap U$ and $A \cap V$ are nonempty, disjoint subsets of A which are open in A. But

$$A = (A \cap U) \cup (A \cap V);$$

thus A is not connected. Therefore either $A \cap U = \phi$ and hence $A \subset V$, or $A \cap V = \phi$ and hence $A \subset U$.

Example 4. Let R be the space of real numbers with the absolute value topology. Then the removal of any point y from R disconnects R into two "rays," open half-lines

$$H^+ = \{x \in R \mid y < x\} \qquad \text{and} \qquad H^- = \{x \in R \mid x < y\}.$$

The removal of y also disconnects any interval which contains y as anything but an endpoint. For if, say, $[a, b]$ contains y in its interior and $[a, b] - \{y\}$ is connected, then $[a, b] - \{y\}$ must lie entirely in either H^+ or H^-, an impossibility.

EXERCISES

1. Find another proof that the space N in Example 3 is connected.
2. Prove that each of the following subspaces of the space of real numbers with the absolute value topology is disconnected.
   a) any finite subset
   b) $(0, 1) \cup (6, 7) \cup (9, 18)$
   c) $\{x \mid x \text{ is irrational}\}$
   d) $\{x \mid x = 1/n, \text{ where } n \text{ is a positive integer or } x = 0\}$
3. Modify the proof of Proposition 1 to show that $(0, 1)$, and hence R, is connected.
4. A space X, τ is said to be *totally disconnected* if X is not connected and the only connected subspaces of X are ϕ and subspaces which consist of only one point. Prove that each of the following spaces are totally disconnected.
   a) the subspace of rational numbers in the usual space of real numbers
   b) any discrete space of more than one point
   c) the set of real numbers with the topology described in Chapter 3, Example 11
5. Suppose A and B are connected subspaces of a connected space X, τ. Show by producing an example that the following need not be connected.
   a) $A \cap B$ b) $A \cup B$ c) $\operatorname{Fr} A$ d) A°
6. Let X be a space with the property that given any $x \in X$ and any neighborhood U of x, there is a neighborhood V of x such that $\operatorname{Cl} V$ is a proper subset of U. Show that X need not be connected. What additional condition on V or U would result in making X connected?

9.2 FURTHER TESTS FOR CONNECTEDNESS

In this section we continue to investigate criteria for determining if a space is connected.

If the reader did Exercise 5 of Section 9.1 and his example for (b) was correct, it must have been that $A \cap B = \phi$, as we see from the next proposition.

Proposition 5. If X, τ is a space and $X = \bigcup_I A_i$, where $\{A_i\}$, $i \in I$, is a collection of connected subspaces of X, then if $\bigcap_I A_i \neq \phi$, X itself is connected.

Proof. Suppose $X = U \cup V$, where U and V are disjoint open subsets of X. Then for each i, either $A_i \subset U$ or $A_i \subset V$ (Proposition 4). If some $A_i \subset U$, then since $\bigcap_I A_i \neq \phi$, some element from each A_i must be in U, and hence every A_i is in U. Therefore we would have $\bigcup_I A_i = X \subset U$ and $V = \phi$. Similarly, if some $A_i \subset V$, then $X \subset V$ and $U = \phi$. We have therefore shown that X could not be expressed as the union of two disjoint, nonempty, open subsets; hence X is connected.

Example 5. We have seen that the space R of real numbers with the absolute value topology is connected (Section 9.1, Exercise 3). Any straight line in Euclidean n-space is homeomorphic to the real line R. This implies that R^n is connected for any n, since R^n is the union of all straight lines in R^n which pass through the origin. The family of such lines therefore fulfills the hypotheses of Proposition 5.

Proposition 6. Let X, τ be a space such that any two elements x and y of X are contained in some connected subspace of X. Then X is connected.

Proof. Let x be a fixed element of X. For any $y \in X$, let $C(x, y)$ be a connected subspace of X which contains x and y. Then $\{C(x, y)\}$, $y \in X$, is a family of connected subspaces of X whose union is X and whose intersection is nonempty (since the intersection at least contains x). Proposition 5 tells us that X is connected.

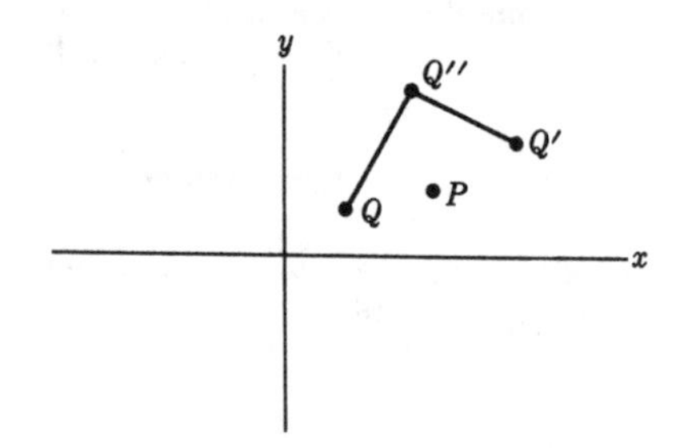

Figure 9.2

Example 6. Any closed line segment in Euclidean n-space R^n is homeomorphic to the closed interval $[0, 1]$, and is hence a connected subspace of R^n. Using this fact, we can show that $R^n - \{P\}$, where P is any point of R^n and $2 \leq n$, is connected. For suppose Q and Q' are any two points of $R^n - \{P\}$. Choose

$$Q'' \in R^n - \{P\}$$

such that $P \notin \overline{QQ''} \cup \overline{Q''Q'}$ (Fig. 9.2). Then $\overline{QQ''} \cup \overline{Q''Q'}$ is connected by Proposition 5, that is, it is the union of the connected subspaces

$$\overline{QQ''}, \qquad \overline{Q''Q'},$$

and

$$\overline{QQ''} \cap \overline{Q''Q'} = \{Q''\} \neq \phi.$$

Therefore Q and Q' are in the same connected subspace of $R^n - \{P\}$. By Proposition 6, then $R^n - \{P\}$ is connected.

We have already seen that since [0, 1] is connected, any homeomorphic image of [0, 1], for example, a closed line segment in R^n, is connected. More generally, of course, any continuous image of [0, 1] is connected. The continuous images of [0, 1] form an important class of spaces known as *paths*. More formally, we make the following definition.

Definition 2. Let X, τ be any space. A subspace Y of X is said to be a *path in* X if there is a continuous function from [0, 1] (with the absolute value topology) onto Y. X is said to be *path connected* if, given any two points x and y in X, there is a path in X containing x and y.

Suppose $X = R^m$, Euclidean m-space. A subset W of R^m is said to be *polygonally connected* if given any two points x and y in W, there are points

$$x_0 = x, x_1, \dots, x_{n-1}, x_n = y$$

such that $\bigcup_{i=1}^n \overline{x_{i-1}x_i} \subset W$, where $\overline{x_{i-1}x_i}$ is the closed segment joining x_{i-1} and x_i (Fig. 9.3).

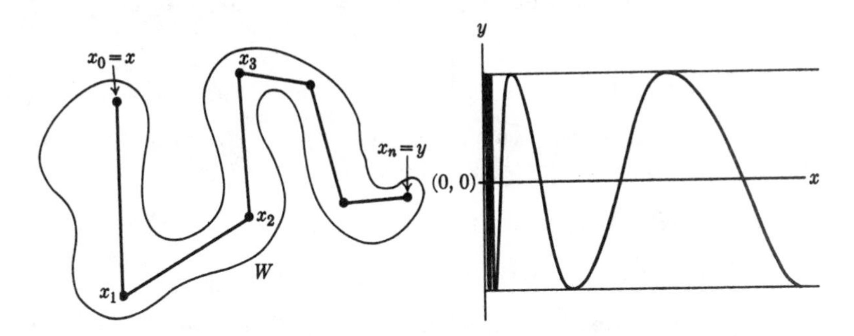

Figure 9.3

Figure 9.4

Any subset of R^n which is path connected is not necessarily polygonally connected. For example, the circle

$$\{(x, y) \mid x^2 + y^2 = 1\} \subset R^2$$

is path connected (it is itself a path), but it is not polygonally connected. On the other hand, any subspace of R^n which is polygonally connected is path connected (Exercise 1). Paths can actually be rather exotic, and may not look anything like [0, 1]. For example, it can be shown that $([0, 1])^n$ is a path for any finite n.

The next example gives a connected subspace of R^2 which is not path connected.

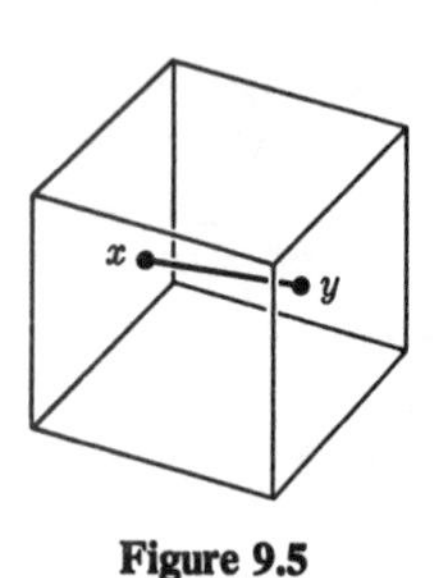

Figure 9.5

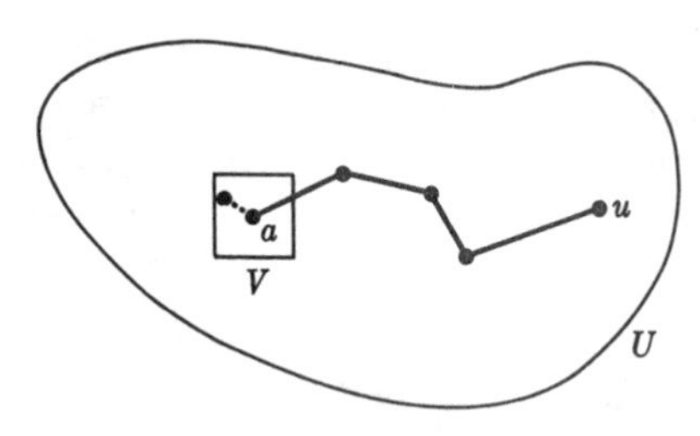

Figure 9.6

Example 7. Let

$$Y = \{(x, y) \mid y = \sin(1/x), x > 0\} \cup \{(0, 0)\} \subset R^2$$

(Fig. 9.4). We shall see from Proposition 11 that Y is connected. If P is any point in Y other than (0, 0), then there is no path in Y which contains (0, 0) and P. For if there were such a path, it would be possible to show that the function f from the space of nonnegative real numbers in R for which Y is the graph is continuous. But f is not continuous (Section 7.4, Exercise 5). Also see Exercise 2 below.

Proposition 7. If X, τ is a path-connected space, then X is connected.

Proof. Suppose X is path connected and $x \in X$. For each $y \in X$, let $P(x, y)$ be a path which contains x and y. Then

$$X = \cup \{P(x, y) \mid y \in X\} \qquad \text{and} \qquad \cap \{P(x, y) \mid y \in X\} \neq \phi.$$

Therefore X is connected by Proposition 6.

Recall that a subset W of Euclidean n-space R^n is *convex* if, given any points x and y of W, the closed segment $\overline{xy}$ is a subset of W. We note that the basic neighborhoods in R^n, considered as the n-fold product of R, are convex subsets of R^n (Fig. 9.5). The "open balls" of the form

$$\{(x_1, \ldots, x_n) \mid x_1^2 + \cdots + x_n^2 < p\},$$

where $p > 0$, are also convex subsets of R^n. Of course any convex subset of R^n is polygonally connected, and hence path connected, and therefore connected (in a very "strong" way).

We now prove a theorem that has wide use in analysis.

Proposition 8. Let U be a connected open subset of R^n. Then U is polygonally connected.

Proof. Choose $u \in U$. Let $A = \{a \in U \mid a$ can be polygonally connected to u in $U\}$ (Fig. 9.6) and $B = U - A$. Then A is open. For since U is open, given any $a \in A$, there is a basic product neighborhood V of a such that $V \subset U$. But V is convex; hence if u can be polygonally connected to a in U, then u can also be polygonally connected in U to any point of V; hence $V \subset A$. Therefore A is open. On the other hand, since B is the set of points of U which cannot be polygonally connected to u in U, a similar argument shows that B is open. Thus $U = A \cup B$, where A and B are disjoint open subsets of U. Since U is connected, it follows that either $A = \phi$ or $B = \phi$. Since $u \in A$, $A \neq \phi$. Therefore $B = \phi$; hence $U = A$ and U is polygonally connected.

We have seen that R^n, Euclidean n-space, is connected for any positive integer n. We intuitively feel, however, that if $n \neq m$, then R^n is not homeomorphic to R^m. In order to confirm our intuitive feeling, we would have to find some topological property of R^n which is not possessed by R^m. In its full generality, this problem is not particularly easy; however, we will prove the following.

Proposition 9. If $n > 1$, then R^n is not homeomorphic to R.

Proof. The space R of real numbers (with the usual topology) has the property that if $x \in R$, then $R - \{x\}$ is disconnected (see Example 4). If R^n were homeomorphic to R, $n > 1$, then R^n would also have the property that if $P \in R^n$, $R^n - \{P\}$ is not connected. But this is not the case (Example 6). Therefore R^n is not homeomorphic to R.

EXERCISES

1. Prove that any polygonally connected subspace of R^n is path connected.
2. Give a detailed proof that Y as described in Example 7 is not path connected.
3. Prove that the only connected subspaces of the space of real numbers with the absolute value topology are intervals (open, closed, or half-open), R, one-point subspaces, rays (open and closed), and ϕ. [*Hint:* First prove that each of the subspaces mentioned is connected, using Propositions 1 and 6 or 7. Suppose W is a connected subspace of R. Break down the discussion into cases where (1) W has neither an upper nor a lower bound; (2) W has one bound, but not both; and (3) W has both an upper and a lower bound.]
4. What are the only compact connected subsets of R? Use this information to prove the following.
   a) If f is a continuous function from a closed interval $[a, b]$ into R and $f(a) < c < f(b)$, then there is $x \in [a, b]$ such that $f(x) = c$.
   b) If f is a continuous function from a closed interval $[a, b]$ into R, then

   $$m = \text{glb}\{f(x) \mid x \in [a, b]\} \qquad \text{and} \qquad M = \text{lub}\{f(x) \mid x \in [a, b]\}$$

   exist; moreover, there are x and x' in $[a, b]$ such that $f(x) = m$ and $f(x') = M$.

5. Let R^2 be the plane with the Pythagorean metric. Prove that $R^2 - C$, where C is any countable set, is polygonally connected. In particular, prove that

$$R^2 - \{(x, y) \mid x \text{ and } y \text{ are rational}\}$$

is polygonally connected. [*Hint:* Through any point in $R^2 - C$, there is a line which does not intersect C.]

6. Which of the following subspaces of R^2 are connected? Indicate clearly how you arrived at your conclusion.

   a) $\{(x, y) \mid y = (1/n)x, n = 1, 2, 3, \ldots\}$
   b) $\{(x, y) \mid \text{either } x \text{ or } y\text{, but not both, is irrational}\}$
   c) $\{(x, y) \mid x \neq 1\}$
   d) $\{(x, y) \mid x \neq 1\} \cup \{(0, 1)\}$

7. Suppose X, τ is a space such that $X = A_1 \cup \cdots \cup A_n$, where each A_i is connected and $A_{i-1} \cap A_i \neq \phi$, $i = 2, \ldots, n$. Is X necessarily connected?

8. Suppose A is a compact subspace of Euclidean n-space R^n, $n \geq 2$. Prove that $R^n - A$ need not be connected.

9. Prove directly (that is, do not refer to Corollary 2 to Proposition 11 of the next section) that if X is connected, then the one-point compactification of X is also connected.

9.3 CONNECTEDNESS AND THE DERIVED SPACES

We have already seen that if X, τ is a connected space, then any identification space derived from X is also connected. In this section we investigate the behavior of connectedness as regards subspaces and product spaces.

It is, of course, false that any subspace of a connected space is connected. The following proposition gives a criterion for determining whether or not a subspace of a given space is connected.

Proposition 10. If A is a subspace of the space X, τ, then A is connected if and only if A cannot be expressed as $S \cup T$, where S and T are nonempty subsets of X and

$$S \cap \text{Cl } T = \text{Cl } S \cap T = \phi.$$

(Note that no demand is made that S and T be open or closed in A.)

Proof. If A is not connected, then $A = S \cup T$, where S and T are disjoint, nonempty subsets of A which are open and closed in A. Suppose

$$x \in S \cap \text{Cl } T.$$

Then since $S \subset A$, $x \in A \cap \text{Cl } T = \text{Cl } T$ in A (Chapter 4, Proposition

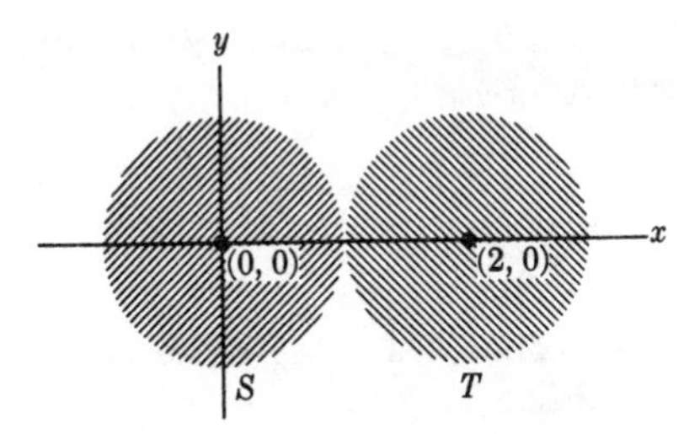

Figure 9.7

4) $= T$. Therefore $x \in S \cap T = \phi$, a contradiction. Then $S \cap \text{Cl } T = \phi$; similarly, $\text{Cl } S \cap T = \phi$.

Suppose that $A = S \cup T$, where $\text{Cl } S \cap T = S \cap \text{Cl } T = \phi$, and S and T are nonempty. Then

$$\begin{aligned} \text{Cl } S \text{ in } A &= A \cap \text{Cl } S = (S \cup T) \cap \text{Cl } S \\ &= (S \cap \text{Cl } S) \cup (T \cap \text{Cl } S) = S \cup \phi = S. \end{aligned}$$

Therefore S is closed in A; similarly, T is closed in A. Hence A is disconnected.

Corollary. Two subsets S and T of a space X, τ are said to be *mutually separated* if

$$\text{Cl } S \cap T = S \cap \text{Cl } T = \phi.$$

Suppose S and T are mutually separated subsets of X and A is a connected subspace of $S \cup T$. Then either $A \subset S$, or $A \subset T$.

The proof of this corollary is left as an exercise.

Example 8. Let R^2 be the plane with the Pythagorean topology,

$$S = \{(x, y) \mid x^2 + y^2 < 1\}$$

and

$$T = \{(x, y) \mid (x - 2)^2 + y^2 < 1\}.$$

(Fig. 9.7). Then $\text{Cl } S \cap T = \text{Cl } T \cap S = \phi$. Therefore $S \cup T$ is a disconnected subspace of R^2. Note, however, that

$$\text{Cl } S \cap \text{Cl } T = \{(1,0)\} \neq \phi.$$

Proposition 11. Suppose A is a connected subspace of X, τ and

$$A \subset Y \subset \text{Cl } A.$$

Then Y is also a connected subspace of X.

Proof. If Y is disconnected, then $Y = S \cup T$, where S and T are mutually separated (Proposition 10). Since A is connected, either $A \subset S$ or $A \subset T$, by the corollary to Proposition 10. Suppose $A \subset S$. Then $\text{Cl } A \subset \text{Cl } S$;

hence $Y \subset \text{Cl } A \subset \text{Cl } S$. But then since $Y = S \cup T$, $T \subset \text{Cl } S$. Since $T \cap \text{Cl } S = \phi$, we have arrived at a contradiction. Therefore Y is connected.

Corollary 1. If a space X, τ contains a connected dense subspace, then X is connected.

Proof. Suppose A is a connected dense subspace of X. Then $\text{Cl } A = X$ is connected by Proposition 11.

Corollary 2. If X, τ is connected, then any compactification Y of X is connected.

Proof. If Y is a compactification of X, then X is a dense connected subspace of Y; therefore Y is connected, by Corollary 1.

Example 9. We see that the space Y in Example 7 is connected as follows: Set $A = Y - \{(0, 0)\}$, and define a function h from $\{x \mid 0 < x\} \subset R$ into R^2 by

$$h(x) = (x, \sin (1/x)).$$

Then h is easily seen to be continuous; in fact, h is a homeomorphism onto its image A. Therefore, since $\{x \mid 0 < x\}$ is connected, A is connected. Now $(0, 0)$ is in $\text{Cl } A$ since every neighborhood of $(0, 0)$ contains infinitely many points of A (cf. Fig. 9.4). Then $A \subset Y \subset \text{Cl } A$; hence, by Proposition 11, Y is connected.

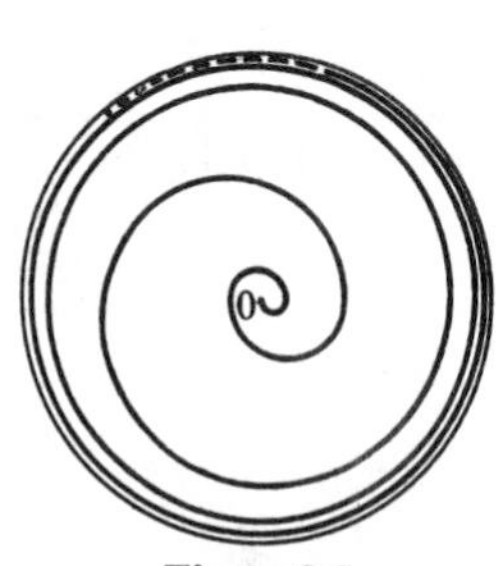

Figure 9.8

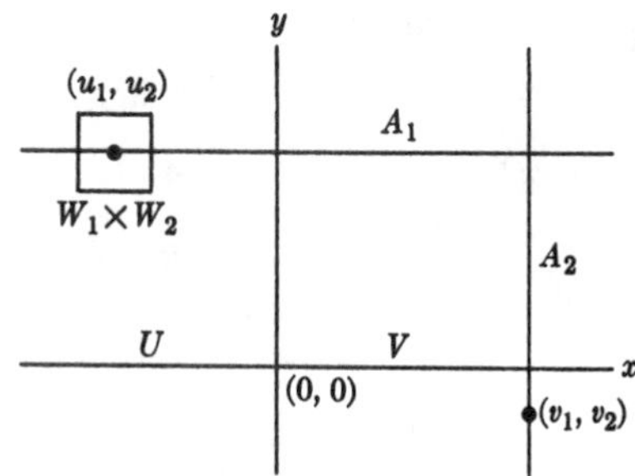

Figure 9.9

Example 10. Consider the graph Y of the equation $r = 1 - 1/A$, $A > 1$, in polar coordinates (considered as a subspace of R^2 with the usual topology) (Fig. 9.8). It may be verified that

$$\text{Cl } Y = Y \cup \{(r, A) \mid r = 1\}$$

(again in polar coordinates.) Thus Y together with any set of points on the unit circle forms a connected subspace of R^2.

We now investigate connectedness and product spaces.

Proposition 12. The product space $\times_I X_i$ of the countable family of nonempty spaces

$$\{X_i, \tau_i\}, \qquad i \in I,$$

is connected if and only if each X_i is connected.

Proof. Suppose each X_i is connected, but $\times_I X_i = U \cup V$, where U and V are disjoint, open, nonempty subsets of $\times_I X_i$. Choose $u \in U$ and $v \in V$ (Fig. 9.9). Since U is open, there is a basic neighborhood $\times_I W_i$ of u such that $\times_I W_i \subset U$, W_i is open in X_i, and $W_i = X_i$, except for $i_1, \ldots, i_n$. Define $c_i^0 = v_i$ (the ith coordinate of v) if $i \neq i_1, \ldots, i_n$, but $c_i^0 = u_i$ if $i = i_1, \ldots, i_n$. Then

$$c^0 = (c_1^0, \ldots, c_i^0, \ldots) \in \underset{I}{\times} W_i \subset U.$$

Define c^1 by letting $c_i^1 = v_i$ if $i \neq i_1, \ldots, i_n$; $c_i^1 = v_i$ if $i = i_1$; and $c_i^1 = u_i$ if $i = i_2, \ldots, i_n$. Generally, define c^m by setting

$$c_i^m = v_i, \quad i \neq i_{m+1}, \ldots, i_n, \qquad \text{and} \qquad c_i^m = u_i, \quad i = i_{m+1}, \ldots, i_n.$$

Then $c^n = v$.

Let

$$A_m = \{x \mid x_{i_m} \text{ is arbitrary}, x_{i_{m+1}} = u_{i_{m+1}}, \ldots, x_{i_n} = u_{i_n}, \text{ and otherwise, } x_i = v_i\}.$$

Then $c^{m-1} \in A_{m-1} \cap A_m$, $1 \leq m \leq n$; $v = c^n \in A_n$. $A_1, \ldots, A_n$ form a "chain" from U to V, for, since $c^m \in A_m \cap A_{m+1}$,

$$A_m \cap A_{m+1} \neq \phi.$$

We now show that each A_i is connected.

Define a function g_m from X_{i_m} onto A_m as follows: $g(x_{i_m})$ is the point of A_m with i_mth coordinate x_{i_m}; i_{m+1}th coordinate $u_{i_{m+1}}, \ldots, i_n$th coordinate u_{i_n}; and for all other i the ith coordinate v_i. Not only is g_m continuous, but also it is a homeomorphism (since it is the inverse of the projection from A_m onto X_{i_m}. Since each X_{i_m} is connected, each A_m is also connected. But then $\bigcup_{i=1}^m A_m$ is connected (Section 9.2, Exercise 7). But $\bigcup_{i=1}^m A_m$ meets both of the disjoint, nonempty, open sets U and V, a contradiction of Proposition 4. Therefore Y must be connected.

If $\times_I X_i$ is connected, then since the projection mapping p_i from $\times_I X_i$ onto X_i is continuous, X_i is also connected.

EXERCISES

1. Prove the corollary to Proposition 10.
2. a) Prove that a space X, τ is connected if and only if given any continuous function f from X into R, the space of real numbers with the absolute value topology, then if a and b are in $f(X)$ and c is a real number such that $a \leq c \leq b$, there is $y \in X$ such that $f(y) = c$.
   b) Use (a) to prove that any connected normal space which contains at least two points contains at least c points, where c is the cardinality of $[0, 1]$.
   c) Prove that any connected separable metric space X, D has either at most one point, or exactly c points. [*Hint:* By (b), if X has 2 points, it has at least c points. Take a countable dense subset $S = \{s_n \mid n \in N\}$. Each point of X is the limit of some sequence of elements of S, and each sequence of elements of S has at most one limit. How many sequences of elements of S are there?]
3. Determine whether each of the spaces below is connected or disconnected.
   a) $\{(x, y) \mid y = 1/x\} \cup \{(x, y) \mid y = 0\} \subset R^2$ with the usual topology
   b) $\{(x, y) \mid x^2 + y^2 < 1\} \cup (x, y) \mid x = 1\} \subset R^2$ with the usual topology
   c) the metric space described in Example 5 of Chapter 2
   d) the plane R^2 with the topology described in Example 5 of Chapter 3
4. A subset A of a space X, τ is said to *disconnect* X if $X - A$ is disconnected. We say that A is a *minimal disconnecting subset* of X if $X - A$ is disconnected, but $X - B$ is not disconnected, where B is any proper subset of A.
   a) Describe some minimal disconnecting subsets for Euclidean n-space, R^n.
   b) If $x \in X$ and $\{x\}$ is a minimal disconnecting subset of X, then x is said to be a *cut point* of X.
      i) Prove that if f is a homeomorphism from a space X, τ onto a space Y, τ', then if x is a cut point of X, $f(x)$ is a cut point of Y.
      ii) Prove that a circle, closed interval, open interval, and half-open interval are not homeomorphic in pairs. [*Hint:* Examine the cut points of each.]
5. A connected subset A of a space X is said to be *irreducibly connected about* $B \subset A$ if A is the smallest connected set which contains B. For example, $[0, 1]$ is irreducibly connected about $\{0, 1\}$ in the usual space R of real numbers. Find sets about which the following subspaces of R^2 are irreducibly connected. If no such set exists, write *none*.
   a) $\{(x, y) \mid 0 \leq x \leq 1, y = 0\}$
   b) $\{(x, y) \mid 0 \leq x \leq 1, 0 \leq y \leq 1\}$
   c) $\{(x, y) \mid x^2 + y^2 = 1\}$
   d) $\{(x, y) \mid x^2 + y^2 > 1\}$

9.4 COMPONENTS. LOCAL CONNECTEDNESS

A disconnected space may, of course, have connected subspaces. The structures of the maximal connected subspaces of any space are indispensable in any description of the entire structure of the space. Unfortunately, even knowing completely the structure of every maximal connected subset of a space does not determine the structure of the space as a whole, as we see from the following.

Example 11. Let Q be the space of rational numbers with the absolute value topology. Then Q is totally disconnected (Section 9.1, Exercise 4); hence the maximal connected subsets of Q are the one-point subsets. If Q were to have the discrete topology, then it would still be true that the maximal connected subsets of Q are the one-point subsets. Note however that relative to the metric topology, each one-point subspace of Q is closed in Q, but not open; but that with respect to the discrete topology, each one-point subset of Q is both open and closed. But clearly we cannot tell which topology Q has just by knowing that each maximal connected subspace of Q is a subspace of exactly one point.

Definition 3. Let X, τ be any space. A maximal connected subspace of X is said to be a *component* of X.

Thus the components of Q as in Example 11 are the one-point subspaces of Q; this is true with respect to either the discrete or the absolute-value topology on Q.

Example 12. In Example 8, S and T are the components of the subspace $S \cup T$ of R^2. In Example 4, H^- and H^+ are the components of $R - \{x\}$.

It has not yet been shown that each element of a space X, τ is actually contained in a component of X; that is, although it is clear that each $x \in X$ is contained in at least one connected subspace of X, i.e., $\{x\}$, we must show that there is a maximal connected subspace of X which contains x. This is easily done, however. For let $\{A_i\}$, $i \in I$, be the family of all connected subspaces which contain x. Then $\bigcup_I A_i$ is a connected subspace which contains x (Proposition 5), and $\bigcup_I A_i$ is clearly a maximal connected subspace of X which contains x.

We saw in Example 11 that a component may or may not be open. The following shows, though, that a component is always closed.

Proposition 13. A component A of a space X, τ is closed.

Proof. By Proposition 11, $\text{Cl}\, A$ is also connected. But $A \subset \text{Cl}\, A$ and A is a maximal connected subspace of X; hence $A = \text{Cl}\, A$. Therefore A is closed.

Proposition 14. Let $\{A_i\}$, $i \in I$, be the set of components of a space X, τ. (We see from Example 11 that $\{A_i\}$, $i \in I$, need not be a finite set.) Then $X = \bigcup_I A_i$, and if $i \neq j$, then $A_i \cap A_j = \phi$.

Proof. Since each $x \in X$ is in some component of X (at least one connected subspace of X contains x, hence a maximal connected subspace of X contains x), $\bigcup_I A_i = X$.

If $i \neq j$, but $A_i \cap A_j \neq \phi$, then $A_i \cup A_j$ is a connected subspace of X which contains both A_i and A_j, a contradiction to the maximality of both A_i and A_j.

Just as we could in a sense localize the notion of compactness, we can also localize connectedness. As we shall see, spaces which are *locally connected* have particularly nice components.

Definition 4. A space X, τ is said to be *locally connected* if there is an open neighborhood system for τ such that for each $x \in X$, $\mathfrak{N}_x$ consists of connected subspaces. Equivalently, X is *locally connected* if given any $x \in X$ and any neighborhood U of x, there is a connected neighborhood V of x such that $V \subset U$ (Exercise 1).

Example 13. If X is any space with the discrete topology, then for each $x \in X$, if we set $\mathfrak{N}_x = \{\{x\}\}$, we obtain an open neighborhood system for the discrete topology. Each subspace $\{x\}$ is, of course, connected. Thus, even though X is totally disconnected, X is still locally connected.

Example 14. Euclidean n-space R^n has a basis for its topology which consists of convex, and therefore connected, subspaces (*cf.* the remarks preceding Proposition 8). But if a space has a basis which consists of connected subspaces, it is certainly locally connected (Proposition 6, Chapter 3).

Euclidean n-space is thus both connected and locally connected. We now give an example of a space which is connected, but not locally connected.

Example 15. The space

$$Y = \{(x, y) \mid y = \sin(1/x), 0 < x\} \cup \{(0, 0)\}$$

is connected as we saw in Example 9. But it is not locally connected. As a matter of fact, if $0 < p < 1$, then $N((0, 0), p)$ has an infinite number of components (Fig. 9.10). The space Y, however, is the image of a locally connected space under a continuous function; hence we see that local connectedness is not generally preserved by a continuous function. Specifically, let

$$A = \{(x, y) \mid y = \sin(1/x), 0 < x\}$$

and

$$B = \{(-1, 0)\}.$$

Let $X = A \cup B \subset R^2$. Define a function f from X onto Y by $f(-1, 0) = (0, 0)$, and $f \mid A$ is the identity mapping. Then f is continuous, since $f \mid A$ and $f \mid B$ are both continuous and A and B are closed subsets of X. However, X is locally connected, since A is homeomorphic to R and since B, the other component of X, is an isolated point.

Since R^2 is locally connected, this example also shows that not every subspace of a locally connected space is locally connected.

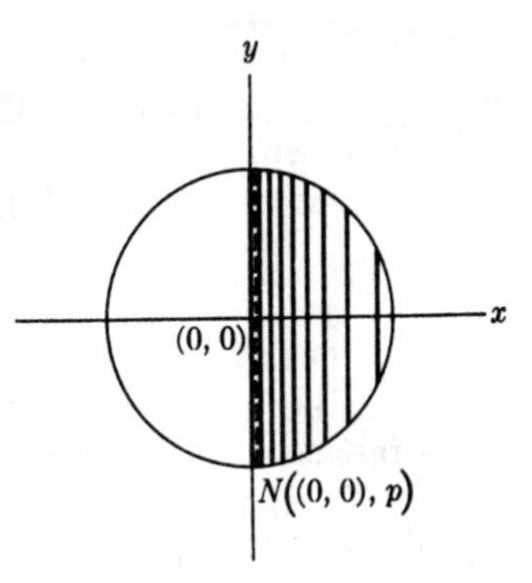

Figure 9.10

Example 16. The previous example can be modified so as to provide an example of a compact, second countable, metric space which is connected, but not locally connected. Define

$$Y' = \mathrm{Cl}\, Y - \{(x, y) \mid x > 3\}.$$

Then Y' is a closed, bounded subspace of R^2, and hence is compact. But if $0 < p < 1$, $N((0, 0), p)$ is not a connected neighborhood of $(0, 0)$. Note too that Y' is locally compact (Proposition 5, Chapter 8), whereas Y is not.

The next proposition gives further criteria for local connectedness.

Proposition 15. The following statements are equivalent for any space X, τ.

a) X is locally connected.
b) Each component of each open subspace of X is open.
c) If C is a component of a subspace Y of X, then $\mathrm{Fr}\, C \subset \mathrm{Fr}\, Y$.

Proof. Statement (a) implies statement (c). Suppose $x \in \mathrm{Fr}\, C$. Then $x \in \mathrm{Cl}\, C \subset \mathrm{Cl}\, Y$. Now

$$\mathrm{Cl}\, Y = Y^\circ \cup \mathrm{Fr}\, Y;$$

thus if $x \notin \mathrm{Fr}\, Y$, then $x \in Y^\circ$. Since X is locally connected, there is a connected neighborhood U of x such that $x \in U \subset Y^\circ \subset Y$. Then

$$C \cap U \neq \phi$$

(since $x \in \mathrm{Fr}\, C$ and $x \in U$), and hence $C \cup U$ is connected and is a subset of Y. But C is a component of Y, and is therefore a maximal connected subspace of Y. It follows then that $C \cup U = C$. Therefore $U \subset C$ and

$x \in C^\circ$, a contradiction, since $C^\circ \cap \operatorname{Fr} C = \phi$. Therefore $x \notin Y^\circ$, and hence $x \in \operatorname{Fr} Y$; thus $\operatorname{Fr} C \subset \operatorname{Fr} Y$.

Statement (c) implies statement (b). Suppose C is a component of U, an open subspace of X. Then

$$C \cap \operatorname{Fr} C \subset U \cap \operatorname{Fr} U;$$

hence $C \cap \operatorname{Fr} C = \phi$ (since $U = U^\circ$ and $U^\circ \cap \operatorname{Fr} U = \phi$). Therefore $C \subset C^\circ$. But $C^\circ \subset C$, and hence $C = C^\circ$. Therefore C is open.

Statement (b) implies statement (a). Suppose U is any neighborhood of $x \in X$. Let V be the component of U which contains x. Then V is open; hence V is a connected neighborhood of x which is a subset of U. Therefore X is locally connected.

Note that in a locally connected space X, τ, the components of X are both open and closed. We immediately have the following corollary.

Corollary. A compact, locally connected space X, τ has at most finitely many components.

Proof. Let $\{A_i\}$, $i \in I$, be the family of components of X. Then $\{A_i\}$, $i \in I$, is an open cover of X, and hence finitely many of the A_i, say $A_{i_1}, \ldots, A_{i_n}$ cover X. But if $i \neq j$, $A_i \cap A_j = \phi$; therefore no A_j can be omitted from $\{A_i\}$, $i \in I$, such that the remaining components still form a cover of X. The components $A_{i_1}, \ldots, A_{i_n}$ then must be all of the components of X.

We have seen that local connectedness is not preserved by continuous functions. Like local compactness, local connectedness is preserved by continuous open mappings.

Proposition 16. If f is a continuous open function from a locally connected space X, τ onto a space Y, τ', then Y is locally connected.

Proof. Suppose $y \in Y$ and U is any neighborhood of y. Since f is onto, there is $x \in X$ such that $f(x) = y$. Then $f^{-1}(U)$ is a neighborhood of x. Since X is locally connected, there is a connected neighborhood V of x. Since f is both continuous and open, $f(V)$ is a connected neighborhood of y with $f(V) \subset U$. Therefore Y is locally connected.

Proposition 17 illustrates another property of local connectedness which is similar to the corresponding property of local compactness (Chapter 8, Proposition 10).

Proposition 17. Let $\{X_i, \tau_i\}$, $i \in I$, be a countable family of nonempty topological spaces. Then the product space $\times_I X_i$ is locally connected if and only if each X_i is locally connected and all but finitely many of the X_i are connected.

Proof. Suppose $\times_I X_i$ is locally connected. Since the projection mapping p_i from $\times_I X_i$ onto X_i is open and continuous, each X_i is locally connected. Let U be any connected neighborhood of any point y of $\times_I X_i$. Then U contains a basic neighborhood of y of the form $\times_I V_i$, where V_i is open in X_i, and $V_i = X_i$ for all but at most finitely many i. Then $p_i(U) = X_i$ for all but at most finitely many i. Since p_i is continuous and connectedness is preserved by continuous functions, X_i is connected for all but at most finitely many i.

Suppose that each X_i is connected and all but finitely many of the X_i are connected. Suppose $y \in \times_I X_i$ and U is a neighborhood of y. Then there is a basic neighborhood $\times_I V_i$ of y which is contained in U, where V_i is open in X_i and $V_i = X_i$ for all but at most finitely many i, say $i_1, \ldots, i_n$. For each i_k, $k = 1, \ldots, n$, there is a connected neighborhood of y_{i_k} (the i_kth coordinate of y) in X_{i_k}, call it W_{i_k}, such that $W_{i_k} \subset V_{i_k}$. There are at most finitely many more i, say $i_{n+1}, \ldots, i_m$, such that

$$X_{i_{n+1}}, \ldots, X_{i_m}$$

are locally connected, but not connected. For each of these i_k,

$$k = n + 1, \ldots, k = m,$$

there is a connected neighborhood W_{i_k} of y_{i_k} with $W_{i_k} \subset V_{i_k}$. For each

$$i \notin \{i_1, \ldots, i_n, i_{n+1}, \ldots, i_m\},$$

set $W_i = X_i$. Set $W = \times_I W_i$. Then W is a neighborhood of y and $W \subset \times_I V_i \subset U$. But W is the product of connected sets and is therefore connected (Proposition 12). Therefore $\times_I X_i$ is locally connected.

EXERCISES

1. Prove that the two definitions of local connectedness given in Definition 4 are equivalent.
2. Prove that every open subspace of a locally connected space is locally connected.
3. Prove that the components of the space Q of rational numbers with the absolute value topology are the one-point subspaces. Prove that Q is not locally connected.
4. Let X, τ be any space. Define an equivalence relation E on X by letting xEy if x and y are contained in some (the same) connected subset of X. Show that the E-equivalence classes are the components of X.
5. Decide which of the following spaces are locally connected. Also describe the components of each space. Both R and R^2 are assumed to have the usual metric topologies.

   a) $R^2 - C$, where C is a compact subset of R^2

b) $\{(x, y) \mid y = x/n, n = 1, 2, 3, \ldots\} \subset R^2$
c) $\{(x, y) \mid y = x/n, n = 1, 2, \ldots ; \text{or } y = 0\} \subset R^2$
d) $\{(x, y) \mid y = x/n, n = 1, 2, \ldots, \text{and } x \neq 0\} \subset R^2$
e) $\{x \mid x = 1/n, n = 1, 2, 3, \ldots\} \cup \{0\} \subset R$
f) the space N in Example 3 of Chapter 7

6. A subset A of a space X, τ is said to be a *path component* of X if A is a maximal path-connected subset of X (see Definition 2).
   a) What is the relation between the path components of a space and the components of the space?
   b) What are the path components of the spaces described in Examples 7 and 10?
   c) Prove that for open subspaces of R^n, a path component is also a component.
   d) What would be meant by saying that a space is locally path connected? Prove that each path component of a locally path-connected space is open.
7. Prove that if a space has only finitely many components, each component is open. Thus it is not sufficient for a space to have only open components in order for the space to be locally connected.
8. If X, τ is locally connected, is the one-point compactification of X necessarily locally connected? Explore conditions under which the one-point compactification of a locally connected space will be locally connected.

9.5 CONNECTEDNESS AND COMPACT T_2- SPACES

Two of the most important properties in topology are compactness and being T_2; hence a compact T_2-space is doubly important. In this section we study the properties of compact T_2-spaces with regard to connectedness and local connectedness. We first introduce some pertinent definitions.

Definition 5. A space X, τ can be *split* between two of its points x and y if there are disjoint open subsets U and V of X such that $x \in U$, $y \in V$, and $X = U \cup V$.

A compact connected T_2-space X, τ is called a *continuum*. The continuum X is said to be *irreducible* about $A \subset X$ if X is a minimal continuum which contains A.

A point x of a space X, τ (not necessarily a continuum) is said to be a *cut point* of X if X is connected but $X - \{x\}$ is not connected. Otherwise, x is said to be a *noncut point*. See Section 9.3, Exercise 4.

Example 17. Evidently if a space X, τ can be split between two points x and y then x and y are in different components of X. It is not true, however, that if x and y are in different components of X that X can necessarily be split between them. Let

$$Y = \{C_n\} \cup \{(x, y) \mid y = 1, \text{ or } y = -1\} \subset R^2$$

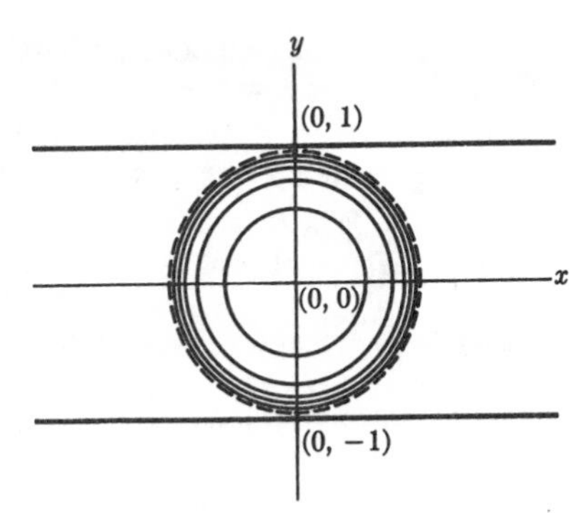

Figure 9.11

(with the usual topology), where $C_n = \{(x, y) \mid x^2 + y^2 = (1 - (1/n))^2\}$, $n = 1, 2, 3, \ldots$ Then the components of Y are each C_n and the two straight lines indicated in Fig. 9.11. It should be intuitively evident that although (0, 1) and (0, −1) are in different components of Y, nevertheless Y cannot be split between these two points.

Example 18. Any closed, bounded, and connected subset of R^n for any n is a continuum. Any path in a T_2-space is a continuum, since any subspace of a T_2-space is T_2 and compactness and connectedness are both preserved by continuous functions.

Every point of the closed interval [0, 1] except 0 or 1 is a cut point of [0, 1]; thus $\{0, 1\}$ is the set of noncut points of [0, 1]. It is easy to show in fact that [0, 1] is a continuum which is irreducible about $\{0, 1\}$. Similarly, if P and Q are any two points of R^n, then the closed segment $\overline{PQ}$ is a continuum irreducible about $\{P, Q\}$.

Proposition 18. Let X, τ be a compact T_2-space and $\{A_i\}$, $i \in I$, be a family of closed subsets of X such that $\{A_i\}$, $i \in I$, is directed by $\leq$ where $A_i \leq A_j$ if $A_j \subset A_i$ (*cf.* Definition 1 and Example 3 of Chapter 6). Suppose that each A_i has the property that it cannot be split between x and y. Then $\bigcap_I A_i$ cannot be split between x and y.

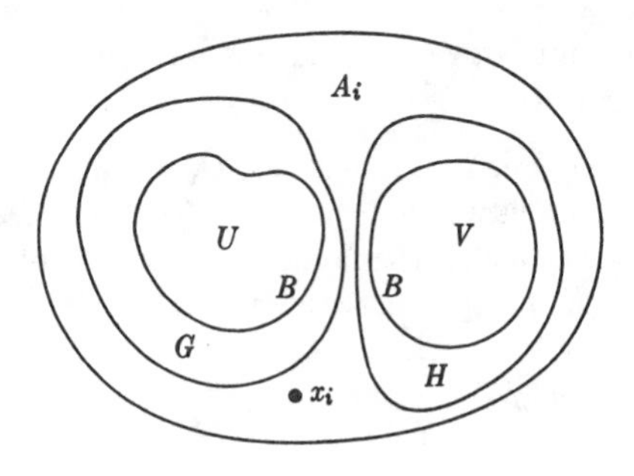

Figure 9.12

Proof. Let $B = \bigcap_I A_i$. Suppose $B = U \cup V$, where U and V are disjoint, open (in B) subsets of B, and $x \in U$ and $y \in V$. Now U and V are also closed in B. But B is closed in X (Exercise 1); hence U and V are closed in X. Since X is compact and T_2, X is T_4, and thus there are open sets G and H such that $U \subset G$, $V \subset H$, and $G \cap H = \phi$ (Fig. 9.12).

For each $i \in I$, $A_i \not\subset G \cup H$, or otherwise we would have

$$x \in A_i \cap G, \qquad y \in A_i \cap H, \qquad \text{and} \qquad (A_i \cap G) \cap (A_i \cap H) = \phi;$$

that is, we could split A_i between x and y. For each A_i, we can therefore find $x_i \in A_i - (G \cup H)$. Since $\{A_i\}$, $i \in I$, is by assumption a directed set, $\{x_i\}$, $i \in I$, is a net. Since X is compact, this net has a limit point z (Proposition 9, Chapter 7). Since any neighborhood N of z contains $\{x_i\}$, $i \in I$, cofinally, any neighborhood N of z meets $\{A_i\}$, $i \in I$, cofinally. But then N meets every A_i. For given A_i, there is $A_j \subset A_i$ such that $N \cap A_j \neq \phi$ (remember that $\{A_i\}$, $i \in I$, is directed); hence $N \cap A_i \neq \phi$. Therefore

$$z \in \text{Cl}\, A_i = A_i$$

for each i, and thus $z \in B = \bigcap_I A_i$. We therefore have that $z \in G \cup H$; hence $G \cup H$ is a neighborhood of z. But $G \cup H$ does not contain any of the x_i, contradicting the choice of z as a limit point of $\{x_i\}$, $i \in I$. It must be then that B cannot be split between x and y.

We have already seen that in a general topological space, the intersection of a family of connected subsets need not be connected. This is not true even in a compact T_2-space, but the following is true.

Proposition 19. Suppose $\{A_i\}$, $i \in I$, is a family of closed connected subsets of a compact T_2-space and $\{A_i\}$, $i \in I$, directed by $\leq$ as in Proposition 18. Then $\bigcap_I A_i$ is connected.

Proof. If x and y are in $\bigcap_I A_i$, then each A_i cannot be split between x and y; therefore, by Proposition 18, $\bigcap_I A_i$ cannot be split between x and y. Proposition 19 then follows at once from the following proposition.

Proposition 20. If X, τ is a compact T_2-space and x and y are in X, then X cannot be split between x and y if and only if x and y are in the same component of X.

Proof. Let C_x be the component of x in X and $Q_x = \{z \mid X \text{ cannot be split between } x \text{ and } z\}$. We already know that $C_x \subset Q_x$. Suppose $y \in Q_x$. We must show that $y \in C_x$. Consider the family $\{A_i\}$, $i \in I$, of closed subsets of X such that A_i cannot be split between x and y. This family in nonempty, since Q_x is a member. Then if $\{A_{i_j}\}$, $j \in J$, is a chain in $\{A_i\}$, $i \in I$, by Proposition 18, $\bigcap_J A_{i_j}$ cannot be split between x and y. Applying Zorn's lemma to the directed (and hence partially ordered) set $\{A_i\}$, $i \in I$, we can find a minimal closed subset B which cannot be split between x and y. We will now show that B is connected.

Suppose B is not connected. Then $B = U \cup V$, where U and V are open and closed in B and nonempty, but $U \cap V = \phi$. If $x \in U$ and $y \in V$, then B splits between x and y, a contradiction. On the other hand, if x and y are in U (or V) and U could be split between x and y, then B could also be split between x and y; if U could not be split between x and y, then B would not be minimal. Therefore B must be connected. But then $B \subset C_x$; hence $y \in C_x$. Then $C_x = Q_x$.

Proposition 21. If X, τ is a continuum and $A \subset X$, then there is a subcontinuum Y of X which is irreducible about A. (As the name indicates, Y is a *subcontinuum* of X if the subspace Y is itself a continuum.)

The proof is left as an exercise.

Example 19. If P and Q are any two distinct points in R^n, then

$$\{P, Q\} \subset \text{Cl}\, N(P, p)$$

for a suitable $p > 0$. Now $\text{Cl}\, N(P, p)$ is a continuum. There is therefore a minimal (or irreducible) subcontinuum of $\text{Cl}\, N(P, p)$ about $\{P, Q\}$. The closed segment $\overline{PQ}$ is an example in this case of such an irreducible subcontinuum.

Proposition 22. A compact T_2-space X, τ is locally connected if and only if every open cover of X has a refinement consisting of a finite number of connected sets.

Proof. Suppose X is locally connected and $\{U_i\}$, $i \in I$, is an open cover of X. Let $\{V_j\}, j \in J$, be the family of components of the U_i. Since X is locally connected, $\{V_j\}, j \in J$, is an open cover of X (Proposition 15); moreover, $\{V_j\}, j \in J$, is a refinement of $\{U_i\}, i \in I$. Since X is compact, there is a finite subcover $\{V_{j_1}, \ldots, V_{j_n}\}$ of $\{V_j\}, j \in J$, and this finite subcover is a refinement of $\{U_i\}, i \in I$, by connected sets.

Conversely, suppose that every open cover of X has a finite refinement by connected sets. Suppose U is a neighborhood of $x \in X$. In order to show that X is locally connected, we must find a connected neighborhood V of x with $V \subset U$. Since X is T_2 and compact, X is T_3; hence there is a neighborhood W of x such that $W \subset \text{Cl}\, W \subset U$. The set $\{U, X - \text{Cl}\, W\}$ is an open cover of X. There is therefore a refinement $\{H_1, \ldots, H_n\}$ of $\{U, X - \text{Cl}\, W\}$ by open, connected subsets of X. Either

$$H_i \subset U \qquad \text{or} \qquad H_i \subset X - \text{Cl}\, W$$

for $i = 1, \ldots, n$ by the definition of a refinement. Let $x \in H_i$. Then $x \in H_i \subset U$. Therefore H_i is a connected neighborhood of x which is contained in U. Hence X is locally connected.

A path is one of the most important types of continua but, as has already been pointed out, paths can be rather peculiar. The purpose of the next proposition is to help establish certain properties of paths. We will then give without proof a proposition which completely characterizes paths.

Proposition 23. If X, τ is a compact T_2-space and if f is a continuous function from X onto a T_2-space Y, then if X is locally connected, Y is also.

The proof of this proposition is outlined in Exercise 4 below. As an immediate consequence of Proposition 23, we have the following.

Corollary 1. Any path in a T_2-space is compact, connected, and locally connected.

Example 20. The space Y' in Example 16 is connected and compact, but is not locally connected, and hence could not be a path.

It can also be proved without much trouble (see Exercise 4) that any path is also second countable; hence, as we shall prove in the next chapter, any path in a T_2-space is a compact, connected, locally connected metric space. As a matter of fact, although we will not prove it in this text, the following is also true.

Proposition 24 (*the Hahn-Mazurkiewicz theorem*). A metric space is compact, connected, and locally connected if and only if it is a path.

This is a somewhat surprising result, since it implies that even spheres, cubes, etc. and their counterparts in R^n for any n, are continuous images of [0, 1]. As a rule, the continuous functions from [0, 1] onto such spaces are not one-one; for if they were, they would be homeomorphisms (Proposition 14, Chapter 7), which, of course, is not generally true.

EXERCISES

1. Prove that the set B in the proof of Proposition 18 is closed. [*Hint:* The proof is extremely simple.]
2. Let X, τ be any space. Define a relation E on X by xEy if X cannot be split between x and y. Prove that E is an equivalence relation on X. An E-equivalence class is said to be a *quasi-component*. Let $x \in X$; denote the component of x by C_x and the quasi-component of x by Q_x. Proposition 20 states that in a compact T_2-space $C_x = Q_x$. Always $C_x \subset Q_x$. Find an example of a space in which $C_x \neq Q_x$.

3. In Example 17, prove that Y cannot be split between $(0, 1)$ and $(0, -1)$.
4. a) Prove that any continuous function from a compact T_2-space onto a T_2-space is *closed*, that is, $f(F)$ is closed if F is closed.
   b) Prove that local connectedness is preserved by continuous closed functions.
   c) Use (b) to prove Proposition 23.
   d) Use (a) to show that any path Y, τ' in a T_2-space is second countable. [*Hint:* Let $\{U_n\}$, $n \in N$, be a countable basis for the usual metric topology on $[0, 1]$, which is known to be second countable. Then $[0, 1] - U_n$ is closed for each n. Prove that $\{Y - f([0, 1] - U_n)\}$, $n \in N$, is a basis for τ'.]
5. Prove Proposition 21.
6. Prove that any continuum is irreducible about its set of noncut points.
7. Find an example of a collection $\{A_i\}$, $i \in I$, of closed, connected subsets of a space X, τ such that $\{A_i\}$, $i \in I$, is directed by $\leq$ as in Proposition 18, but $\bigcap_I A_i$ is not connected.
8. Find an example of a continuum in R^2 (with the usual topology) which is irreducible about each of the following subsets of R^2.

   a) $\{(0, 0), (1, 1), (0, 1)\}$
   b) $\{(x, y) \mid x \text{ and } y \text{ are rational and } x^2 + y^2 < 1\}$
   c) $\{(x, y) \mid x = 0; 1 < y < 2\}$

   Even before they were found, how could we be certain that such continua existed?
9. Suppose that X is a compact T_2-space which is irreducibly connected about two points a and b. Prove that if A and B are connected subsets of X each of which contains a, then $A \subset B$ or $B \subset A$.

10
METRIZABILITY. COMPLETE METRIC SPACE

10.1 METRIZABLE SPACES

This chapter is concerned with two topics pertaining to metric spaces. The first question to be considered is, When is a space a metric space? This of course is a very poor statement of the real issue, since we have already defined what we mean by a metric space, that is, a set with a metric D. We have seen, however, that metric spaces have a certain topology induced by their metric. Suppose that instead of starting with a set with a metric, we begin with a topological space X, τ. We might then ask, Is there a metric D which can be defined on X such that the topology induced by D is τ? This is in fact a very profound question, and satisfactory answers to it were not provided until comparatively recently. We shall only partially answer the question in this text.

Definition 1. A space X, τ is said to be *metrizable* if a metric D can be defined on X such that the topology induced by D is τ. Otherwise, X is said to be *nonmetrizable.*

The question now is, When is a space metrizable?

Example 1. As we have seen, a topology can be defined on the space R of real numbers using the open intervals as a basis. This topology is the same topology as is induced by the absolute value metric; hence the open interval topology on R is metrizable.

Example 2. If X, D is any metric space, then the product space $X \times X$ can be defined without direct reference to metric. The product topology on $X \times X$ turns out to be the same topology, however, as is induced by the metric D', defined by

$$D'((x_1, x_2), (y_1, y_2)) = D(x_1, y_1) + D(x_2, y_2).$$

Therefore if X is metrizable, then $X \times X$ is also metrizable. (See Proposition 1, Chapter 8.)

Example 3. If N is the set of positive integers with the topology defined by calling a set U open if $U = N - F$, where F is finite, then N is not metrizable. For any metrizable space must be normal, but N is not even T_2.

A theorem which tells us when a space is metrizable is called a metrization theorem. One of the most important metrization theorems is *Urysohn's metrization theorem.*

Proposition 1. Let X, τ be a T_1-space. Then the following statements are equivalent:

a) X is regular and second countable.
b) X is separable and metrizable.
c) X is homeomorphic to a subspace of the product space $\times_N [0, 1]$, that is, the product of $[0, 1]$ with itself countably infinitely many times, where $[0, 1]$ has the absolute value topology. The product space $\times_N [0, 1]$ is known as the *Hilbert cube.*

As a first step toward proving Proposition 1, we prove the following.

Proposition 2. Let $\{X_n, D_n\}$, $n \in N$, be a countable family of second countable metric spaces. Then the product space $\times_N X_n$ is second countable and metrizable.

Proof. Since each X_n is second countable, $\times_N X_n$ is second countable by Proposition 4, Chapter 7. We define a metric on $\times_N X_n$ as follows: Let x and y be any points of $\times_N X_n$; denote the nth coordinates of x and y by x_n and y_n, respectively. Define

$$D'_n(x_n, y_n) = \min(D_n(x_n, y_n), 1).$$

It is easily proved that the metric D'_n thus defined on X_n is equivalent to the metric D_n (*cf.* Section 2.3, Exercise 6). Define

$$D(x, y) = \textstyle\sum_N \frac{D'_n(x_n, y_n)}{2^n}.$$

By comparison with the series $\sum_N(\frac{1}{2})^n$, we see that $D(x, y)$ is defined for each x and y in $\times_N X_n$. It can be verified in a straightforward fashion that D is in fact a metric for $\times_N X_n$. What must now be shown is that the topology induced on $\times_N X_n$ by D is the same as the product topology. We will use Corollary 1, Proposition 9, Chapter 3.

Let $y \in \times_N X_n$, and let U be a basic neighborhood of y in the product topology. Then $U = \times_N W_n$, where W_n is open in X_n and $W_n = X_n$ for all but at most finitely many n, say $n_1, \ldots, n_t$. Since W_n is open in X_n, we can find positive numbers $p_1, \ldots, p_t$ such that

$$N(y_{n_i}, p_i) \subset W_{n_i}, \qquad i = 1, \ldots, t.$$

Choose $p = \min(p_1, \ldots, p_t)$. Then if $z \in N(y, p)$, $z_n \in W_n$ for each n; hence $z \in U$, and therefore $N(y, p) \subset U$.

On the other hand, suppose $N(y, p)$ is a D-p-neighborhood of y. Choose $q \in N$ such that

$$\sum_{n=q}^{\infty} (1/2)^n < p/2$$

and choose positive numbers $p_1, \ldots, p_{q-1}$ such that

$$p_1 + \cdots + p_{q-1} < p/2.$$

Let $V = \times_N H_n$, where $H_n = N(y_n, p_n)$, $n = 1, \ldots, q-1$, and $H_n = X_n$ otherwise. Then V is a basic neighborhood of y in the product topology, and $V \subset N(y, p)$. Therefore the product and metric topologies on $\times_N X_n$ are equivalent; hence $\times_N X_n$ is metrizable.

Corollary. The Hilbert cube $\times_N [0, 1]$ is second countable and metrizable.

It should be clear to the reader than any subspace of a metrizable space is metrizable (using the same metric which makes the original space into a metric space).

We now proceed to the proof of Proposition 1.

Proof (Proposition 1)

Statement (c) implies statement (b). This is true since any subspace of a second countable metric space is both second countable (and hence separable) and itself a metric space. Any metric space is regular and, in metric spaces, separability and second countability are equivalent notions (Proposition 5, Chapter 7); hence (b) implies (a). The difficult part of this proof then is to show that (a) implies (c).

Statement (a) implies statement (c). Since X is second countable, X is Lindelöf (Chapter 7, Proposition 3). Since X is a regular Lindelöf space, X is normal (Proposition 6, Chapter 7). Let

$$\mathcal{B} = \{B_n \mid n = 1, 2, 3, \ldots\}$$

be a countable basis for τ. There are a countable number of ordered pairs of the form (B_m, B_n) such that $\operatorname{Cl} B_m \subset B_n$ (since X is regular, there is at least one such pair). Since there are countably many such pairs, we will enumerate them

$$\{(U_k, V_k) \mid k = 1, 2, 3, \ldots, \text{ where } U_k = B_{m_k} \text{ and } V_k = B_{n_k}\}.$$

Then $\operatorname{Cl} U_k \subset V_k$. Therefore $\operatorname{Cl} U_k$ and $X - V_k$ are disjoint closed subsets of X. Applying Urysohn's lemma (Proposition 10, Chapter 5) to $\operatorname{Cl} U_k$ and $X - V_k$, we can find a continuous function f_k from X into $[0, 1]$ such that $f_k(x) = 0$ for $x \in \operatorname{Cl} U_k$ and $f_k(x) = 1$ if $x \in X - V_k$.

We now define a function F from X into the Hilbert cube as follows: For each $x \in X$, let $F(x)$ be that point of the Hilbert cube whose kth coordinate is $f_k(x)$. Then F is a function from X onto a subspace Z of $\times_N [0, 1]$. We now prove that F is a homeomorphism of X onto Z.

F is continuous, since each f_k is continuous (Proposition 21, Chapter 4).

F is one-one: Suppose $x \neq y$. Since X is T_1, $X - \{y\}$ is a neighborhood of x. Therefore there is $B_m \in \mathcal{B}$ such that $x \in B_m \subset X - \{y\}$. X is regular; hence there is an open set U such that $U \subset \text{Cl } U \subset B_m$. Again, there is $B_n \in \mathcal{B}$ such that $x \in B_n \subset U$. Then

$$\text{Cl } B_n \subset \text{Cl } U \subset B_m,$$

and thus $(B_n, B_m) = (U_k, V_k)$ for some k. For this k, $x \in B_n = U_k$, and hence $f_k(x) = 0$. But

$$y \in X - B_m = X - V_k,$$

and then $f_k(y) = 1$. Therefore $F(x)$ and $F(y)$ differ at least in the kth coordinate; thus $F(x) \neq F(y)$.

It remains to be shown that F^{-1} is continuous. Let $z \in Z$, and suppose $F^{-1}(z) = x$. Now Z is a metric space with the metric described in Proposition 2. Let G be any neighborhood of x. We will show that $F(G) = (F^{-1})^{-1}(G)$ is an open subset of Z. In order to accomplish this, it is sufficient to prove that for some positive number p,

$$Z \cap N(z, p) \subset F(G).$$

As in showing that F was one-one, we can again find members of $\mathcal{B}$, say B_m and B_n, such that

$$x \in B_m \subset G \qquad \text{and} \qquad x \in B_n \subset \text{Cl } B_n \subset B_m.$$

Again for some k, $(B_n, B_m) = (U_k, V_k)$. Set $p = (\frac{1}{2})^k$. Suppose

$$z' \in Z \cap N(z, p),$$

and let $F^{-1}(z') = x'$. We will prove $x' \in G$. If $x' \notin G$, then

$$x' \in X - B_m = X - V_k;$$

therefore $f_k(x') = 1$. (That is, the kth coordinate of $F(x')$ is 1; we may denote this by $z'_k = 1$.) But

$$F^{-1}(z) = x \in B_m = U_k,$$

and hence $f_k(x) = 0$; that is, $z_k = 0$. Then

$$D(z, z') \geq |z_k - z'_k|/2^k = (1/2)^k = p,$$

a contradiction to $z' \in N(z, p)$. Therefore $x' \in G$; hence $F(G)$ is an open subset of Z, which completes the proof.

Proposition 1 is one of the most important metrization theorems in topology, although there are many others. It does not, however, completely characterize metrizable spaces, since there are nonseparable metric spaces (Example 6, Chapter 7); hence there are metrizable spaces which cannot be homeomorphic to a subspace of the Hilbert cube.

Corollary. Any separable metric space X, D has a metrizable compactification.

Proof. Since X is a separable metric space, X is homeomorphic to a subspace Z of the Hilbert cube. The Hilbert cube is, however, compact (since it is the product of compact spaces). Therefore $\operatorname{Cl} Z$ is a closed subset of a compact metric space, and is therefore itself a compact metric space. But $\operatorname{Cl} Z$ is a compactification of X; hence the desired result.

Example 4. Note that Euclidean n-space R^n is a separable metric space for each n, and that R^n is hence embeddable as a subspace of the Hilbert cube. Since R is homeomorphic to the open interval $(0, 1)$ by some homeomorphism h, a specific homeomorphism f of R^n in the Hilbert cube might be defined as follows: Let $x = (x_1, \ldots, x_n)$ be any point of R^n. Set

$$f_k(x) = h(x_k), \quad k = 1, \ldots, n,$$

and

$$f_k(x) = 0 \quad \text{for } k > n.$$

Define

$$f(x) = \big(f_1(x), f_2(x), \ldots, f_k(x), \ldots\big).$$

It is easily verified that f is a homeomorphism from R^n onto a subspace Z of the Hilbert cube. It can also be confirmed that $\operatorname{Cl} Z$ is homeomorphic to $([0, 1])^n$.

We can see from Example 7 of Chapter 8 that there can be more than one metrizable compactification of R, and that hence there may be a number of distinct embeddings of R (or R^n) as a subspace of the Hilbert cube.

Example 5. We stated after Proposition 22 of the last chapter that we would prove that any path in a T_2-space is a metric space. We do so now. Any path in a T_2-space is a compact subspace of a T_2-space and hence is regular (in fact, it is normal). We also saw in the last chapter that any path was second countable. By Proposition 1, then, any path is homeomorphic to a subspace of the Hilbert cube, which is a metric space; hence any path is a metric space.

EXERCISES

1. Which of the following spaces are not homeomorphic to a subspace of the Hilbert cube? If a given space is homeomorphic to a subspace of the Hilbert cube, produce a homeomorphism.
   a) the open interval (8, 11) with the absolute value topology
   b) N, the space of positive integers with the discrete topology
   c) $\{0, 1\}$ with the trivial topology
   d) $\{(x, y) \mid x^2 + y^2 = 4\} \subset R^2$ with the Pythagorean topology
2. The following refer to Proposition 2.
   a) Prove that D_n' is a metric on X_n which is equivalent to the metric D_n.
   b) Prove that D is a metric for $\times_N X_n$.
3. Find necessary and sufficient conditions on a space X, τ which make the one-point compactification of X a separable metric space.
4. Prove or disprove each of the following statements.
   a) If X and Y are each homeomorphic to a subspace of the Hilbert cube, then the product space $X \times Y$ is also.
   b) If Z is a subspace of X, τ and X is separable and metrizable, then Z is regular and second countable.
   c) A locally compact T_2-space is metrizable if and only if it is second countable.
   d) The continuous image of a separable metric space is a separable metric space.
   e) The product space of a countable family of compact separable spaces is homeomorphic to a subspace of the Hilbert cube.
5. Show that there is no finite n such that $([0, 1])^n$ could replace the Hilbert cube in the statement of Proposition 1.
6. Define a sequence $\{s_n\}$, $n \in N$, in the Hilbert cube by letting $s_n^n = 1$, and $s_n^k = 0$ if $k \neq n$, that is, by letting the nth coordinate of s_n be 1 and every other coordinate of s_n be 0. Since the Hilbert cube is compact, this sequence has a limit point. Find a limit point of this sequence. Does the sequence converge?
7. Suppose a space X has a dense metrizable subspace Y. Is X necessarily metrizable? For example, suppose D is a metric on Y. For each $x, y \in X$, let $a_n \to x$ and $b_n \to y$, $a_n, b_n \in Y$. Define

$$D(x, y) = \lim D(a_n, b_n).$$

10.2 CAUCHY SEQUENCES

Preparatory to a discussion of complete metric spaces, we will investigate a notion which the reader may have encountered as early as freshman calculus, that of a *Cauchy sequence* (although, as was the case with *metric*, the terminology might have been different).

Suppose X, D is any metric space and $\{s_n\}$, $n \in N$, is any sequence in X which converges to some point y of X. Then the following proposition is true.

Proposition 3. Given any number $p > 0$, there is a positive integer M such that if k and m are any two integers greater than M, then $D(s_k, s_m) < p$.

Proof. Since $s_n \to y$, we may find a positive integer M such that if $n > M$, $D(s_n, y) < p/2$. Then if k and m are both integers greater than M, we have

$$D(s_k, s_m) \leq D(s_k, y) + D(y, s_m) < p/2 + p/2 = p.$$

This result inspires the following definition.

Definition 1. Let X, D be a metric space. Then a sequence $\{s_n\}, n \in N$, in X is said to be a *Cauchy sequence* if given any positive number p, there is a positive integer M such that if m and k are integers greater than M, then $D(s_k, s_m) < p$.

Proposition 3 states that if a sequence in a metric space converges, then that sequence is a Cauchy sequence. It is not true, however, that every Cauchy sequence in any metric space converges.

Example 6. Let $\{s_n\}$, $n \in N$, be the sequence in the space R of real numbers (with the absolute value metric) defined by $s_n = 1/n$. Then $s_n \to 0$; hence $\{s_n\}$, $n \in N$, is a Cauchy sequence in R, or any subspace of R which contains it. But $\{s_n\}$, $n \in N$, is therefore a Cauchy sequence in $R - \{0\}$; however, it does not converge in $R - \{0\}$.

When the reader first studied the structure of the space R of real numbers, he may have taken as a basic axiom any one of the following:

A. Any nonempty subset W of R which has an upper bound has a least upper bound.

B. Any nonempty subset W of R which has a lower bound has a greatest lower bound.

C. Every Cauchy sequence in R converges.

We will now show that all three of these statements are equivalent relative to R. In Exercise 2, the reader is asked to show the equivalence of A and B. Propositions 4 and 5 now prove that A and C are equivalent.

Proposition 4. Assume property A of the space R of real numbers. Then a sequence $\{s_n\}$, $n \in N$, in R converges if and only if it is a Cauchy sequence.

Proof. If $\{s_n\}, n \in N$, converges, it is a Cauchy sequence by Proposition 3.

Suppose that $\{s_n\}$, $n \in N$, is a Cauchy sequence. Set $p = 1$. Then there is a positive integer M such that if k and m are integers greater than M, $|s_k - s_m| < 1$. Let

$$T = \max(|s_1|, |s_2|, \ldots, |s_{M+1}|).$$

Then for any positive integer n, $|s_n| < T + 1$; hence

$$-(T+1) < s_n < T + 1.$$

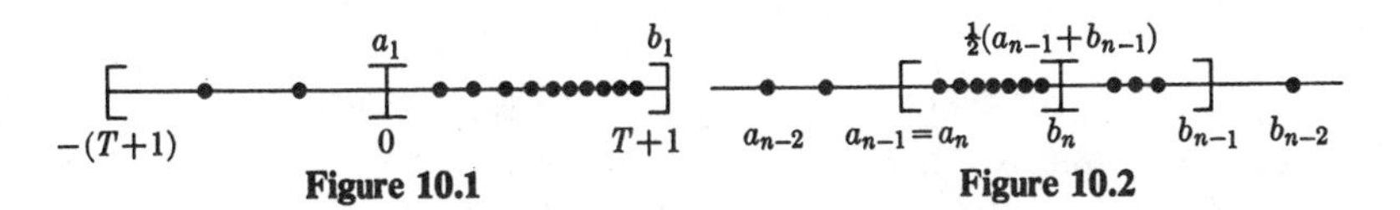

Figure 10.1 **Figure 10.2**

Divide $[-(T+1), T+1]$ into two intervals $[-(T+1), 0]$ and $[0, T+1]$. One of these intervals must contain infinitely many of the s_n. Set the left-hand endpoint of that interval equal to a_1 and the right-hand endpoint of that interval equal to b_1 (Fig. 10.1). Divide $[a_1, b_1]$ into

$$[a_1, (a_1 + b_1)/2] \qquad \text{and} \qquad [(a_1 + b_1)/2, b_1].$$

One of these intervals contains infinitely many of the s_n. Set the left-hand endpoint of that interval equal to a_2 and the right-hand endpoint of that interval equal to b_2. Continuing in like manner, suppose we have found that the interval $[a_{n-1}, b_{n-1}]$ contains infinitely many s_n. Divide

$$[a_{n-1}, b_{n-1}]$$

into

$$[a_{n-1}, (a_{n-1} + b_{n-1})/2] \qquad \text{and} \qquad [(a_{n-1} + b_{n-1})/2, b_{n-1}]$$

(Fig. 10.2). One of these intervals contains infinitely many of the s_n. Set the left-hand endpoint of that interval equal to a_n and the right-hand endpoint equal to b_n. By construction the following statements hold:

a) $a_{n-1} \leq a_n$, $n = 1, 2, 3, \ldots$;
b) $b_n \leq b_{n-1}$, $n = 1, 2, 3, \ldots$;
c) $b_n - a_n = (\frac{1}{2})^n((T+1) - (-(T+1))) = (\frac{1}{2})^{n-1}(T+1)$, $n = 1, 2, 3, \ldots$

Let $A = \{a_n \mid n \in N\}$ and $B = \{b_n \mid n \in N\}$. Since A is nonempty and has an upper bound $T + 1$, A has a least upper bound, say a. Since B is nonempty and has a lower bound, B has a greatest lower bound, say b. Then $a_n \leq a \leq b \leq b_n$ for each $n \in N$. If $a \neq b$, then $b - a > 0$. But

$$b - a \leq b_n - a_n = (\tfrac{1}{2})^{n-1}(T+1)$$

for each n. If then $b - a > 0$, there is a positive integer M' such that if $n > M'$,

$$(\tfrac{1}{2})^{n-1}(T + 1) < b - a,$$

which is impossible; therefore $a = b$.

We now show that $s_n \to a$. Let $p > 0$. Then there is an integer M_1 such that $n > M_1$ implies $b_n - a_n < p/2$. Consequently

$$[a_n, b_n] \subset N(a, p/2) = (a - p/2, a + p/2).$$

Therefore $N(a, p/2)$ contains infinitely many of the s_n. Since $\{s_n\}$, $n \in N$, is a Cauchy sequence, there is a positive integer M such that if k and m are greater than M, $|s_k - s_m| < p/2$. Since $N(a, p/2)$ contains infinitely many s_n, there is at least one integer $m' > M$ such that $s_{m'} \in N(a, p/2)$. Therefore if $n > M$, then

$$|s_n - a| \leq |s_n - s_{m'}| + |s_{m'} - a| < p/2 + p/2 = p.$$

Hence if $n > M$, then $s_n \in N(a, p)$. Therefore $s_n \to a$.

Proposition 5. If property C is assumed for the space R of real numbers, then every nonempty subset W of R which has an upper bound has a least upper bound; that is, property C implies property A.

Proof. Let W be a nonempty subset of R such that W has an upper bound T. We can find a sequence $\{s_n\}$, $n \in N$, such that (1) $s_n \leq s_{n+1}$ for each n; (2) $s_n \in W$ for each n; and (3) given any $w \in W$, there is a positive integer M such that $n > M$ implies $w \leq s_n$. The construction of such a sequence is left as an exercise. We now prove that this sequence must be a Cauchy sequence.

Suppose $\{s_n\}$, $n \in N$, is not a Cauchy sequence. Then for some $p > 0$ there is no integer M' such that if m and k are integers greater than M', then $|s_k - s_m| < p$. By construction, if $k < m$, then $s_k \leq s_m$; hence

$$|s_k - s_m| = s_m - s_k.$$

Therefore there is an integer m_1 such that $s_{m_1} - s_1 > p$. There is an integer $m_2 > m_1$ such that $s_{m_2} - s_{m_1} > p$; therefore $s_{m_2} - s_1 > 2p$. There is an integer $m_3 > m_2$ such that $s_{m_3} - s_{m_2} > p$, and hence $s_{m_3} - s_1 > 3p$. Continuing in like fashion, we see that the set of s_n, and hence W, could not have an upper bound, contradicting the assumption that W has an upper bound. Therefore $\{s_n\}$, $n \in N$, is a Cauchy sequence.

Since $\{s_n\}$, $n \in N$, is a Cauchy sequence, it converges to some limit y. It is left as an exercise to prove that y is the least upper bound of W.

EXERCISES

1. The following refer to the proof of Proposition 5.
   a) Indicate how to construct a sequence $\{s_n\}$, $n \in N$, having the properties (1) through (3) required for the proof of Proposition 5.
   b) Prove that y, the limit of $\{s_n\}$, $n \in N$, is the least upper bound of W.
2. Prove that properties A and B are equivalent relative to the space of real numbers. [*Hint:* Assume A is true, $W \neq \phi$, and a a lower bound for W. Prove that $\{a - w \mid w \in W\}$ has a least upper bound b. Prove that the greatest lower bound of W is $a - b$.]
3. Prove that a sequence in R^n (with metric D as defined in Proposition 1 of Chapter 8) converges if and only if it is a Cauchy sequence. [*Hint:* Use Proposition 12, Chapter 6.]
4. Provide an example which shows that the following is *not* an adequate characterization of a Cauchy sequence: Given any $p > 0$, there is an integer M such that for any $n > M$, $D(s_n, s_{n+1}) < p$.
5. Examine the proof of Proposition 4 and tell how this proof might be used to prove that every closed bounded subset of R is sequentially compact.
6. Describe all the Cauchy sequences in the space N of positive integers with the absolute value metric.
7. Does the subspace Z of integers in R satisfy axioms A through C for the real numbers? What properties does Z have which R does not have? Does the subspace Q of rational numbers in R satisfy A through C? Are any differences between Q and R due entirely to the fact that Q does not satisfy A, but R does?
8. Prove that a Cauchy sequence with respect to one metric may fail to be a Cauchy sequence with respect to an equivalent metric. Thus show that the image of a Cauchy sequence under a homeomorphism between metric spaces may not be a Cauchy sequence. Is the image of a convergent sequence under a homeomorphism convergent? Can a Cauchy sequence whose image under a homeomorphism is not Cauchy be convergent?

10.3 COMPLETE METRIC SPACES

The space of real numbers with the absolute value metric is only one of a special class of metric spaces in which a sequence converges if and only if it is a Cauchy sequence. Intuitively, one might feel that those spaces in which not every Cauchy sequence converges are in some sense incomplete; they have gaps which should be filled in, for example, as we "fill in" the rational numbers to get the real numbers. We therefore say the following.

Definition 2. A metric D on a set X is said to be *complete* if every Cauchy sequence in X, D converges to a point of X. If D is a complete metric on X, then X, D is said to be a *complete metric space.*

The absolute value metric D on the space R of real numbers is a complete metric on R; thus R, D is a complete metric space. Note that D is not a complete metric for the set Q of rational numbers.

We said that a metric space is complete if its metric is complete. Two metric spaces may be homeomorphic as topological spaces, but one might be a complete metric space and the other not. The following example illustrates this point.

Example 7. Let N be the set of positive integers. Let D be the usual absolute value metric on N, i.e., $D(m, n) = |m - n|$ for all $m, n \in N$. Then N, D is a complete metric space, since only Cauchy sequences in N are those sequences which are constant from some point on (see Section 10.2, Exercise 6). The topology induced on N by D is the discrete topology.

Now define a metric D' on N by

$$D'(m, n) = |1/m - 1/n|.$$

It is easily verified that D' is a metric on N which also induces the discrete topology. Therefore, considered as topological spaces, N, D and N, D' are homeomorphic (the identity mapping being an explicit homeomorphism). However, D' is not a complete metric space, since the sequence $\{s_n\}, n \in N$, in N defined by $s_n = n$ for each $n \in N$ is a Cauchy sequence, but does not converge.

The following terminology proves useful in the discussion of complete metric spaces.

Definition 3. Let A be a subset of a metric space X, D. The *diameter* of A is defined to be the least upper bound of

$$\{D(x, y) \mid x, y \in A\}.$$

We denote the diameter of A by $d(A)$.

Example 8. If $A = \{(x, y) \mid x^2 + y^2 = 1\} \subset R^2$ with the Pythagorean metric, then $d(A) = 2$ (Fig. 10.3). If $B = \{(x, y) \mid |x| < 2, |y| \leq 1\} \subset R^2$, then $d(B) = \sqrt{20}$ (Fig. 10.4).

Proposition 6. If A is a subset of the metric space X, D then

$$d(A) = d(\text{Cl } A).$$

Proof. Let p be any positive number. We will show that

$$d(\text{Cl } A) < d(A) + p.$$

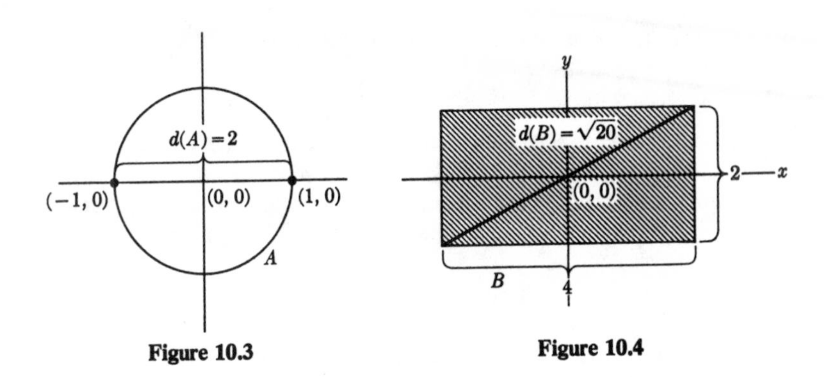

Figure 10.3

Figure 10.4

Suppose x and y are in Cl A. Then both $N(x, p/2)$ and $N(y, p/2)$ meet A. Choose

$$x' \in N(x, p/2) \cap A$$

and

$$y' \in N(y, p/2) \cap A.$$

Then

$$\begin{aligned} D(x, y) &\leq D(x, x') + D(x', y) \leq D(x, x') + D(x', y') + D(y, y') \\ &< p/2 + d(A) + p/2 = d(A) + p. \end{aligned}$$

Therefore $d(\text{Cl } A) \leq d(A)$. But since $A \subset \text{Cl } A$, $d(A) \leq d(\text{Cl } A)$; hence $d(A) = d(\text{Cl } A)$.

We now use Proposition 6 to prove an important criterion for completeness.

Proposition 7. A metric space X, D is complete if and only if given a countable family $\{A_n\}$, $n \in N$, of closed, nonempty subsets of X such that

$$A_1 \supset A_2 \supset \cdots \supset A_n \supset \cdots \qquad \text{and} \qquad d(A_n) \to 0,$$

$\bigcap_N A_n \neq \phi$.

Proof. Suppose X, D is a complete metric space. For each $n \in N$, choose $a_n \in A_n$. Suppose $p > 0$. Then there is an integer M such that $n > M$ implies $d(A_n) < p/2$. If k and m are both integers greater than M, then a_k and a_m are both elements of A_{M+1}; hence, since $d(A_{M+1}) < p/2$,

$$D(a_k, a_m) < p.$$

We therefore see that $\{a_n\}$, $n \in N$, is a Cauchy sequence. Since X, D is assumed to be a complete metric space, the sequence $\{a_n\}$, $n \in N$, converges to some point y. Then for any n, $\{a_n, a_{n+1}, \ldots\}$ is also a sequence which converges to y. But A_n is closed and

$$\{a_n, a_{n+1}, \ldots\} \subset A_n,$$

for each n; therefore $y \in A_n$ (since $y \in \text{Cl } A_n$ and $\text{Cl } A_n = A_n$). Since n was arbitrary, $y \in \bigcap_N A_n$; therefore $\bigcap_N A_n \neq \phi$.

Conversely, suppose that given any decreasing sequence $A_1 \supset A_2 \supset \cdots$ of closed, nonempty subsets of X such that

$$d(A_n) \to 0, \qquad \bigcap_N A_n \neq \phi.$$

Let $\{s_n\}$, $n \in N$, be a Cauchy sequence in X. Set

$$B_n = \{s_k \mid k \geq n\} \qquad \text{and} \qquad A_n = \text{Cl } B_n$$

for all $n \in N$. Then $\{A_n\}$, $n \in N$, fulfills the necessary conditions, and hence $\bigcap_N A_n \neq \phi$. Choose $y \in \bigcap_N A_n$. We now show that $s_n \to y$.

Let $p > 0$. Then there is an integer M such that if $n > M$, $d(B_n) < p$. By Proposition 6, $d(A_n) < p$ as well. Then $D(s_n, y) < p$ for all $n > M$; that is, if $n > M$, then $s_n \in N(y, p)$. Therefore $s_n \to y$.

As we have seen, not every metric space is complete. The next proposition enables us to say that any compact metric space is complete.

Proposition 8. If a metric space X, D is compact, then D is complete.

Proof. Suppose we have a decreasing sequence $A_1 \supset A_2 \supset \cdots$ of closed, nonempty subsets of X. Then, by Proposition 8, Chapter 7, $\bigcap_N A_n \neq \phi$. Therefore, by Proposition 7, D is complete.

Corollary. Any separable metric space X, D is homeomorphic to a dense subspace of a complete metric space.

Proof. By the corollary to Proposition 1, X has a metrizable compactification Z, D'. But Z, D' is a complete metric space by Proposition 8.

It is not true that any subspace of a complete metric space is necessarily complete (e.g., the rational numbers form an incomplete subspace of the real numbers). We do, however, have the following.

Proposition 9. Any closed subspace Y of a complete metric space X, D is a complete metric space.

Proof. Suppose $\{s_n\}$, $n \in N$, is a Cauchy sequence in Y. Then $\{s_n\}$, $n \in N$, is also a Cauchy sequence in X; hence $s_n \to y$, for some $y \in X$. But then $y \in \text{Cl } Y = Y$, and thus $\{s_n\}$, $n \in N$, converges in Y.

Suppose $\{X_n, D_n\}$, $n \in N$, is a countable family of nonempty metric spaces. Then a metric D can be defined for $\times_N X_n$ as in Proposition 2.

The following proposition gives a necessary and sufficient condition for $\times_N X_n, D$, to be complete.

Proposition 10. $\times_N X_n, D$, is a complete metric space if and only if each component space X_n, D_n is a complete metric space.

Proof. First suppose that X_n, D_n is a complete metric space for each n, and that $\{s_n\}, n \in N$, is a Cauchy sequence in $\times_N X_n$. If we denote the kth coordinate of s_n by $s_n(k)$, then it is easily verified that $\{s_n(k)\}, n \in N$, is a Cauchy sequence in X_k. Therefore $\{s_n(k)\}, n \in N$, converges in X_k. The convergence of $\{s_n\}, n \in N$, then follows from Proposition 12, Chapter 6.

Suppose now that one of the X_n, D_n, say X_1, D_1, is not complete. Then there is a Cauchy sequence $\{s_n(1)\}, n \in N$, in X_1 which does not converge. Select a point a_n from each $X_n, n \geq 2$. Then the sequence

$$\{s_n\}, \qquad n \in N, \qquad \text{in } \underset{N}{\times} X_n,$$

defined by setting the first coordinate of s_n equal to $s_n(1)$ and the nth coordinate of s_n equal to $a_n, n \neq 1$, (that is, $\{s_n\}, n \in N$, is a constant sequence in all but the first coordinate) is a Cauchy sequence which could not converge in $\times_N X_n$, since it does not converge in the first coordinate.

Corollary. The Hilbert cube is a complete metric space.

Proof 1. Since [0, 1] with the absolute value metric is compact, it is a complete metric space; hence the Hilbert cube $\times_N [0, 1]$ is a complete metric space by Proposition 10.

Proof 2. The Hilbert space is a compact metric space, and hence is complete by Proposition 8.

We have already seen that any separable metric space is homeomorphic to a dense subspace of a complete metric space. Actually, the following stronger result is true.

Proposition 11. Let X, D be any metric space. Then X may be embedded as a dense subspace of a complete metric space Y, D' by an embedding which preserves distances [that is, if h is the embedding of X, then $D(x, y) = D'(h(x), h(y))$ for any x and y in X].

We saw earlier that any topological space was homeomorphic to a dense subset of a compact space. Here we have a somewhat analogous theorem for metric spaces, compact spaces being the most important type of topological space, and complete metric spaces being the most important type of metric space. A space Y, D' as described in Proposition 11 is called a *completion* of X, D.

The proof of Proposition 11 in its entirety is long and cumbersome, and little is to be gained by going through all the sordid details. Therefore only an outline of the proof is presented here.

Outline of proof (Proposition 11). Let Y' be the set of all Cauchy sequences in X. Two Cauchy sequences $\{s_n\}$, $n \in N$, and $\{t_n\}$, $n \in N$, will be considered to be equivalent if $D(s_n, t_n) \to 0$. It must be verified that we have thus defined a genuine equivalence relation on Y'. Denote the equivalence class of a sequence $\{s_n\}$, $n \in N$, by $\{s_n\}''$; denote the set of equivalence classes by Y. We define a metric D' on Y by setting

$$D'(\{s_n\}'', \{t_n\}'') = \lim D(s_n, t_n).$$

It must be shown that the required limit always exists and is independent of the representatives of the equivalence classes. Moreover, it must be shown that D' is actually a metric. If $x \in X$, then x can be identified with the equivalence class of the constant sequence $\{s_n = x\}$, $n \in N$. The mapping thus defined in distance preserving is hence a homeomorphism. By a method of diagonalization, it can be shown that each Cauchy sequence in Y converges, and that each element of Y is the limit of a Cauchy sequence each member of which is the class of a constant sequence. For more details, the reader might see Theorem 2–72 in Hocking and Young, *Topology* (Addison-Wesley, 1961).

EXERCISES

1. In Example 7, confirm that D' is a metric on N and that D' induces the discrete topology on N. Describe a completion of N, D; of N, D'.
2. Find a necessary and sufficient condition for a completion of a space X, D to be compact. [*Hint:* Let R be the space of real numbers and let D be the absolute value metric. Set $D'(x, y) = \min(D(x, y), 1)$ for all $x, y \in R$. Then D' is a metric on R which is equivalent to D. Is D' a complete metric on R? Describe the completion of R, D'.]
3. Prove Proposition 8 by means of Proposition 9, Chapter 7 and the definition of a Cauchy sequence.
4. In the proof of Proposition 10, verify that each $\{s_n(k)\}$, $n \in N$, is really a Cauchy sequence in X_k.
5. Let f be a continuous function from a metric space X, D into itself with the property that
$$D(f(x), f(y)) \leq kD(x, y),$$
where $0 \leq k < 1$, for any $x, y \in X$. That is, f "contracts" distances. Set $f^1 = f, f^2 = f \circ f$, and, in general, $f^n = f \circ f^{n-1}$. Choose $y \in X$. Consider the sequence $\{s_n\}$, $n \in N$, defined by $s_n = f^n(y)$.

   a) Prove that $\{s_n\}$, $n \in N$, is a Cauchy sequence.

b) Prove that if D is complete, then f has a unique fixed point, that is, that there is one and only one $z \in X$ such that $f(z) = z$. [*Hint:* Let z be the unique limit of $\{s_n\}$, $n \in N$. The sequence $\{f(s_n)\}$, $n \in N$, must have the same limit as $\{s_n\}$, $n \in N$, but $f(z)$ is also the limit of $\{f(s_n)\}$, $n \in N$. Suppose z and z' are both fixed points of f, and show that this contradicts the assumption that f is a contracting function.]

c) Show by example that if D is not complete, a contracting function may not have any fixed points. If D is not complete, might a contracting function have more than one fixed point?

6. A distance-preserving function is called an *isometry*. Prove that it is not possible to isometrically embed a complete metric space X, D as a dense proper subspace of another complete metric space Y, D'. Prove, however, that it might be possible to embed X as a dense proper subspace of a complete metric space Y, D' if the embedding is not required to be an isometry.

7. A subset A of a metric space X, D is said to be *totally bounded* if given any positive number p, the open cover $\{N(x, p)\}$, $x \in A$, of A has a finite subcover. Prove that a subspace of a complete metric space is compact if and only if it is closed and totally bounded. (Cf. Exercise 6 of Section 8.1.)

10.4 BAIRE CATEGORY THEOREM

We now come to a theorem of great importance in mathematics, particularly in the construction of existence proofs in analysis. An existence proof is a proof which shows that something exists, or can be found at least *in theory*, even if we cannot actually come up with a specific example of what exists. For example, we may wish to know that such and such an equation has a solution even if we cannot find the solution, or that such and such a function exists even if we cannot at the moment construct an example of the function. To know that something either can or cannot be done either encourages us to try to do it, or saves us the time and effort of trying. Unfortunately, there will always be intrepid unbelievers who will insist upon trying to trisect angles with ruler and compass, square circles, and solve quintic equations by radicals.

We now state and prove the famous *Baire Category Theorem.*

Proposition 12. Let X, D be a nonempty complete metric space. Then the following hold:

a) If X is expressed as the union of countably many subsets A_1, $A_2, \ldots, A_n, \ldots$, then at least one of the A_n is somewhere dense. That is, for one of the A_n, $\text{Cl } A_n$ contains an open subset of X.

b) If $U_1, U_2, \ldots$ are countably many dense open subsets of X, then $\bigcap_N U_n$ is dense in X, that is, $\text{Cl}(\bigcap_N U_n) = X$.

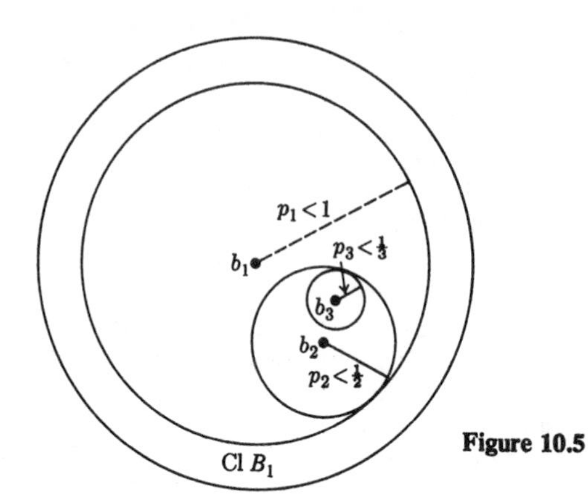

Figure 10.5

Proof

a) If (a) is false, then there is a countable family $\{A_n\}$, $n \in N$, of subsets of X such that $X = \bigcup_N A_n$, but $(\text{Cl } A_n)^\circ = \phi$ for each $n \in N$. For each n then, $\text{Cl } A_n \neq X$. Select $b_1 \in X - \text{Cl } A_1$. Since $X - \text{Cl } A_1$ is open, there is a positive number $p_1 < 1$ such that

$$N(b_1, p_1) \subset X - \text{Cl } A_1.$$

Set $B_1 = N(b_1, p_1/2)$ (Fig. 10.5). Then $\text{Cl } B_1 \subset N(b_1, p_1)$; hence

$$\text{Cl } B_1 \cap \text{Cl } A_1 = \phi.$$

Now B_1 is a nonempty open subset of X, and therefore $B_1 \not\subset \text{Cl } A_2$. Choose $b_2 \in B_1 - \text{Cl } A_2$. Since $B_1 - \text{Cl } A_2$ is open, there is $p_2 > 0$ such that $N(b_2, p_2) \subset B_1 - \text{Cl } A_2$. We lose no generality in further requiring that $p_2 < \frac{1}{2}$. Set $B_2 = N(b_2, p_2/2)$. Then

$$B_2 \subset B_1 \qquad \text{and} \qquad \text{Cl } B_2 \cap \text{Cl } A_2 = \phi.$$

Proceeding in like fashion, we can obtain a decreasing sequence of open p_n-neighborhoods $B_1 \supset B_2 \supset \cdots \supset B_n \supset \cdots$ such that $\text{Cl } B_n \cap \text{Cl } A_n = \phi$ and $p_n < 1/n$. Then

$$\text{Cl } B_1 \supset \text{Cl } B_2 \supset \cdots \supset \text{Cl } B_n \supset \cdots \qquad \text{and} \qquad d(B_n) \to 0.$$

Therefore by Proposition 7, $\bigcap_N \text{Cl } B_n \neq \phi$. Pick $x \in \bigcap_N B_n$. Then $x \in A_n$ for some n, since $\bigcup_N A_n = X$. But then

$$x \in \text{Cl } A_n \cap \text{Cl } B_n,$$

which is impossible, since $\text{Cl } A_n$ and $\text{Cl } B_n$ are disjoint. Therefore (a) is proved.

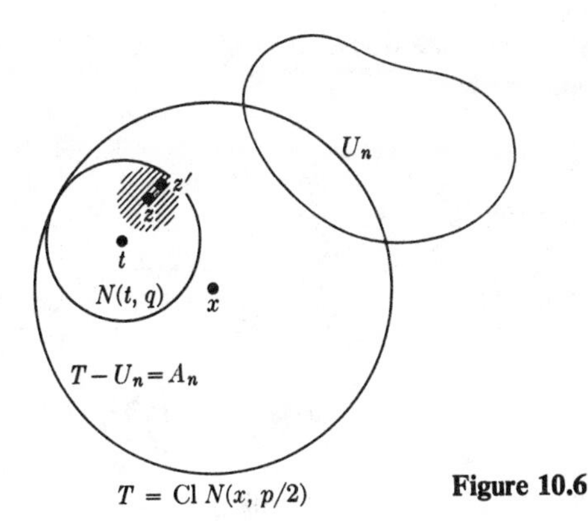

Figure 10.6

b) Suppose $\{U_n\}$, $n \in N$, is a countable family of dense open subsets of X. In order to prove that $\bigcap_N U_n$ is dense, it is sufficient to prove that each neighborhood of any point of X meets $\bigcap_N U_n$. Choose any $x \in X$ and any $p > 0$; we will show that

$$N(x, p) \cap \left(\bigcap_N U_n\right) \neq \phi.$$

(This suffices to prove statement (b), since the collection of p-neighborhoods is a basis for the topology induced by D.) Set

$$T = \text{Cl}\, N(x, p/2);$$

then $T \subset N(x, p)$. We now show that $T \cap (\bigcap_N U_n) \neq \phi$. Since T is closed, the subspace T is itself a complete metric space (Proposition 9). Set $A_n = T - U_n$. Since

$$A_n = T - U_n = T \cap (X - U_n),$$

the intersection of two closed subsets of X, A_n is closed in both X and T.

Suppose A_n is somewhere dense. Then there is $t \in T$ and $q > 0$ such that

$$N(t, q) \cap T \subset \text{Cl}\, A_n \cap T = A_n.$$

Therefore $N(t, q) \cap (T - A_n) = \phi$. Now $t \in T = \text{Cl}\, N(x, p/2)$ (Fig. 10.6); hence $N(t, q)$ meets $N(x, p/2)$ in some point z. We may choose $q' > 0$ such that

$$N(z, q') \subset N(t, q) \cap N(x, p/2).$$

But since U_n is dense, $N(z, q')$ intersects U_n, say, in z'. Then

$$z' \in T \cap N(t, q) \subset A_n.$$

But $A_n = T - U_n$, and hence $z' \in T - U_n$; that is, $z' \notin U_n$, a contradiction. Therefore A_n must be nowhere dense in T.

By (a) then, $T \neq \bigcup_N A_n$ (remember that T is a complete metric space); thus there is $s \in T - \bigcup_N A_n$. Therefore, since $A_n = T - U_n$, $s \in T \cap (\bigcap_N U_n)$. Then $T \cap (\bigcap_N U_n) \neq \phi$, and hence

$$N(x, p) \cap \left(\bigcap_N U_n\right) \neq \phi.$$

This completes the proof of (b).

A topological space X, τ which is the union of countably many subsets each of which is nowhere dense in X is said to be of the *first category*. Otherwise, X is said to be of the *second category*. Proposition 12 thus states that every complete metric space is of the second category.

Example 9. Assign a positive integer n to each real number. Set

$$A_n = \{x \in R \mid \text{such that the positive integer } n \text{ has been assigned to } x\}.$$

Then $R = \cup A_n$. Since R with the absolute value metric is a complete metric space, at least one of the A_n must be somewhere dense in R. Actually, we can show that some A_n is dense in any closed interval.

Note that since the subspace Q of rational numbers is countable, we could assign a different positive integer to each rational number, and thus Q could be expressed as the union of countably many subsets of Q each of which is nowhere dense in Q.

Example 10. The plane R^2 with the Pythagorean metric is a complete metric space. Any straight line L in R^2 is a closed subset of R^2; moreover, $R^2 - L$ is an open dense subset of R^2. If L_1 and L_2 are any two lines in R^2, then

$$(R^2 - L_1) \cap (R^2 - L_2) = R^2 - (L_1 \cup L_2).$$

By Proposition 11, however, $(R^2 - L_1) \cap (R^2 - L_2)$ is a dense subset of R^2. In general, we may remove countably many straight lines from the plane and still have what remains a dense subset of R^2 (though it may not be open).

Proposition 12 is used in existence proofs in the following ways. A suitable complete metric space is first constructed. Suppose we wish to prove that something exists which does not have a certain property P. We express the set of elements of X which have P as the union of countably many nowhere dense subsets of X. Since this union could not be all of X by Proposition 12(a), there must be some element of X which does not have P. This approach is used to show that there is a continuous function from the space of real numbers to the space of real numbers which is nowhere differentiable. If we wish to show that some element of X has a given property Q, we find countably many conditions such that if an element of X satisfies all of the conditions, then that element has Q. If the countable family of conditions is such that the set of elements of X which satisfy any one of the given conditions is an open dense subset of X, then there is an element which satisfies them all, and hence has Q, by Proposition 12(b). For examples and details of some existence theorems from analysis which use Baire's theorem, the reader is referred to Chapter 13, Section 4.2 of Dugundji, *Topology* (Allyn & Bacon, 1964).

EXERCISES

1. Use Proposition 12 to show that a complete metric space which is connected must contain uncountably many points if it contains more than one.
2. Suppose at each point of R^2 (with the Pythagorean metric) we draw a circle with integral radius. Is it necessarily true that the set of circles with radius n is somewhere dense in R^2 for at least one positive integer n?
3. Prove that the union of countably many nowhere dense, closed subsets of a complete metric space can still be dense. Why is this not a contradiction to Proposition 11(b)? [*Hint:* Consider the rationals in the space of reals.]
4. Prove that if X, D is a complete metric space, then the removal of countably many closed, nondense subsets of X still leaves a dense subset of X. Does this remain true if the subsets removed are not required to be closed?
5. The results obtained in Exercise 5 of Section 10.3 are also used in some important existence proofs. Indicate how an existence proof which uses these results might be constructed.
6. Which of the following subspaces of the usual space R of real numbers are of the second category?
   a) The irrational numbers
   b) $\{x \mid 0 < x \leq 1\}$
   c) $\{x \mid x = 1/n,\ n$ a positive integer, or $x = 0\}$
   d) $(0, 1) \cup (3, 4)$
7. Suppose Y and Z are subspaces of some space X, τ and Y and Z are both of the second category. Decide whether each of the following must be of the second category.
   a) $X \cap Y$ b) $X \cup Y$ c) the product space $X \times Y$ d) $X - Y$
8. Prove that every locally compact metric space X, D can be given a metric D' such that D' is equivalent to D and X, D' is complete. Hence each locally compact metric space is of the second category.

10.5 PARACOMPACTNESS. COMPLETE REGULARITY

Paracompactness and *complete regularity* are topological notions of relatively recent origin, but because they express properties of substantial significance in important areas of modern mathematics, they are widely used in advanced mathematical literature today. It is beyond the scope of this text to discuss these concepts in depth or give much insight into why they are important, but since the reader is likely to encounter them in more advanced work in analysis or topology, we are including their definitions and some of their elementary properties in this section. We first treat paracompactness.

Definition 4. An open covering $\{U_i\}$, $i \in I$, of a topological space X, is said to be *locally finite* if each point of X has a neighborhood which meets only finitely many of the U_i.

The space X, τ is *paracompact* if X is T_2 and if each open cover of X has a locally finite open refinement. (For the definitions of *open cover* and *refinement*, see Definition 1 of Chapter 7.)

Since a finite cover is necessarily locally finite, it follows that any open cover of a compact space has a locally finite open refinement (any finite subcover of the given open cover will do). Since T_2 is also necessary for paracompactness, we have:

Proposition 13. Any compact T_2-space is paracompact.

It is also true that any metrizable space is paracompact though we will not offer any proof of this fact. In fact, paracompactness and metrizability are very closely related as we will see shortly.

It is not necessarily true that the product of even two paracompact spaces is paracompact, nor is it true that every subspace of a paracompact space need be paracompact. We do, however, have the following.

Proposition 14. Every closed subspace of a paracompact space is paracompact.

Proof. Suppose that A is a closed subspace of the paracompact space X and $\{U_i\}$, $i \in I$, is an open cover of A. Then since each U_i is open in A, we have $U_i = A \cap V_i$, where V_i is an open subset of X, for each $i \in I$. Also

$$\{V_i \mid i \in I\} \cup \{X - A\}$$

forms an open cover of X, and, hence has a locally finite open refinement $\{W_k\}$, $k \in K$. It follows now that $\{A \cap W_k\}$, $k \in K$, is a locally finite open refinement of $\{U_i\}$, $i \in I$. Therefore A is paracompact.

We proved earlier that every compact T_2-space is normal. In fact, every paracompact space is normal. We prove this important result in two stages: we first show any paracompact space is regular and then prove that any paracompact regular space is normal.

Proposition 15. Any paracompact space is regular.

Proof. Let X be a paracompact space. Since X is T_2, X is also T_1. We must show that X is T_3, that is, given any closed set F of X and any point x of $X - F$, there are disjoint open sets U and V such that $F \subset V$ and $x \in U$.

Let F be a closed set in X and $x \in X - F$. Because X is T_2, for each $y \in F$, we can find disjoint open sets U_y and V_y with $x \in U_y$ and $y \in V_y$. Now the V_y, together with $X - F$, form an open cover of X, and this open cover has a locally finite open refinement $\{W_k\}$, $k \in K$. Let

$$V = \cup \{W_k \mid W_k \cap F \neq \emptyset\}.$$

Then V is an open set which contains F. Since the open cover $\{W_k\}$, $k \in K$, is locally finite, x has a neighborhood U' which meets only finitely many W_k, say $W_{k_1}, \ldots, W_{k_n}$. If some W_{k_i}, $i = 1, \ldots, n$, meets F, then it must be a subset of some V_y, say V_{y_i}. Set

$$U = U' \cap (U_{y_1} \cap \cdots \cap U_{y_n}).$$

Then $x \in U$, U is open, and $U \cap V = \emptyset$. Therefore X is regular.

Proposition 16. Any paracompact space X is normal.

Proof. Since X is T_2, and hence T_1, it remains to show that X is T_4, that is, given two disjoint closed sets F and F' in X, there are disjoint open sets U and V such that $F \subset U$ and $F' \subset V$.

Let F and F' be disjoint closed subsets of X. By Proposition 15 for each $x \in F$, we can find disjoint open sets U_x and V_x such that $x \in U_x$ and $F' \subset V_x$. Then the U_x for all $x \in F$, together with $X - F$, form an open cover of X; this open cover has a locally finite open refinement $\{W_k\}$, $k \in K$. Let

$$U = \cup \{W_k \mid W_k \cap F \neq \emptyset\};$$

then U is an open set which contains F. For each $y \in F'$ we can find a neighborhood H_y which meets only finitely many W_k, say $W_{k_1}(y), \ldots,$ $W_{k_n}(y)$ (the value of n also depending on y). Each $W_{k_i}(y)$ which meets F is contained in some U_x, say U_{x_i}, for $x_i \in F$. Set

$$G_y = H_y \cap (V_{x_1} \cap \cdots \cap V_{x_n})$$

(where V_{x_i} is the set corresponding to U_{x_i}). Then G_y is an open set which contains y but does not meet U. Let

$$V = \bigcup_{y \in F'} G_y.$$

Then V is an open set which contains F' but does not meet U. Therefore X is normal.

Local finiteness and paracompactness are both strongly related to metrizability. We will not review the various metrizability theorems, but will content ourselves with presenting without proof a theorem due to the Russian mathematician Smirnov, who is renowned for his work on metrizability. Two eminent American topologists, John Hocking and Gail Young, cite this theorem as "the most natural metrization theorem" they have seen. We first introduce a definition.

Definition 5. A space X is *locally metrizable* if each point $x \in X$ has a neighborhood which is metrizable (as a subspace).

Proposition 17. A locally metrizable T_2-space is metrizable if and only if it is paracompact.

The topological concept of *completely regular* has particular importance in that branch of mathematics known as *functional analysis.* As with paracompactness, we will content ourselves with presenting the definition and certain basic facts pertinent to this concept.

Definition 6. A space X is said to be *completely regular* (sometimes $T_{3\frac{1}{2}}$) if given any $x \in X$ and any closed subset F of X, $x \notin F$, there is a continuous function $f: X \to [0, 1]$ such that $f(x) = 0$ and $f(y) = 1$ for all $y \in F$.

The space X is said to be *Tychonoff* if X is T_1 and completely regular.

Tychonoff spaces lie between regular and normal spaces in that normal implies Tychonoff, and Tychonoff implies regular. Regular does not necessarily imply completely regular, nor does Tychonoff necessarily imply normal. We will soon see, though, that Tychonoff and normal spaces are rather closely related.

Proposition 17. A completely regular space X is T_3.

Proof. Suppose F is a closed subset of X and $x \in X - F$. Then there is a continuous function $f: X \to [0, 1]$ such that $f(x) = 0$ and $f(y) = 1$ for all $y \in F$. Now $[0, \frac{1}{2})$ and $(\frac{1}{2}, 1]$ are disjoint open subsets of $[0, 1]$ which contain 0 and 1, respectively; hence by the continuity of f, $f^{-1}([0, \frac{1}{2}))$ and $f^{-1}((\frac{1}{2}, 1])$ are disjoint open subsets of X which contain x and F, respectively. Therefore X is T_3.

We have already seen that every locally compact T_2-space is regular (Proposition 6 of Chapter 8). We now prove the following stronger result.

Proposition 18. Every locally compact T_2-space X is Tychonoff.

Proof. Since X is T_2, X is T_1. We now show that X is completely regular. Let Y be the one-point compactification of X. Then Y is a compact T_2-space (Proposition 12 of Chapter 8) and hence is normal. Suppose F is a closed subset of X and $x \in X - F$. Then $\{x\}$ and F are both disjoint closed subsets of the normal space Y; hence there is a continuous function $f: Y \to [0, 1]$ such that $f(x) = 0$ and $f(y) = 1$ for all $y \in F$ (by Urysohn's lemma). By taking $f \mid X: X \to [0, 1]$ we obtain a function which proves the complete regularity of X.

We leave the proof of the following proposition to the reader.

Proposition 19. Any normal space is a Tychonoff space.

Since a compact T_2-space is normal, we also have

Corollary. Any compact T_2-space is Tychonoff.

In fact, compact T_2-spaces can be shown to characterize Tychonoff spaces in the following sense.

Proposition 20. A space X is Tychonoff if and only if it is homeomorphic to a subspace of a compact T_2-space.

We make no attempt to prove Proposition 20. We note, however, that if we found a Tychonoff space which is not normal, then Proposition 20 tells us that we have found a nonnormal subspace of a normal space.

Proposition 21.

a) Every subspace of a completely regular space is completely regular; hence every subspace of a Tychonoff space is Tychonoff.
b) If $\{X_n\}$, $n \in N$, is a countable family of completely regular spaces, then the product space $\times_n X_n$ is also completely regular; hence the product of a countable family of Tychonoff spaces is Tychonoff.

Proof. We prove (b) and leave (a) as an exercise. Let F be a closed subset of $\times_n X_n$ and $x \in \times_n X_n - F$. Since $\times_n X_n - F$ is open and contains x, there is a basic neighborhood $\times_n W_n$ which contains x and fails to meet F. All but finitely many W_n are equal to X_n. Suppose $W_{n_1}, \ldots, W_{n_t}$ are those $W_n \neq X_n$. For $i = 1, \ldots, t$, $X_{n_i} - W_{n_i}$ is a closed subset of X_{n_i} which does not contain x_{n_i}, then n_ith coordinate of x. For each $i = 1, \ldots, t$, we therefore have a function

$$g_i: X_{n_i} \to [0, 1]$$

such that $g_i(x_{n_i}) = 0$ and $g_i(y) = 1$ for each $y \in X_{n_i} - W_{n_i}$. Define $f: \mathsf{X}_n X_n \to [0, 1]$ by setting

$$f(w) = \max\{g_i \circ p_{n_i}(w) \mid i = 1, \ldots, t\},$$

where p_{n_i} is the projection into the n_ith component. We leave it to the reader to demonstrate that f is continuous, $f(x) = 0$ and $f(y) = 1$ for all $y \in F$ (in fact, for all $y \in \mathsf{X}_n X_n - \mathsf{X}_n W_n$). This establishes that $\mathsf{X}_n X_n$ is completely regular.

The importance of completely regular spaces rests partly on the following property.

Definition 7. Let X be a topological space and let $C(X, R)$ denote the set of continuous functions from X into R, the usual space of real numbers. We say that $C(X, R)$ *separates points* if given any distinct real numbers r and s and two distinct points x and y of X, there is an element f of $C(X, R)$ such that $f(x) = r$ and $f(y) = s$.

Proposition 22. If X is Tychonoff, then $C(X, R)$ as described in Definition 7 above separates points.

Complete regularity is also related to metrizability, but we will not develop this aspect of the concept.

EXERCISES

1. Prove Proposition 19.
2. Prove Proposition 22.
3. Prove (a) of Proposition 21.
4. Prove that both paracompactness and complete regularity are topological properties, that is, they are preserved by homeomorphisms.
5. Prove that if X is completely regular and F and F' are disjoint subsets of X such that F is closed and F' is compact, then there is a continuous function $f: X \to [0, 1]$ such that $f(F) = 0$ and $f(F') = 1$.
6. Prove that any locally compact T_2-space that is the union of a countable number of compact sets is paracompact.
7. As a corollary of Exercise 6 show that any locally compact T_2-space is paracompact if it is second countable.
8. Prove that the product of a compact T_2-space and a paracompact space is paracompact.

11
INTRODUCTION TO HOMOTOPY THEORY

11.1 HOMOTOPIC FUNCTIONS

Good mathematical terminology is generally intuitively appealing. For example, consider this statement: The unit disk

$$Y = \{(x, y) \mid x^2 + y^2 \leq 1\} \subset R^2$$

(with the usual topology) can be *contracted* to $(0, 0)$; that is, Y is a *contractible space* (Fig. 11.1). Most likely, the reader has not encountered the notion of a contractible space before, but having studied the topology of Y, he might feel that such a statement fits in with his notions about Y.

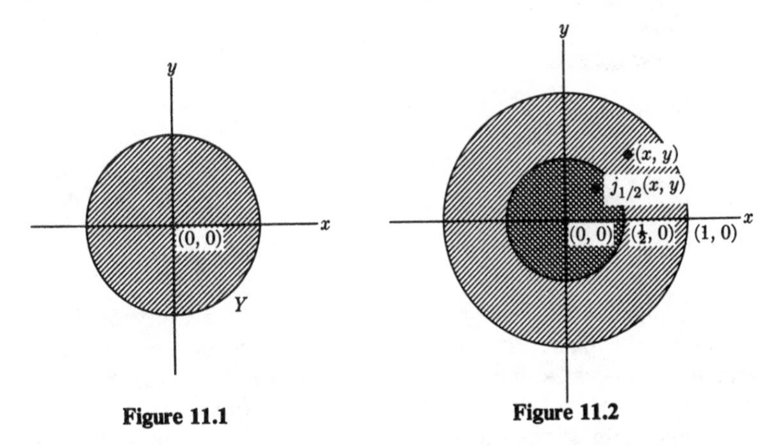

Figure 11.1 **Figure 11.2**

The word *contractible* implies the idea of shrinkability, that somehow we can reduce Y to something smaller; in particular, contractible to $(0, 0)$ implies that Y can in some reasonable way be shrunk down to $(0, 0)$. One of the most obvious ways to contract the disk Y is to slide its points along radii toward the center $(0, 0)$. One such "contracting function," call it $j_{1/2}$, of the disk into itself could be described as follows: For each $(x, y) \in Y$, let $j_{1/2}(x, y)$ be the point of Y on the segment $\overline{(0, 0)(x, y)}$ midway between (x, y) and $(0, 0)$ (Fig. 11.2). It is easily seen that $j_{1/2}$ is continuous. We may further note that if $0 \leq r \leq 1$, we may define $j_r \colon Y \to Y$ by letting

$j_r(x, y)$ be the point on $\overline{(0,0)(x,y)}$ which is $1/r^{\text{th}}$ of the distance from $(0, 0)$ to (x, y). Then j_1 is the identity mapping on Y, and j_0 maps all of Y into $(0, 0)$.

For each $r \in [0, 1]$, we have a continuous function j_r from Y into Y. We can therefore define a function j from $Y \times [0, 1]$ into Y by

$$j\big((x, y), r\big) = j_r(x, y)$$

for each $(x, y) \in Y$ and $r \in [0, 1]$. Thus

$$j_1 = j \mid (Y \times \{1\}) \qquad \text{and} \qquad j_0 = j \mid (Y \times \{0\}).$$

If we look at the images of j_r in Y for each $r \in [0, 1]$, we note that as r proceeds from 1 to 0, we gradually shrink Y to $(0, 0)$. Looking at it from the point of view of mappings, we are transforming the identity function on Y into a constant function in a natural sort of way. An informal three-dimensional representation of what is happening is given in Fig. 11.3.

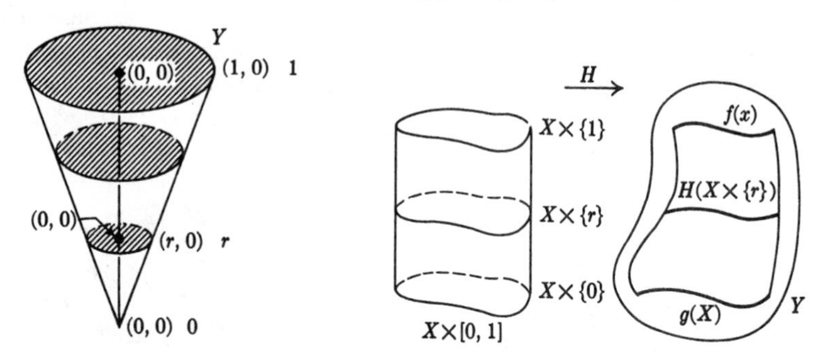

Figure 11.3 **Figure 11.4**

Throughout this chapter we will assume that the space R of real numbers, the plane R^2, and Euclidean n-space in general, have their standard topologies unless it is specified otherwise. Any subsets of R^n will be assumed to have the subspace topology from R^n.

Look again at the function $j: Y \times [0, 1] \to Y$. It is intuitively clear that since j_r is "close to" $j_{r'}$ provided r is close to r', j is continuous. Of course, the continuity of j can also be demonstrated quite rigorously.

To sum it all up then, for each $r \in [0, 1]$, we have a function j_r from Y into Y such that j_1 is the identity function on Y and j_0 is a constant function. As r varies smoothly from 1 to 0, j_r varies smoothly from j_1 to j_0. This enables us to define a continuous function $j: Y \times [0, 1] \to Y$ such that $j \mid (Y \times \{r\}) = j_r$. We have in a sense smoothly transformed the identity function on Y into a constant function. Thus we now have more justification for calling Y a contractible space.

This notion of transforming one function continuously into another is one of the most important ideas in topology. We express it formally in the following definition.

Definition 1. Let f and g be continuous functions from a space X, τ into a space Y, τ'. Then f is said to be *homotopic* to g if there is a continuous function H from $X \times [0, 1]$ into Y such that

$$H \mid (X \times \{1\}) = f \quad \text{and} \quad H \mid (X \times \{0\}) = g.$$

The function H is said to be a *homotopy between f and g* (Fig. 11.4).

Intuitively again, f and g are homotopic if f can be continuously transformed into g. We see that the identity function on the unit disk is homotopic to a constant function, that is, the function which maps the entire unit disk into $(0, 0)$.

Example 1. Reexamine Example 17 of Chapter 5. Note that any two continuous functions from $[0, 1]$ into R^2 are homotopic. More generally, we can say that any two continuous functions from $[0, 1]$ into any absolute retract (Section 5.5, Exercise 2) are homotopic.

The fact that two functions f and g are homotopic is generally far easier to see intuitively than to express analytically; that is, it is often evident that f and g are homotopic even when the actual writing out of an explicit homotopy between f and g would require considerable labor.

Example 2. We have already seen that the identity function $j_1 = i$ on the unit disk Y is homotopic to the function j_0 which maps all of Y into $(0, 0)$. We could also show that i is homotopic to the function $k: Y \to Y$ defined by

$$k(x, y) = (0, \tfrac{1}{2})$$

for any $(x, y) \in Y$. One argument to show that i is homotopic to k could use functions similar to the j_r previously used to show that i was homotopic to j_0. Still another method to produce a homotopy between i and k would be to break $[0, 1]$ up into $[0, \frac{1}{2}] \cup [\frac{1}{2}, 1]$. Let j be the homotopy between $j_1 = i$ and j_0 as defined earlier. Define

$$H((x, y), r) = \begin{cases} j((x, y), 1 - 2r), & \text{if } 0 \le r \le \frac{1}{2}, \\ (0, r - \frac{1}{2}), & \text{if } \frac{1}{2} \le r \le 1. \end{cases}$$

Geometrically, H first contracts Y to $(0, 0)$ and then slides it along the segment $\overline{(0, 0)(0, \frac{1}{2})}$ from $(0, 0)$ to $(0, \frac{1}{2})$ (Fig. 11.5). Since

$$H \mid (Y \times [0, \tfrac{1}{2}]) \quad \text{and} \quad H \mid (Y \times [\tfrac{1}{2}, 1])$$

are each continuous, and since $H \mid (Y \times \{\frac{1}{2}\})$ is well-defined, H is con-

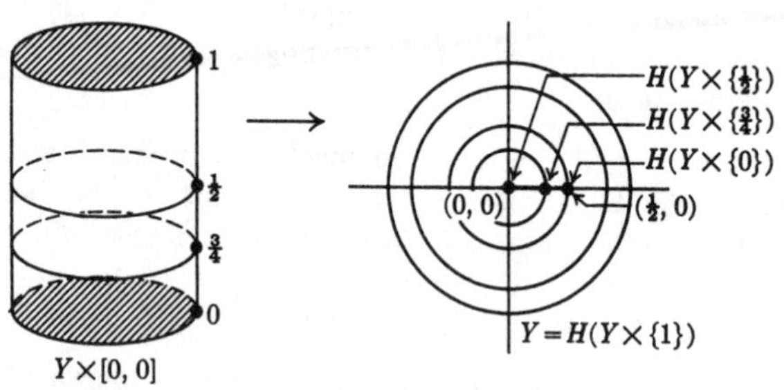

Figure 11.5

tinuous by Proposition 11, Chapter 4. Since

$$H \mid (Y \times \{1\}) = i \qquad \text{and} \qquad H \mid (Y \times \{0\}) = k,$$

H is a suitable homotopy between i and k.

Example 3. An *arc* is a homeomorphism from [0, 1] into any space; thus any arc is also a path. Since any two paths in R^2 are homotopic, it is certainly true that any two arcs in R^2 are homotopic. In particular, if a_1 and a_2 are distinct arcs in R^2 such that $a_1(0) = a_2(0)$ and $a_1(1) = a_2(1)$, then a_1 and a_2 are homotopic. Suppose that P is some point in the area bounded by the images of a_1 and a_2 (Fig. 11.6), and that a_1 and a_2 are considered as arcs in $R^2 - \{P\}$. Then a_1 and a_2 are not homotopic in $R^2 - \{P\}$, since the missing point would prevent us from transforming a_1 continuously into a_2. Intuitively, in order to transform a_1 into a_2 we would have to have a break at some stage of the transformation to get past the barrier posed by the missing point.

In like manner, removing a point from the interior of the unit disk Y prevents the identity map on Y from being homotopic to a constant map. We therefore see that any two given functions from one space into another are not necessarily homotopic. The next proposition shows that the notion of homotopy enables us to classify functions from one space into another according to the functions to which they are homotopic.

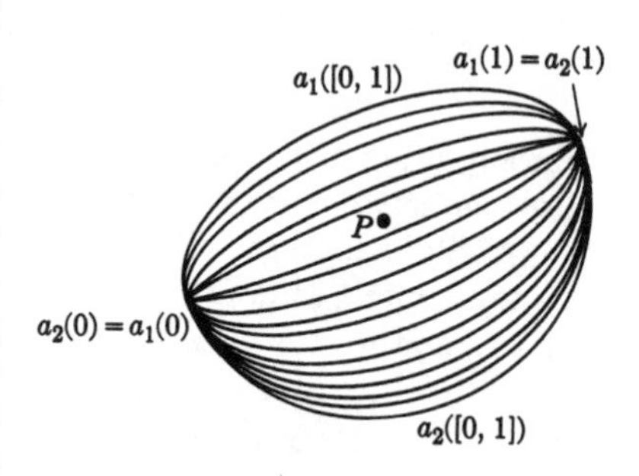

Figure 11.6

Proposition 1. Let X, τ and Y, τ' be topological spaces, and let Y^X denote the set of all continuous functions from X into Y. (The reason for the notation Y^X will be made clearer in the appendix.) Let $\sim$ denote "is homotopic to," that is, $f \sim g$ will mean that f is homotopic to g. Then $\sim$ is an equivalence relation on Y^X.

Proof. If $f \in Y^X$, then $f \sim f$. The explicit homotopy $H: X \times [0, 1] \to Y$ between f and f is given by $H(x, r) = f(x)$ for all $x \in X$ and $r \in [0, 1]$. In order to verify that H is a suitable homotopy, we must show that

$$H \mid (X \times \{1\}) = H \mid (X \times \{0\}) = f$$

and that H is continuous. Now

$$H \mid (X \times \{1\})(x, 1) = H(x, 1) = f(x)$$

for any $x \in X$, and hence $H \mid (X \times \{1\}) = f$; similarly, $H \mid (X \times \{0\}) = f$. Suppose U is any open subset of Y. Then

$$H^{-1}(U) = f^{-1}(U) \times [0, 1].$$

But since f is continuous, $f^{-1}(U)$ is an open subset of X; thus $f^{-1}(U) \times [0, 1]$ is an open subset of $X \times [0, 1]$. Therefore H is continuous.

If $f \sim g$, then $g \sim f$. Since $f \sim g$, there is a homotopy

$$H: X \times [0, 1] \to Y$$

such that

$$H \mid (X \times \{1\}) = f \qquad \text{and} \qquad H \mid (X \times \{0\}) = g.$$

Define $H': X \times [0, 1] \to Y$ by

$$H'(x, r) = H(x, 1 - r) \qquad \text{for any} \quad x \in X \quad \text{and} \quad r \in [0, 1].$$

Then $H'(X \times \{0\}) = f$ and $H' \mid (X \times \{1\}) = g$. All we have done is to turn H upside down, or reverse its direction, if you prefer. Then H' is a suitable homotopy between g and f.

If $f \sim g$ and $g \sim k$, then $f \sim k$. Since $f \sim g$, there is a homotopy $H_1: X \times [0, 1] \to Y$ such that

$$H_1 \mid (X \times \{1\}) = f \qquad \text{and} \qquad H_1 \mid (X \times \{0\}) = g.$$

Since $g \sim k$, there is a homotopy $H_2: X \times [0, 1] \to Y$ such that

$$H_2 \mid (X \times \{1\}) = g \qquad \text{and} \qquad H_2 \mid (X \times \{0\}) = k.$$

Define $H: X \times [0, 1] \to Y$ by

$$H(x, r) = \begin{cases} H_2(x, 2r), & 0 \le r \le \frac{1}{2}, \\ H_1(x, 2r - 1), & \frac{1}{2} \le r \le 1. \end{cases}$$

Note that, for all $x \in X$,

$$\begin{aligned} H(x, 1) &= H_1(x, 2 - 1) = H_1(x, 1) = f(x); \\ H(x, 0) &= H_2(x, 0) = k(x); \\ H(x, \tfrac{1}{2}) &= H_1(x, 0) = H_2(x, 1) = g(x). \end{aligned}$$

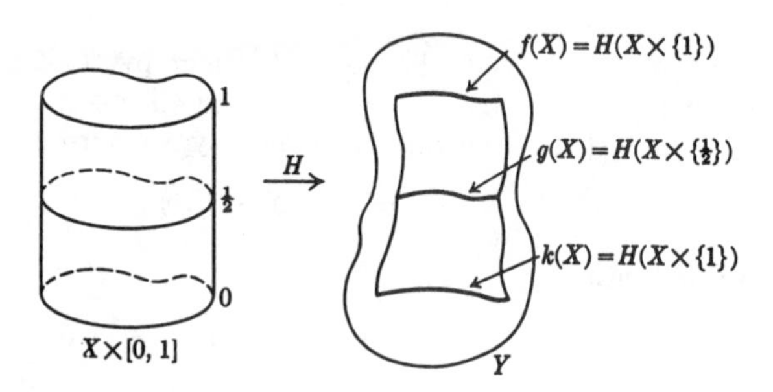

Figure 11.7

Therefore H is well-defined and will be a homotopy between f and k if it is continuous. The proof of the continuity of H is left as an exercise. Essentially what we have done is to "paste" H_1 and H_2 together to get a new homotopy between f and k (Fig. 11.7).

We have therefore shown that $\sim$ is an equivalence relation on Y^X.

Definition 2. If $f \in Y^X$, then the family of continuous functions from X into Y which are homotopic to f is called the *homotopy class* of f.

Because of Proposition 1, we can say the homotopy classes of Y^X form a partition of Y^X.

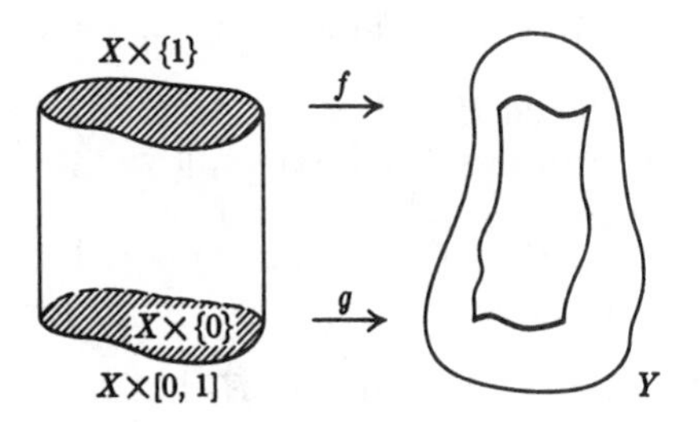

Figure 11.8

We close this section by noting that the question of whether two functions in Y^X are homotopic is really a question of whether or not a certain function can be extended. In particular, we may define

$$h: X \times \{0\} \cup X \times \{1\} \to Y$$

by $h(x, 0) = f(x)$ and $h(x, 1) = g(x)$ for all $x \in X$ (Fig. 11.8). Then f and g are homotopic if and only if h can be extended to a continuous function H from $X \times [0, 1] \to Y$ such that

$$H \mid (X \times \{0\}) \cup (X \times \{1\}) = h.$$

This observation shows the connection between two of the most difficult questions in topology: the extension of functions and the questions of when two functions are homotopic. Because of Proposition 11, Chapter 5, this connection is helpful in many important instances.

EXERCISES

1. In the proof of Proposition 1, prove that the homotopies constructed to prove

$$f \sim g \quad \text{implies} \quad g \sim f$$

and

$$f \sim g \text{ and } g \sim k \quad \text{imply} \quad f \sim k$$

are continuous.

2. Suppose Y, τ' is an absolute retract. Prove that any two continuous functions from a compact T_2-space X, τ into Y are homotopic. Is it necessarily true that any two continuous functions from any space Z, τ'' into Y are homotopic? Explain.

3. A space X, τ is said to be *contractible* to one of its points x_0 if the identity function on X is homotopic to the function which maps all of X into x_0.
   a) Which of the following spaces are contractible to at least one of their points? (Remember that subspaces of R^n are being given their usual topology.) An intuitive argument is all that is expected.
      i) $\{(x, y) \mid x^2 + y^2 = 1\} \subset R^2$
      ii) $\{(x, y) \mid |x| \leq 1, |y| < 1\} \subset R^2$
      iii) $\{(x, y, z) \mid x^2 + y^2 = 1; -1 \leq z \leq 1\} \subset R^3$
      iv) $\{x \mid 3 < x < 7\} \subset R$
   b) A space X, τ is said to be *arc-connected* if any two distinct points of X can be joined by the image of an arc, that is, X is arc-connected if given any distinct points x and y of X, there is an arc a (see Example 3) in X such that $a(0) = x$ and $a(1) = y$. Prove that if X is arc-connected and contractible to one of its points, then X is contractible to any of its points. Show that this is not true if X is not arc-connected.
   c) Prove that if f is a homeomorphism from a contractible space X, τ onto a space Y, τ', then Y is also contractible. [*Hint:* If H is the homotopy between the identity map i_X on X and a constant function k on X, consider $H': Y \times [0, 1] \to Y$ defined by $H'(y, r) = f(H(x, r))$, where $f(x) = y$.
   d) Discuss the following statement and try to determine its truth or falsity: If f is a continuous function from a contractible space X, τ onto a space Y, τ', then Y is contractible.
   e) Try to determine the truth or falsity of the following statement: Any absolute retract is contractible.

4. Suppose f and g are continuous and homotopic functions from a space X, τ into itself. Let k be any continuous function from X into X. Prove that $f \circ k$ is homotopic to $g \circ k$. More generally, prove that if k is homotopic to k', another continuous function from X into X, then $f \circ k \sim g \circ k'$ and $k \circ f \sim k' \circ g$.

5. Explain intuitively why the identity function on the unit circle $X = \{(x, y) \mid x^2 + y^2 = 1\}$ is not homotopic to $f: X \to X$ where $f(z) = (1, 0)$ for all $z \in X$. Take a rubber band and pin the rubber band at one point to a table. Then try to push the rubber band back along itself to the point at which it is pinned.

11.2 LOOPS

It should be well known to the reader that one branch of mathematics can often be used to give results in another branch. Algebra and geometry are wedded in algebraic geometry, and calculus is a tool in differential geometry. There is no branch of mathematics today which is really self-contained. Even logic has started to draw heavily in recent years on topological methods to obtain some of its most significant discoveries. It must also have occurred to the reader that topology is highly geometric; at the same time, topology often has its inspiration and applications in real and complex analysis. What we are going to do now is to begin to develop a method by which algebra can be used to express topological properties. The method we will study is only one of several applications of algebra to topology, and, in fact, we will study only a small part of the method at that. Algebraic topology is both one of the oldest and one of the newest areas of topological studies; the *fundamental group* dates back to the early days of topology (*c.* 1900, which really was not so long ago), while much of homology theory only dates back a few years, or less.

If X, τ and Y, τ' are arbitrary spaces, there is not much algebraic structure that might be given to either Y^X, the family of continuous functions from X into Y, or the homotopy equivalence classes of Y^X. We therefore would like to find a suitable space X, τ so that either Y^X or the homotopy classes could be given an algebraic structure which could help us to study Y. The following definition proves useful.

Definition 3. Let Y, τ' be any space and $y_0 \in Y$. Then a continuous function a from $[0, 1]$ into Y such that

$$a(0) = a(1) = y_0$$

is said to be a *loop* in Y with *base point* y_0 (Fig. 11.9). Two loops a_0 and a_1 in Y with base point y_0 are said to be *homotopic relative to* y_0 if there is a homotopy H between a_0 and a_1 such that for each $r \in [0, 1]$,

$$H(0, r) = H(1, r) = y_0;$$

that is, for each $r \in [0, 1]$, $H \mid [0, 1] \times \{r\}$ is a loop in Y with base point y_0 (Fig. 11.10). We will denote the set of all loops in Y with base point y_0 by $L(Y, y_0)$.

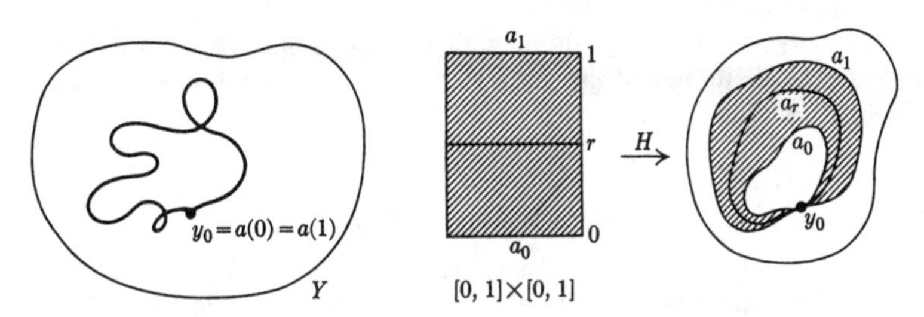

Figure 11.9

Figure 11.10

The following notation will occasionally prove useful: Let X, τ and Y, τ' be spaces, and let A and B be subspaces of X and Y, respectively. Then $f: X, A \to Y, B$ will denote that the function f from X to Y has the property that $f(A) \subset B$. Thus $f: [0, 1], \{0, 1\} \to Y, y_0$ would denote a loop in Y based on y_0 if f were continuous.

Now $L(Y, y_0)$ is a subset of the family $Y^{[0,1]}$ of all continuous functions from $[0, 1]$ into Y. The relation on $L(Y, y_0)$ defined by "is homotopic relative y_0 to," which we will again denote by $\sim$, is seen to be an equivalence relation on $L(Y, y_0)$ by the same argument as was used in Proposition 1. We will denote the set of homotopy (relative to y_0) classes of $L(Y, y_0)$ by $\pi_1(Y, y_0)$. If a is any loop in Y with base point y_0, we will denote the equivalence class of a by $|a|$. We shall now determine an algebraic structure on $\pi_1(Y, y_0)$; in particular, we shall make $\pi_1(Y, y_0)$ into a group.

We define an operation # on $\pi_1(Y, y_0)$ as follows: Suppose $|a_1|$ and $|a_2|$ are elements of $\pi_1(Y, y_0)$, that is, $|a_1|$ and $|a_2|$ are the homotopy relative y_0 equivalence classes of the loops a_1 and a_2. Define $|a_1| \# |a_2|$ to be the equivalence class of the loop $a_1 \# a_2$ defined by

$$(a_1 \# a_2)(r) = \begin{cases} a_1(2r), & 0 \le r \le \frac{1}{2}, \\ a_2(2r-1), & \frac{1}{2} \le r \le 1. \end{cases}$$

We first verify that $a_1 \# a_2$ is a *bona fide* element of $L(Y, y_0)$. Now $a_1 \# a_2$ is at least a function from $[0, 1]$ into Y. Note that $a_1 \# a_2$ is formed by "going around" a_1 once and then going around a_2 once. Also $a_1 \# a_2$ is continuous because $a_1 \# a_2$ is continuous on both $[0, \frac{1}{2}]$ and $[\frac{1}{2}, 1]$, and is well-defined at $r = \frac{1}{2}$ since $(a_1 \# a_2)(\frac{1}{2}) = a_1(1) = a_2(0) = y_0$. [One of the principal reasons that we restricted ourselves to $L(Y, y_0)$ was so that we could "add" functions in this way and have them well-defined. If a_1 and a_2 both did not begin and end at y_0, we would have no assurance that $a_1 \# a_2$ was well-defined at $r = \frac{1}{2}$.] Since

$$(a_1 \# a_2)(0) = a_1(0) = (a_1 \# a_2)(1) = a_2(1) = y_0,$$

we see that $a_1 \# a_2$ is really an element of $L(Y, y_0)$. We are therefore justified in taking its homotopy equivalence class, which is an element of $\pi_1(Y, y_0)$. Defining

$$|a_1| \# |a_2| = |a_1 \# a_2|,$$

we have a beginning on an operation on $\pi_1(Y, y_0)$.

We are not sure yet, though, that # is really an operation. In defining $|a_1| \# |a_2|$, we made use of particular representatives of the equivalence classes $|a_1|$ and $|a_2|$. In order to have a valid operation on $\pi_1(Y, y_0)$, however, $|a_1 \# a_2|$ must be independent of the representatives we pick. That is, the "sum" of two equivalence classes must depend only on the equivalence classes we are adding and not on which loops we pick from each class to compute the sum. Proposition 2 shows that # is a well-defined operation on $\pi_1(Y, y_0)$.

Proposition 2. If a_1, a_2, a_3, and a_4 are any elements of $L(Y, y_0)$ and $a_1 \sim a_3$, then

$$a_1 \# a_2 \sim a_3 \# a_2.$$

Similarly, if $a_2 \sim a_4$, then

$$a_1 \# a_2 \sim a_1 \# a_4.$$

Therefore

$$|a_1| \# |a_2| = |a_1 \# a_2| = |a_3 \# a_2| = |a_3| \# |a_2|$$

and

$$|a_1| \# |a_2| = |a_1| \# |a_4|.$$

Proof. Since $a_1 \sim a_3$, there is a homotopy (relative to y_0)

$$H: [0, 1] \times [0, 1] \to Y$$

between a_1 and a_3. Define H': $[0, 1] \times [0, 1] \to Y$ by

$$H'(r, s) = \begin{cases} H(2r, s), & \text{if } 0 \leq r \leq \frac{1}{2}, \\ a_2(2r - 1), & \text{if } \frac{1}{2} \leq r \leq 1. \end{cases}$$

Then

$$H'(r, 0) = \begin{cases} a_1(2r), & \text{if } 0 \leq r \leq \frac{1}{2}, \\ a_2(2r - 1), & \text{if } \frac{1}{2} \leq r \leq 1, \end{cases}$$

$$= (a_1 \# a_2)(r) \qquad \text{for any} \quad r \in [0, 1].$$

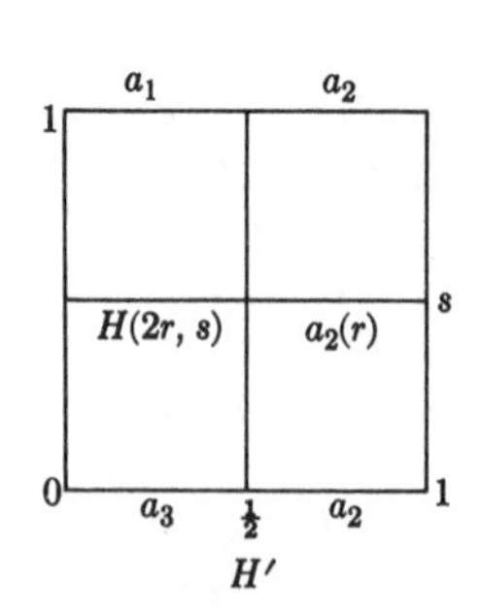

Figure 11.11

Similarly, $H'(r, 1) = (a_3 \# a_2)(r)$ for any $r \in [0, 1]$. Also,

$$H'(0, s) = H(0, s) = y_0 = H'(1, s) = a_2(1).$$

Direct computation further shows that H' is well-defined for $r = \frac{1}{2}$. We have yet to show that H' is continuous.

The continuity of H' can be demonstrated in a formal argument, and the reader is urged to provide such an argument in this case. In general, however, an appeal to a "picture" of the homotopy is much easier, and usually just as convincing. For example, Fig. 11.11 gives a picture of the homotopy H'. Note how we continuously deform a_1 into a_3 while keeping a_2 fixed. The homotopy is continuous on $[0, \frac{1}{2}] \times [0, 1]$ and on $[\frac{1}{2}, 1] \times [0, 1]$, and hence is continuous.

It is left as an exercise to prove that $a_1 \# a_2 \sim a_1 \# a_4$.

Corollary. If a_1, a_2, a_3, and a_4 are in $L(Y, y_0)$, and if $a_1 \sim a_3$ and $a_2 \sim a_4$, then

$$a_1 \# a_2 \sim a_3 \# a_4.$$

Therefore if $|a_1| = |a_3|$ and $|a_2| = |a_4|$ in $\pi_1(Y, y_0)$, then

$$|a_1| \# |a_2| = |a_3| \# |a_4|.$$

Proof. $a_1 \# a_2 \sim a_3 \# a_2 \sim a_3 \# a_4$.

Although it is not obvious at first glance, the operation # defined on $\pi_1(Y, y_0)$ is not necessarily commutative. In other words, there is no particular reason for $a_1 \# a_2$ to always be homotopic to $a_2 \# a_1$.

We now have a set $\pi_1(Y, y_0)$ with an operation #, and we claim that $\pi_1(Y, y_0)$, # is a group. In order to prove this assertion, we must show that # is an associative operation, that there is an identity in $\pi_1(Y, y_0)$ with respect to #, and that each element of $\pi_1(Y, y_0)$ has an inverse with respect to #.

We first prove the associativity of #.

Proposition 3. If $|a_1|$, $|a_2|$, and $|a_3|$ are any three elements of $\pi_1(Y, y_0)$, then

$$(|a_1| \# |a_2|) \# |a_3| = |a_1| \# (|a_2| \# |a_3|).$$

Proof. It suffices to show that $(a_1 \# a_2) \# a_3 \sim a_1 \# (a_2 \# a_3)$. By definition,

$$\big((a_1 \# a_2) \# a_3\big)(r) = \begin{cases} (a_1 \# a_2)(2r), & \text{if } 0 \le r \le \frac{1}{2}, \\ a_3(2r - 1), & \text{if } \frac{1}{2} \le r \le 1. \end{cases}$$

Therefore

$$\big((a_1 \# a_2) \# a_3\big)(r) = \begin{cases} a_1(4r), & \text{if } 0 \le r \le \frac{1}{4}, \\ a_2(4r - 1), & \text{if } \frac{1}{4} \le r \le \frac{1}{2}, \\ a_3(2r - 1), & \text{if } \frac{1}{2} \le r \le 1. \end{cases}$$

Similarly,

$$(a_1 \# (a_2 \# a_3))(r) = \begin{cases} a_1(2r), & \text{if } 0 \le r \le \frac{1}{2}, \\ a_2(4r - 2), & \text{if } \frac{1}{2} \le r \le \frac{3}{4}, \\ a_3(4r - 3), & \text{if } \frac{3}{4} \le r \le 1. \end{cases}$$

Define

$$H(r, s) = \begin{cases} a_1(4r/(1 + s)), & \text{if } 0 \le r \le \frac{1}{4}(1 + s), \\ a_2(4r - 1 - s), & \text{if } \frac{1}{4}(1 + s) \le r \le \frac{1}{4}(2 + s), \\ a_3(1 - 4(1 - r)/(2 - s)), & \text{if } \frac{1}{4}(2 + s) \le r \le 1. \end{cases}$$

Direct computation shows that H is a suitable homotopy between $(a_1 \# a_2) \# a_3$ and $a_1 \# (a_2 \# a_3)$. Actually, a convincing argument for the suitability of H can be made from its picture alone (Fig. 11.12).

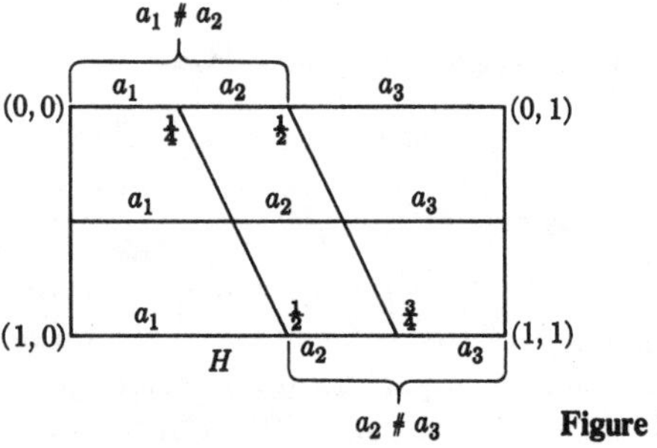

Figure 11.12

The reader should have begun to realize by now that pictures are a very useful way of arriving at homotopies. Sometimes a picture alone is sufficient to convince one that a homotopy exists. Almost always, at any rate, a picture helps to obtain an analytic expression of the homotopy. Perhaps the reader has been brought up to believe that arguing from pictures is a cardinal sin in mathematics; certainly there is no doubt that pictures can be misleading if improperly used. Nevertheless, diagrams intelligently employed can be indispensable tools in a mathematical argument. This is particularly true with regard to homotopy arguments for two reasons. First, the notion of a homotopy is highly geometric (as is much of topology); hence we should expect pictures to be apropos. Second, actually producing a homotopy analytically may be so cumbersome, while, at the same time, producing a picture may be so easy and convincing, that no reasonable person would demand the explicit analytic expression. Let the reader then accustom himself to the use of pictures in homotopy theory, while, at the same time, being sure that he understands the theory and meaning behind any picture he uses, and would be able to produce an analytic expression if necessary.

EXERCISES

1. In Proposition 2, prove $a_1 \# a_2 \sim a_1 \# a_4$.
2. Suppose the space Y, τ' is contractible to the point y_0; let k denote the function which maps Y onto y_0. Let f be any continuous function from a space X, τ into Y. Prove that f is homotopic to k', where k' is the function which maps all of X onto y_0. Prove that $\pi_1(Y, y_0)$ contains only one element.
3. Let Y, τ' be any space, $y_0 \in Y$, and k the function which maps all of $[0, 1]$ onto y_0. Prove that $|k|$ is an identity for $\pi_1(Y, y_0)$ with respect to #. That is, prove that if $|a_1| \in \pi_1(Y, y_0)$, then

$$|k| \# |a_1| = |a_1| \# |k| = |a_1|.$$

[*Hint:* Figures 11.13 and 11.14 give pictures of the desired homotopies. Express analytically what these pictures say graphically.]

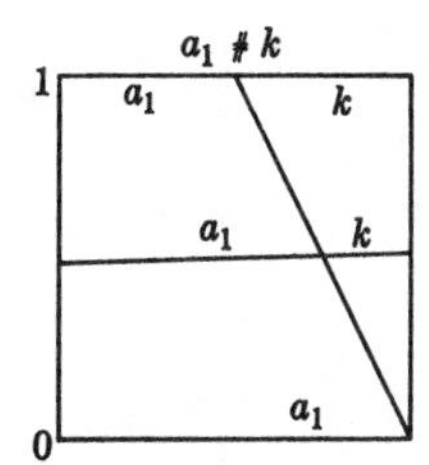

Figure 11.13

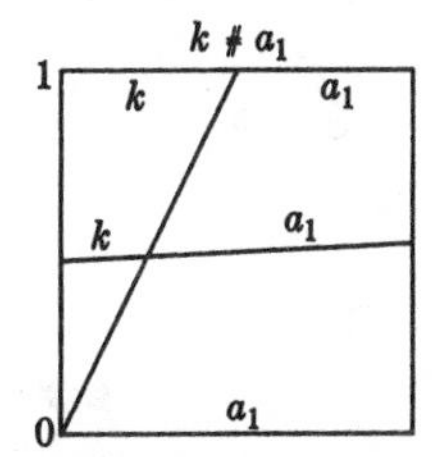

Figure 11.14

4. Give an argument to show that if $Y = \{(x, y) \mid x^2 + y^2 = 1\} \subset R^2$ and P is any point of Y, then $\pi_1(Y, P)$ contains infinitely many elements. [*Hint:* Let k be the loop which maps all of $[0, 1]$ onto P, and let a_1 be the loop which "wraps" $[0, 1]$ once around Y (the identification of 0 and 1, if you prefer). Show that a_1 is not homotopic to k. Can $a_1 \# a_1$ be homotopic to a_1? For any positive integer n, define $na_1 = a_1 \# \cdots \#$ (n times) $\# a_1$. Try to show that for any positive integers n and m, na_1 is homotopic to ma_1 if and only if $m = n$.]
5. Find an explicit homotopy between the loop α on the disk $\{(x, y) \mid x^2 + y^2 \leq 1\}$ with base point $(0, 1)$ defined by

$$\alpha(x) = \begin{cases} (4x, \sqrt{1 - (4x)^2}), & 0 \leq x \leq \frac{1}{4}, \\ (2 - 4x, -\sqrt{1 - (2 - 4x)^2}), & \frac{1}{4} \leq x \leq \frac{1}{2}, \\ (-4x + 2, -\sqrt{1 - (2 - 4x)^2}), & \frac{1}{2} \leq x \leq \frac{3}{4}, \\ (4x - 4, \sqrt{1 - (4x - 4)^2}), & \frac{3}{4} \leq x \leq 1, \end{cases}$$

with the loop which maps $[0, 1]$ entirely into $(0, 1)$. There are in fact infinitely many such homotopies.

11.3 THE FUNDAMENTAL GROUP

Thus far we have seen that $\pi_1(Y, y_0)$, # is a semigroup (that is, a set with an associative operation) which also has an identity which we will denote by $|k|$ (Section 11.2, Exercise 3). It remains to be shown that each element of $\pi_1(Y, y_0)$ has an inverse with respect to #. If $a \in L(Y, y_0)$, we define a^{-1} by letting $a^{-1}(r) = a(1 - r)$ for each $r \in [0, 1]$. Then

$$a^{-1}(0) = a(1) = y_0 \qquad \text{and} \qquad a^{-1}(1) = a(0) = y_0.$$

Since a is continuous, a^{-1} is also; hence a^{-1} is an element of $L(Y, y_0)$. Geometrically, a^{-1} is a going around in the opposite direction (Fig. 11.15).

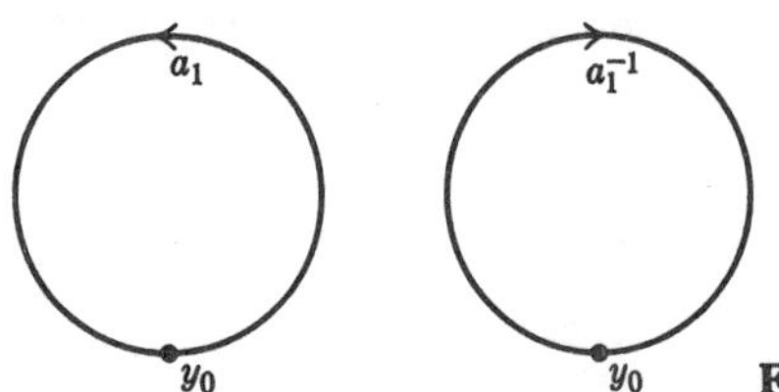

Figure 11.15

Proposition 4. If $a \in L(Y, y_0)$, then

$$a \# a^{-1} \sim a^{-1} \# a \sim k.$$

Therefore the inverse of $|a|$ in $\pi_1(Y, y_0)$ is $|a^{-1}|$.

Proof. An explicit homotopy between k and $a \# a^{-1}$ is given by

$$H(r, s) = \begin{cases} a(2r(1 - s)), & \text{if } 0 \leq r \leq \frac{1}{2}, \\ a(2(1 - r)(1 - s)), & \text{if } \frac{1}{2} \leq r \leq 1. \end{cases}$$

The reader should confirm that this is indeed a suitable homotopy. The geometric idea behind it is that we are starting at y_0, then going out along a to a certain point and finally coming back along a^{-1}, each time shortening the distance we go until have pulled $a \# a^{-1}$ entirely back into y_0. The reader should also produce a homotopy to show $a^{-1} \# a \sim k$.

We therefore conclude that $\pi_1(Y, y_0)$, # is a group with identity $|k|$ in which $|a| \in \pi_1(Y, y_0)$ has as its inverse $|a^{-1}|$. We are therefore justified in making the following definition.

Definition 4. Let Y, τ' be any space, and let $y_0 \in Y$ and $\pi_1(Y, y_0)$, # be as described above. Then

$$\pi_1(Y, y_0), \#$$

is called the *fundamental group based on* y_0 of the space Y.

Of course it is all well and good to know that $\pi_1(Y, y_0)$, # is a group. This is a good beginning, but hardly any more than that. For to be of much use in the study of topological spaces, $\pi_1(Y, y_0)$ must have certain properties. First, $\pi_1(Y, y_0)$ should be computable, at least for the majority of spaces which might be of interest. Second, $\pi_1(Y, y_0)$ should tell us something about the space Y, τ'; if it gives no information about the structure of Y as a topological space, it clearly has no value in the study of topology. Third, it is rather repugnant that $\pi_1(Y, y_0)$ should depend on the point of Y on which it is based; in other words, $\pi_1(Y, y_0)$ should depend on Y and τ' rather than on Y, τ', and y_0. Otherwise, we might get different fundamental groups for the same space without any clear idea of how to choose among them, and we would also be given the repugnant implication that some point of Y was better, or at least in some way significantly different than, other points of Y. Fourth, if there is a continuous function f from Y onto a space Z, τ'', we would hope that there is naturally associated with f a homomorphism from the fundamental group of Y into the fundamental group of Z. This would enable us to attack the difficult problem of whether there is a continuous function from one space onto another; most of all, however, we would expect that if groups are to be associated with spaces, then homomorphisms of groups will be associated with continuous functions from one space to the other.

All of the above considerations are very basic and very important. It will be the goal of the remainder of this chapter to at least partly settle each one of them. We first consider the question of the computation of fundamental groups.

There are a number of theorems and techniques for computing fundamental groups of spaces. Many of these are beyond the scope of this book. For most of the simpler spaces, however, it is not very difficult to compute the fundamental group.

Example 4. Let Y, τ' be a space which is contractible to one of its points y_0. Then from Section 11.2, Exercise 2, we see that $\pi_1(Y, y_0)$, # is a group of precisely one element.

Example 5. Let

$$Y = \{(x, y) \mid x^2 + y^2 = 1\} \subset R^2$$

and let y_0 be any point of Y. Let a be the loop which goes once around the circle in the counterclockwise direction. Then a cannot be homotopic to k (the function which takes all of $[0, 1]$ onto y_0), since a cannot be "pulled back" into y_0 without breaking (see Section 11.2, Exercise 4). For each $a \in L(Y, y_0)$ and each integer n, define

$$na = \begin{cases} a \# \cdots \# (n \text{ times}) \# a, & \text{if } n \text{ is positive,} \\ k, & \text{if } n = 0, \\ a^{-1} \# \cdots \# (-n) \# a^{-1}, & \text{if } n \text{ is negative.} \end{cases}$$

(See Fig. 11.16.)

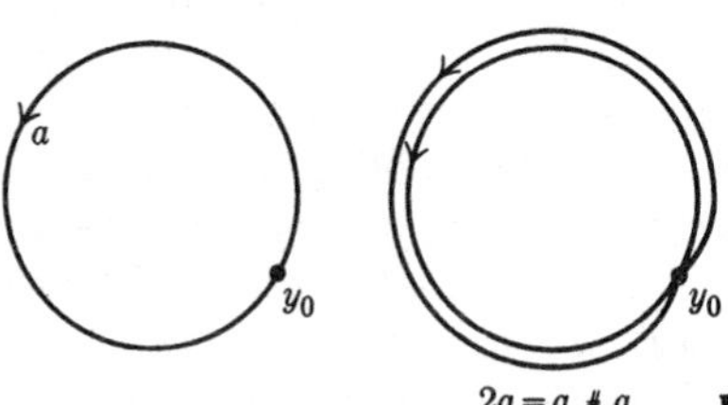

Figure 11.16

It can rather readily be shown that $(m + n)a \sim ma + na$ and that $na \sim ma$ if and only if $m = n$, where m and n are any two integers. On the other hand, every loop in Y based on y_0 is homotopic to ma for some integer m. There is therefore an isomorphism between

$$\pi_1(Y, y_0), \# \qquad \text{and} \qquad Z, +,$$

the additive group of integers; the isomorphism, call it j, is given explicitly by $j(|ma|) = m$. Note that $Z, +$ and $\pi_1\ (Y, y_0)$ are both groups which are generated by one of their elements; $Z, +$ is generated by 1, and $\pi_1(Y, y_0), \#$ is generated by $|a|$.

Example 6. Since R^2 (Fig. 11.17) is homeomorphic to

$$\{(x, y) \mid x^2 + y^2 < 1\} \subset R^2,$$

a contractible space, R^2 is a contractible space, and thus has a trivial fundamental group. Let $Y' = R^2 - \{(0, 0)\}$, and let y_0' be any point of Y'. Let a be a loop in Y' based on y_0' which goes once around (0, 0). The same reasoning used in Example 5 shows that $\pi_1(Y', y_0')$ is generated by $|a|$ and is isomorphic to the additive group of integers. Note that Y' and Y from Example 5 are not homeomorphic, but that they have isomorphic fundamental groups.

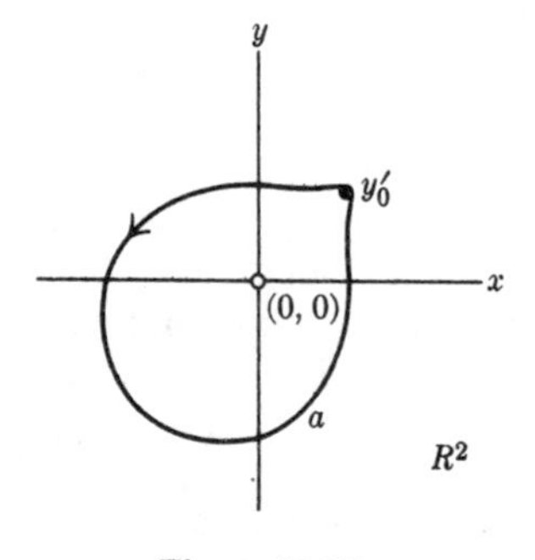

Figure 11.17

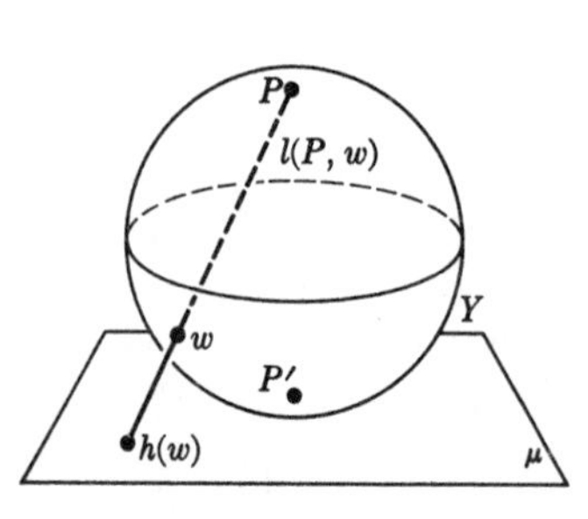

Figure 11.18

Example 7. Let

$$Y = \{(x, y, z) \mid x^2 + y^2 + z^2 = 1\} \subset R^3 \qquad \text{and} \qquad y_0 \in Y.$$

If P is any point of Y, we can show that $Y - \{P\}$ is homeomorphic to R^2 and hence is a contractible space. For let P' be the point of Y antipodal to P, and let μ be the plane in R^3 tangent to Y at P' (Fig. 11.18). For each $w \in Y - \{P\}$, the line $l(P, w)$ determined by P and w intersects μ in a unique point u. Define $h: Y - \{P\} \to \mu$ by

$$\{h(w)\} = l(P, w) \cap \mu$$

for each $w \in Y - \{P\}$. It is not hard to show that h is a homeomorphism.

Now Y is not contractible to any of its points (the intuitive argument runs, "You can't peel an orange without breaking its skin"), but if a is any loop in Y and P is any point in $Y - a([0, 1])$, then a is essentially a loop in R^2; hence a is homotopic to k, where $|k|$ is the identity of $\pi_1(Y, y_0)$ (Fig. 11.19). Thus any loop in Y based on y_0 is homotopic to k, and therefore $\pi_1(Y, y_0)$ consists only of $|k|$.

Y is hence an example of a noncontractible space which has a trivial fundamental group. Note that if P and Q are any two points of Y, then $Y - \{P, Q\}$ is homeomorphic to the space Y' in Example 6 and thus has a fundamental group isomorphic to the additive group of integers.

The following proposition aids in the computation of many fundamental groups.

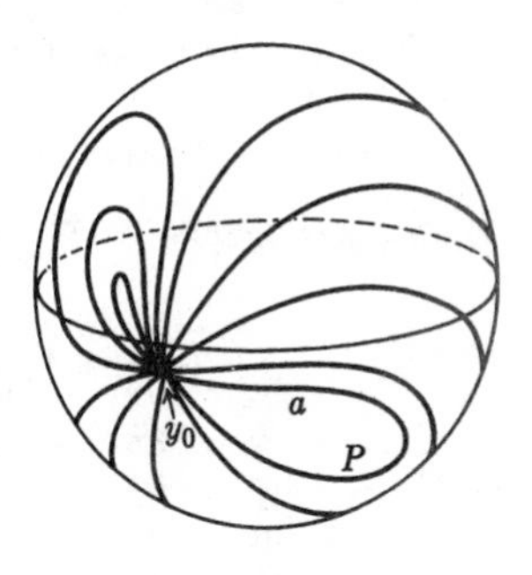

Figure 11.19

Proposition 5. Let X, τ and Y, τ' be spaces with base points x_0 and y_0, respectively. Then

$$\pi_1(X \times Y, (x_0, y_0))$$

is isomorphic to the direct sum of $\pi_1(X, x_0)$ and $\pi_1(Y, y_0)$,

$$\pi_1(X, x_0) \oplus \pi_1(Y, y_0).$$

That is, the fundamental group of the product space of two spaces is the direct sum of the fundamental groups of the component spaces.

Proof. Suppose $a \in L(X \times Y, (x_0, y_0))$. Let p_X and p_Y be the projections of $X \times Y$ into X and Y, respectively. Then

$$p_X \circ a \in L(X, x_0)$$

and

$$p_Y \circ a \in L(Y, y_0).$$

Define

$$f: \pi_1(X \times Y, (x_0, y_0)) \to \pi_1(X, x_0) \oplus \pi_1(Y, y_0)$$

by

$$f(|a|) = (|p_X \circ a|, |p_Y \circ a|).$$

We will prove that f is the desired isomorphism.

We first show that f is well-defined, that is, that $f(|a|)$ is independent of the representative of $|a|$ that is used. Suppose $a \sim a'$. Then there is a homotopy

$$H: [0, 1] \times [0, 1], \{0, 1\} \times [0, 1] \to X \times Y, (x_0, y_0)$$

between a and a'. It can, however, then be verified by straightforward computation that $p_X \circ H$ and $p_Y \circ H$ are suitable homotopies between $p_X \circ a$ and $p_X \circ a'$, and $p_Y \circ a$ and $p_Y \circ a'$, respectively. (Compare this to Section 11.1, Exercise 4.) Therefore f is well-defined.

We now show that f is onto: Suppose

$$(|a_1|, |a_2|) \in \pi_1(X, x_0) \oplus \pi_1(Y, y_0).$$

Define $a \in L(X \times Y, (x_0, y_0))$ by

$$a(r) = \begin{cases} (a_1(2r), y_0), & \text{if } 0 \leq r \leq \frac{1}{2}, \\ (x_0, a_2(2r - 1)), & \text{if } \frac{1}{2} \leq r \leq 1. \end{cases}$$

Then $p_X \circ a \sim a_1$ and $p_Y \circ a \sim a_2$ (Exercise 1). Moreover, a is easily seen to be continuous (a is continuous on both $[0, \frac{1}{2}]$ and $[\frac{1}{2}, 1]$ and is well-defined at $r = \frac{1}{2}$); also

$$a(0) = (a_1(0), y_0) = (x_0, y_0) = a(1),$$

and hence a is an element of $L(X \times Y, (x_0, y_0))$. But $f(|a|)$ then is $(|a_1|, |a_2|)$; therefore f is onto.

Now we show that f is one-one: Suppose $f(|a|) = f(|a'|)$. Then

$$(|p_X \circ a|, |p_Y \circ a|) = (|p_X \circ a'|, |p_Y \circ a'|).$$

Therefore there is a homotopy H_1 between $p_X \circ a$ and $p_X \circ a'$ and a homotopy H_2 between $p_Y \circ a$ and $p_Y \circ a'$. Define a homotopy

$$H: [0, 1] \times [0, 1] \to X \times Y$$

by

$$H(r, s) = (H_1(r, s), H_2(r, s)).$$

Direct computation shows that H is a homotopy between a and a'; hence $|a| = |a'|$. Therefore f is one-one.

It remains to be shown that f is a homomorphism. Suppose $|a|$ and $|a'|$ are elements of $\pi_1(X \times Y, (x_0, y_0))$. Then

$$f(|a| \# |a'|) = f(|a \# a'|) = (|p_X \circ (a \# a')|, |p_Y \circ (a \# a')|)$$
$$= (|p_X \circ a \# p_X \circ a'|, |p_Y \circ a \# p_Y \circ a'|)$$ (this latter equality follows at once from the manner in which the addition of loops has been defined)
$$= (|p_X \circ a| \# |p_X \circ a'|, |p_Y \circ a| \# |p_Y \circ a'|)$$
$$= (|p_X \circ a|, |p_Y \circ a|) \# (|p_X \circ a'|, |p_Y \circ a'|)$$ [where # is here "addition" in $\pi_1(X, x_0) \oplus \pi_1(Y, y_0)$]
$$= f(|a|) \# f(|a'|).$$

Therefore f is a homomorphism, and, consequently, an isomorphism.

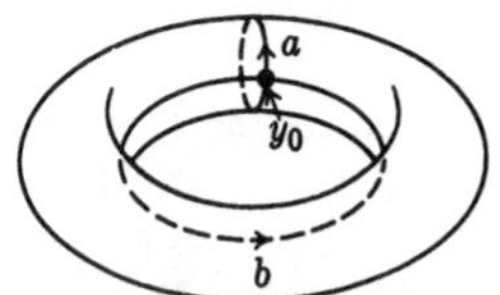

Figure 11.20

Example 8. We have seen that the fundamental group of the circle Y in Example 5 (relative to any base point) is isomorphic to the additive group of integers $Z, +$. The torus $Y \times Y$ then has a fundamental group which is isomorphic to the direct sum of the additive group of integers with itself, that is, $Z \oplus Z, +$. Note that $Z \oplus Z, +$ has two generators, $(0, 1)$ and $(1, 0)$. These correspond to the homotopy classes of the loops a and b (actually the images of loops) in $Y \times Y$ shown in Fig. 11.20. Note that since the fundamental group of Y does not depend on which base point is used, neither does the fundamental group of $Y \times Y$.

EXERCISES

1. The following refer to the proof of Proposition 5.
   a) Prove that $p_X \circ a \sim a_1$ and $p_Y \circ a \sim a_2$.
   b) Prove each of the equalities in the chain of equalities used to show that f is a homomorphism.
2. A space X, τ is said to be *simply connected* if its fundamental group (with respect to some base point) is trivial. The circle is an example of a space which is connected but not simply connected. Which of the following spaces are

simply connected? In all cases, compute the fundamental groups using the given base point.

a) $(0, 1) \subset R$, using any base point
b) $(0, 1) \times Y$, with Y as in Example 5 and using any base point
c) $[0, 1] \times Y'$, with Y' as in Example 7 and using any base point
d) $\{(x, y) \mid x^2 + y^2 < 1\} \cup \{(0, 1)\} \subset R^2$, with $(0, 1)$ as base point
e) Y^n, where n is any positive integer, Y is as in Example 5, and any base point is used

3. Suppose X, τ and Y, τ' are contractible to x_0 and y_0, respectively (Section 11.1, Exercise 3). Prove that the product space $X \times Y$ is contractible and simply connected.

4. Prove that the function H defined in Proposition 4 is a genuine homotopy between $a \# a^{-1}$ and k. Find a homotopy between $a^{-1} \# a$ and k.

5. In the examples given in this section, the fundamental group has not depended on the base point which was used to compute it. Actually the following important proposition is true:

If Y, τ' is arc-connected, and if y_0 and y_1 are any two points of Y, then $\pi_1(Y, y_0)$ is isomorphic to $\pi_1(Y, y_1)$.

[Recall that an arc in Y is a homeomorphism from $[0, 1]$ into Y, and that Y is said to be arc-connected if given any two distinct points x and y in Y, there is an arc h in Y such that $h(0) = x$ and $h(1) = y$.] Prove this proposition. [*Hint:* An isomorphism can be defined as follows: Since Y is arc-connected, there is a homeomorphism j from $[0, 1]$ into Y such that $j(0) = y_1$ and $j(1) = y_0$. Define $j^{-1}: [0, 1] \to Y$ by

$$j^{-1}(r) = j(1 - r)$$

for each $r \in [0, 1]$. Defining $j \# j^{-1}$ in the "natural" way, it can be shown that $j \# j^{-1}$ is homotopic to k, where $|k|$ is the identity of $\pi_1(Y, y_0)$. Suppose $|a| \in \pi_1(Y, y_0)$. Define $f: \pi_1(Y, y_0) \to \pi_1(Y, y_1)$ by

$$f(|a|) = |j \# a \# j^{-1}|$$

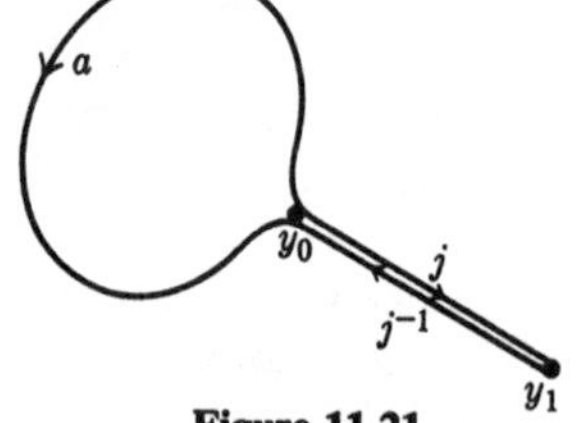

Figure 11.21

(Fig. 11.21). We know that $f(|a|)$ is a well-defined element of $\pi_1(Y, y_1)$. The details showing that f is an isomorphism are straightforward, but should be supplied carefully by the reader to check his understanding of homotopies and the fundamental group.]

Note that if a space is not arc-connected, then the fundamental group might depend on the base point. For example, let

$$Y = \{(x, y) \mid x = 3\} \cup \{(x, y) \mid x^2 + y^2 = 1\} \subset R^2.$$

If a base point y_0 in $\{(x, y) \mid x = 3\}$ is chosen, then all of the loops based on y_0 are also in $\{(x, y) \mid x = 3\}$ since otherwise we would have a continuous image of a connected space $[0, 1]$ which was not connected. Therefore if

$y_0 \in \{(x, y) \mid x = 3\}$, then $\pi_1(Y, y_0)$ is trivial. If a base point in the other component of Y were picked, then the fundamental group would turn out to be isomorphic to the additive group of integers. We therefore conclude that the fundamental group will only reflect the topology of the arc-component of the base point, and that the remainder of the space will be ignored. It is for this reason that one generally restricts one's attention to arc-connected spaces when discussing the fundamental group.

6. Divide the loops in the following figure according to homotopy class. Each loop goes around just once in the direction indicated. The shaded areas are "holes" in the space.

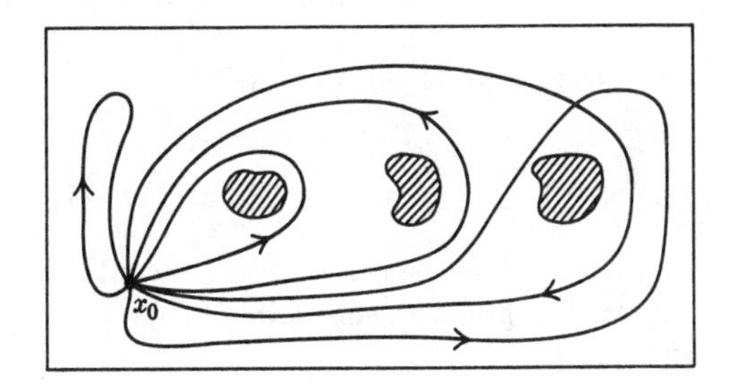

Figure 11.22

11.4 THE FUNDAMENTAL GROUP AND CONTINUOUS FUNCTIONS

We now have some knowledge about computing fundamental groups, and we have established the fact that, in arc-connected spaces, the fundamental group does not depend on the base point. In this section we will therefore restrict our attention to arc-connected spaces; we will thus be able to speak of *the* fundamental group of a space without regard to a base point. This section will be devoted to studying the relation between fundamental groups and continuous functions, and we shall also try to draw some conclusions as to what topological information the fundamental group conveys.

We first show that continuous functions do indeed induce homomorphisms on the fundamental groups.

Proposition 6. Let X, τ and Y, τ' be spaces, and suppose

$$f: X, x_0 \to Y, y_0$$

is a continuous function. Then there is associated with f a homomorphism f_* from $\pi_1(X, x_0)$ into $\pi_1(Y, y_0)$.

Proof. We first define a function $\bar{f}$ from $L(X, x_0)$ into $L(Y, y_0)$ as follows: For each $a \in L(X, x_0)$, set $\bar{f}(a) = f \circ a$. Since $\bar{f}(a)$ is the composition of two continuous functions, it is continuous. Furthermore, $\bar{f}(a)$ is a function from $[0, 1]$ into Y such that

$$\bar{f}(a)(0) = f(a(0)) = f(x_0) = y_0 = \bar{f}(a)(1);$$

hence $\bar{f}(a)$ is indeed an element of $L(Y, y_0)$. For any $|a| \in \pi_1(X, x_0)$, set

$$f_*(|a|) = |\bar{f}(a)|.$$

We first show that f_* is well-defined, that is, that $f_*(|a|)$ does not depend on the representative of $|a|$ used to compute it, but only on the equivalence class. Suppose $a \sim a'$, that is, $|a| = |a'|$. Then there is a homotopy

$$H: [0, 1] \times [0, 1], \{0, 1\} \times [0, 1] \to X, x_0$$

such that

$$H \mid [0, 1] \times \{0\} = a \qquad \text{and} \qquad H \mid [0, 1] \times \{1\} = a'.$$

Define $H' = f \circ H$. Since H' is the composition of continuous functions, it is continuous. Direct computation shows that H' is a homotopy between $f \circ a'$ and $f \circ a$, and hence

$$|f \circ a| = f_*(|a|) = |f \circ a'| = f_*(|a'|).$$

Therefore f_* is well-defined.

We now prove that f_* is a homomorphism. Suppose $|a|$ and $|a'|$ are any elements of $\pi_1(X, x_0)$. Now $a \# a'$ is defined by

$$(a \# a')(r) = \begin{cases} a(2r), & \text{if } 0 \leq r \leq \frac{1}{2}, \\ a'(2r - 1), & \text{if } \frac{1}{2} \leq r \leq 1. \end{cases}$$

Therefore $f \circ (a \# a')$ is the loop defined by

$$f \circ (a \# a')(r) = \begin{cases} f(a(2r)), & \text{if } 0 \leq r \leq \frac{1}{2}, \\ f(a'(2r - 1)), & \text{if } \frac{1}{2} \leq r \leq 1. \end{cases}$$

But this is precisely the definition of $f \circ a \# f \circ a'$. Then

$$\bar{f}(a \# a') = \bar{f}(a) \# \bar{f}(a').$$

Therefore $f_*(|a \# a'|) = f_*(|a|) \# f_*(|a'|)$. Since $|a \# a'| = |a| \# |a'|$, we have

$$f_*(|a| \# |a'|) = f_*(|a|) \# f_*(|a'|).$$

Hence f_* is a homomorphism.

The following example shows that f_* may not be onto even when f is.

Example 9. We have already seen that there is a continuous function f from the interval $[0, 1]$ onto a circle Y. But $[0, 1]$ is a contractible space and thus has a trivial homotopy group, whereas the homotopy group of Y is isomorphic to the additive group of integers. Therefore f_* in this instance is merely a function which takes the sole element (the identity) of $\pi_1([0, 1])$ onto the identity $|k|$ of $\pi_1(Y)$; hence f_* is clearly not onto.

The next two propositions show that the homomorphisms induced by continuous functions behave fairly respectably in relation to the functions that induce them.

Proposition 7. Suppose f is a continuous function from X, τ into Y, τ' and g is a continuous function from Y, τ' into Z, τ''. Also suppose that $f(x_0) = y_0$ and $g(y_0) = z_0$. Then

$$(g \circ f)_*: \pi_1(X, x_0) \to \pi_1(Z, z_0) \qquad \text{is the same as} \quad g_* \circ f_*.$$

That is, the composition of continuous functions gives a corresponding composition of the homomorphisms that these functions induce.

Proof. Suppose $a \in L(X, x_0)$. Then

$$(\overline{g \circ f})(a) = (g \circ f) \circ a = g \circ (f \circ a) = \bar{g}(f \circ a) = \bar{g}(\bar{f}(a)) = (\bar{g} \circ \bar{f})(a).$$

Therefore

$$(g \circ f)_*(|a|) = (g_* \circ f_*)(|a|).$$

Proposition 8. Suppose f and g are continuous functions from X, τ into Y, τ' and $f(x_0) = g(x_0) = y_0$. Then if f is homotopic to g,

$$f_* = g_*.$$

Proof. If f is homotopic to g, let H be a homotopy between f and g. Define

$$(H_* a)(r, s) = H(a(r), s)$$

for each $(r, s) \in [0, 1] \times [0, 1]$. Then $H_* a$ is easily verified to be a homotopy between $f \circ a$ and $g \circ a$ for any $a \in L(X, x_0)$. Then $\bar{f}(a) \sim \bar{g}(a)$ for any $a \in L(X, x_0)$; hence

$$f_*(|a|) = g_*(|a|) \qquad \text{for any} \quad |a| \in \pi_1(X, x_0).$$

If fundamental groups are to have much meaning topologically, then homeomorphic spaces should have isomorphic fundamental groups. As we have already seen, however, it is quite possible for two spaces which are not homeomorphic to have isomorphic fundamental groups. We may therefore suspect that there is a condition even weaker than homeomorphism which assures that two spaces have isomorphic fundamental groups. Experience has shown that this suspicion is indeed quite correct. The following definition proves to be appropriate.

Definition 5. Let X, τ and Y, τ' be (arc-connected) spaces. Then X and Y are said to have the *same homotopy type*, or to be *homotopically equivalent*, if there are continuous functions

$$f: X \to Y \qquad \text{and} \qquad g: Y \to X$$

such that $f \circ g$ is homotopic to the identity function i_Y on Y and $g \circ f$ is homotopic to the identity function i_X on X.

Being of the same homotopy type is a weaker condition than being homeomorphic, since X and Y would be homeomorphic if and only if there were continuous functions $f: X \to Y$ and $g: Y \to X$ such that $f \circ g = i_Y$ and $g \circ f = i_X$ (hence $f = g^{-1}$). Of course, if two spaces are homeomorphic, they are also of the same homotopy type.

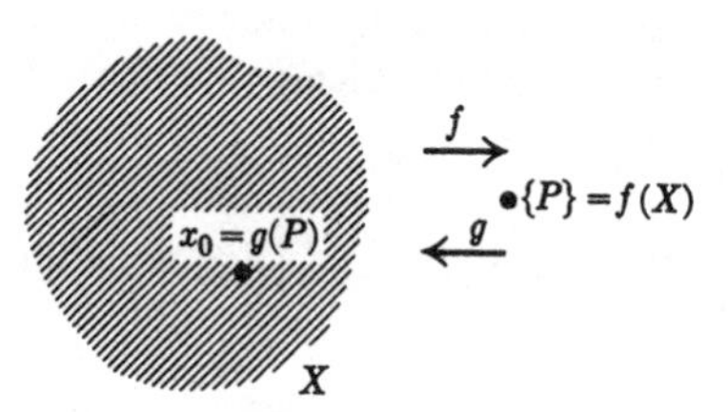

Figure 11.23

Example 10. Let X, τ be any contractible space and let Y, τ' be a space consisting of a single point P. Then X has the same homotopy type as Y. Let x_0 be a point of X to which X can be contracted and let k be the function on X which takes all of X into x_0 (Fig. 11.23). Then $i_X \sim k$. Let $f: X \to Y$ be defined by $f(x) = P$ for all $x \in X$, and $g: Y \to X$ be defined by $g(P) = x_0$. Then

$$f \circ g = k \sim i_Y$$

and

$$g \circ f = i_X.$$

Therefore X and Y have the same homotopy type.

Note how apparently different two spaces of the same homotopy type can be.

Example 11. Suppose W is a subspace of a space X, τ. Then a continuous function $f: X \to W$ is said to be a *retraction* of X onto W if $f \mid W = i_W$. We call W a *deformation retract* of X if i_X is homotopic to a retraction of X onto W. For example, the letter O is a deformation retract of the letter Q; here the retraction f could be described by saying that f takes any point in the tail of the Q into the point where the tail crosses the O part of the Q, and leaves all other points fixed.

If W is a deformation retract of X, then W and X have the same homotopy type. For convenience, set $i_X \mid W = i_W$. Let f be the continuous function from X into W which is homotopic to i_X. Then

$$f \circ i_W = f \mid W \sim i_W \qquad \text{and} \qquad i_W \circ f = f \sim i_X.$$

Therefore W and X have the same homotopy type.

The following proposition is pure set theory and will be stated without proof. It is of sufficiently wide application that the reader should already be familiar with it; if such is not the case, he should supply a proof.

Proposition 9. Let f be a function from a set S into a set T. Then f is one-one and onto if and only if there is a function g from T into S such that $f \circ g$ is the identity mapping on T and $g \circ f$ is the identity mapping on S, that is, f is one-one and onto if and only if it has a two-sided inverse.

We use this immediately to prove the following.

Proposition 10. Suppose two spaces X, τ and Y, τ' have the same homotopy type. Then $\pi_1(X)$ is isomorphic to $\pi_1(Y)$. (Recall that we are assuming X and Y arc-connected; hence their fundamental groups are independent of the base point.)

Proof. Since X and Y are of the same homotopy type, there are continuous functions $f: X \to Y$ and $g: Y \to X$ such that $f \circ g \sim i_Y$ and $g \circ f \sim i_X$. Applying Propositions 7 and 8, we have

$$f_* \circ g_* = i_{Y*} \qquad \text{and} \qquad g_* \circ f_* = i_{X*}.$$

But i_{X*} and i_{Y*} are the identity functions on $\pi_1(X)$ and $\pi_1(Y)$, respectively. It follows then from Proposition 9 that f_* is one-one and onto, and hence is an isomorphism.

It is not true that if two spaces have isomorphic fundamental groups they are then of the same homotopy type (e.g., see Example 7). Nevertheless, homotopy equivalence does give a partition of the family of topological spaces, just as homotopy gave a partition of the family of continuous functions from one space into another.

Proposition 11. Let the phrase "is of the same homotopy type as" be denoted by $\simeq$, and let $\mathcal{T}$ denote the family of all topological spaces. Then $\simeq$ is an equivalence relation on $\mathcal{T}$.

Proof. If $X \in \mathcal{T}$, then $X \simeq X$. Let $f = g = i_X$. Then $f \circ g = g \circ f = i_X$; therefore $X \simeq X$.

If $X \simeq Y$, then $Y \simeq X$. Since $X \simeq Y$, there are continuous functions $f: X \to Y$ and $g: Y \to X$ such that

$$f \circ g \sim i_Y \qquad \text{and} \qquad g \circ f \sim i_X.$$

The same f and g, however, give us $Y \simeq X$.

If $X \simeq Y$ and $Y \simeq Z$, then $X \simeq Z$. Since $X \simeq Y$, there are continuous functions $f: X \to Y$ and $g: Y \to X$ such that

$$g \circ f \sim i_X \qquad \text{and} \qquad f \circ g \sim i_Y.$$

Since $Y \simeq Z$, there are continuous functions $q: Y \to Z$ and $h: Z \to Y$ such that

$$q \circ h \sim i_Z \qquad \text{and} \qquad h \circ q \sim i_Y.$$

Set $u = q \circ f: X \to Z$ and $v = g \circ h: Z \to X$. Since u and v are the composition of continuous functions, they are continuous. Also

$$u \circ v = (q \circ f) \circ (g \circ h) = q \circ \big((f \circ g) \circ h\big) \sim q \circ (i_Y \circ h) = q \circ h \sim i_Z.$$

Similarly, $v \circ u \sim i_X$; hence $X \simeq Z$. Therefore $\simeq$ is an equivalence relation.

Corollary. Any two contractible spaces are of the same homotopy type.

Proof. Any two contractible spaces are each homotopically equivalent to the space consisting of exactly one point (Example 10); hence they are homotopically equivalent to each other.

Homeomorphism also defines an equivalence relation on $\mathcal{T}$. Since any two homeomorphic spaces are also of the same homotopy type, but not conversely, the homotopy equivalence classes are generally larger than the classes of homeomorphic spaces.

We now come to the question of what the fundamental group of a space can tell us about the space. First, although we cannot be sure that two spaces are not of the same homotopy type if their fundamental groups are isomorphic, we can be certain that they are not of the same homotopy type if their fundamental groups are not isomorphic. For example, we can conclude that the circle and closed line interval are not of the same homotopy type (and hence could not be homeomorphic) because the segment is contractible and hence has a trivial fundamental group, whereas the fundamental group of a circle is nontrivial.

In a rather imperfect way, the fundamental group measures the "holes" in a space. A contractible space, geometrically speaking, is a space without any holes. On the other hand, it should be noted that although the sphere also has a trivial fundamental group, it could hardly be said not to have any holes. The hole in a sphere, however, is a higher-dimensional hole, and is not one which can be registered by the fundamental group. Note that a torus has two types of holes and that its fundamental group has two generators. We should not, however, try to push this point too far, since it is only approximately true.

The fact that there are "higher-dimensional holes," as well as the fact that the fundamental group can give but rather limited information, leads us to hope that there are other algebraic structures which can be associated with a space to supplement the information given by the fundamental group. Such is indeed the case. There are *higher homotopy groups* [previously implied by using the notation $\pi_1(X)$ instead of merely $\pi(X)$], *homology groups*, *cohomology groups*, and a long list of others, but these will not be explored in this text.

EXERCISES

1. Supply an argument to prove more fully that O is a deformation retract of Q (Example 11).
2. Classify each of the diagrams in Fig. 11.24 (considered as subspaces of R^2) according to homotopy type.

Figure 11.24

3. In each of the following, decide whether or not the two spaces given are of the same homotopy type.
   a) a circle in R^2 and $R^2 - \{(0, 0)\}$
   b) the sphere in R^3 and a circle in R^2
   c) a triangle and a circle in R^2
   d) R^3 and R^2
4. In which of the following is the second space a deformation retract of the first space? An intuitive argument may be all the reader will be able to give.
   a) $\{(x, y) \mid x^2 + y^2 \leq 1\}$ and $\{(x, y) \mid x^2 + y^2 = 1\}$
   b) R^2 and $\{(0, 0)\}$
   c) $\{(x, y) \mid x^2 + y^2 \leq 1\}$ and $\{(x, y) \mid x^2 + y^2 < 1\}$
5. Suppose W is a subspace of X, r such that $i_W = i_X \mid W$ can be extended to a continuous function from X into W. Discuss the relation between $\pi_1(W)$ and

$\pi_1(X)$. Formulate a proposition which tells us when i_W cannot be extended to a continuous function from X into W. Let

$$W = \{(x, y) \mid x^2 + y^2 = 1\} \subset X = \{(x, y) \mid x^2 + y^2 \leq 1\}.$$

In this case can i_W be extended to a continuous function from X into W? How could we interpret this result geometrically?

6. Suppose X, τ is a space with a contractible subspace A. Let X/A be the identification space obtained by identifying all the points of A (and leaving the other points of X as they are). What can be said about the relation of $\pi_1(X)$ to $\pi_1(X/A)$? Give an argument for your assertion. Can you interpret your statement geometrically?

7. We will call a continuous function f from a space X, τ into a space Y, τ' *trivial* if f_* takes all of $\pi_1(X)$ onto the identity of $\pi_1(Y)$. Examine each of the following and decide whether or not there is a nontrivial function from the first space into the second. If there is a nontrivial function, describe one.

   a) a circle in R^2 and the open interval $(0, 1)$
   b) a circle in R^2 and $R^2 - \{(0, 0)\}$
   c) a circle C in R^2 and the torus $C \times C$
   d) a torus and a circle in R^2
   e) a space with fundamental group Z_5 and a space with fundamental group Z_3, where Z_5 and Z_3 represent the additive groups of integers modulo 5 and 3, respectively

8. If X and Y have the same homotopy type, which of the following situations cannot occur?

   a) X compact, but Y not compact
   b) X connected, but Y disconnected
   c) X arcwise connected, but Y not arcwise connected

APPENDIX ON INFINITE PRODUCTS

Let $\{X_i, \tau_i\}$, $i \in I$, be any family of topological spaces indexed by some set I. If I is countable, then we already have a definition of the product space of this family. But I need not always be countable; thus if we are to consider product spaces in their full generality, we need to have a product of uncountably many spaces also.

Let R be the set of real numbers. Then, as we know, $R \times R = R^2$ is the set of all ordered pairs (x, y) where x and y are elements of R. To each ordered pair (x, y), we can associate a function c from the set $\{1, 2\}$ into R defined by $c(1) = x$ and $c(2) = y$. For each $(x, y) \in R$, there is a distinct function from $\{1, 2\}$ into R, and for each function $c: \{1, 2\} \to R$, there is a unique point $(c(1), c(2))$ of R^2.

If X_1 and X_2 are any two sets, then

$$X_1 \times X_2 = \{(x_1, x_2) \mid x_1 \in X_1 \text{ and } x_2 \in X_2\}.$$

But we are able to show that $X_1 \times X_2$ can be associated in a natural way with the set of all functions c from $\{1, 2\}$ into $X_1 \cup X_2$ which have the property that $c(1) \in X_1$ and $c(2) \in X_2$. Note that $\{1, 2\}$ is the index set for the family of sets $\{X_1, X_2\}$ of which we are taking the product. We therefore make the following definition.

Definition 1. Let $\{X_i\}$, $i \in I$, be any family of sets. Define the *product* of the family $\{X_i\}$, $i \in I$, to be the collection of all functions c from I into $\bigcup_I X_i$ such that

$$c(i) \in X_i \qquad \text{for each} \quad i \in I.$$

The product of $\{X_i\}$, $i \in I$, is denoted by $\times_I X_i$. If $c \in \times_I X_i$, then $c(i)$, usually denoted by c_i, is called the *ith coordinate of c*. X_i is called the *ith component* of the set $\times_I X_i$.

Note that this definition of the product does not depend on the cardinality of I. It should not cause the reader much work to show that where I is countable between the old definition of $\times_I X_i$ and the new, there is a natural equivalence. We now use the considerations put forth in Chapter 4 concerning what properties the product topology should have to define a topology on $\times_I X_i$ if each X_i is also a topological space.

Definition 2. Let $\{X_i, \tau_i\}$, $i \in I$, be a family of spaces, and let $\times_I X_i$ be the product set of the family of X_i as defined in Definition 1. Let

$$\mathcal{S} = \{\times_I V_i \mid V_i = X_i \text{ for all but at most one } i \in I \text{ and each } V_i \text{ is an open subset of } X_i\}.$$

Then $\mathcal{S}$ is a subbasis for a topology τ on $\times_I X_i$ called the *product topology*. The space $\times_I X_i, \tau$ is called the *product space* of $\{X_i, \tau_i\}$, $i \in I$. (Compare this to Definition 7, Chapter 4.)

There is a natural function

$$p_i: \underset{I}{\times} X_i \to X_i$$

defined by $p_i(c) = c_i$ for each $i \in I$. We call p_i the *projection into the* ith *component.*

The reader should promptly prove that the product topology is the coarsest topology which makes each projection p_i continuous.

The propositions and proofs in this text which deal with product spaces have purposely been designed so that it would be easy to adapt them to the more general notion of a product space (provided that they are valid when generalized). We now give two examples of propositions and their proofs which generalize and one example of a proposition which is not true when stated for the product of uncountably many spaces.

Proposition 1. Let $Y = \times_I X_i$ be the product space of the family of non-empty spaces $\{X_i, \tau_i\}$, $i \in I$. Then Y is T_2 if and only if each X_i is T_2. (See Proposition 3, Chapter 5.)

Proof. Suppose each X_i is T_2, and let x and y be distinct points of Y. We will use x_i and y_i to denote the ith coordinate of x and y, respectively. Since $x \neq y$, $x_i \neq y_i$ for at least one $i \in I$, say for i'. Therefore there are open sets $U_{i'}$ and $V_{i'}$ in $X_{i'}$ such that

$$x_{i'} \in U_{i'}, \qquad y_{i'} \in V_{i'}, \qquad \text{and} \qquad U_{i'} \cap V_{i'} = \phi.$$

Set $U = \times_I H_i$, where $H_i = X_i$, $i \neq i'$, and $H_{i'} = U_{i'}$; and set $V = \times_I G_i$, where $G_i = X_i$, $i \neq i'$, and $G_{i'} = V_{i'}$. Then U and V are neighborhoods of x and y, respectively, and since any point of U differs from any point of V at least in the ith coordinate, $U \cap V = \phi$. Therefore Y is T_2.

By the generalization of Proposition 20, Chapter 4, each X_i, τ_i is homeomorphic to a subspace of Y. Thus if Y is T_2, then each X_i is also T_2.

Proposition 2. Suppose f is a function from a space X, τ into the product space $\times_I X_i, \tau'$. Define $f_i: X, \tau \to X_i, \tau_i$ by

$$f_i(x) = p_i \circ f(x) \qquad \text{for each} \quad x \in X,$$

where p_i is the projection into the ith component. Then f is continuous if and only if f_i is continuous for each $i \in I$. (*Cf.* Proposition 21, Chapter 4.)

Proof. If f is continuous, then $f_i = f \circ p_i$ is the composition of two continuous functions and therefore is continuous.

Suppose now that f_i is continuous for each $i \in I$. We first note that the ith coordinate of $f(x)$ is $f_i(x)$ for each $x \in X$. A basis $\mathcal{B}$ for τ' consists of all sets of the form $X_I V_i$, where V_i is open in X_i for each $i \in I$ and $V_i = X_i$ for all but finitely many i. Suppose $X_I V_i$ is any member of $\mathcal{B}$, and $V_i = X_i$ for each $i \in I$ except $i_1, \ldots, i_m$. Now $f^{-1}(X_I V_i)$ is the set of all points x of X such that

$$f(x) \in \underset{I}{X} V_i.$$

But this is easily seen to be $\bigcap_I f^{-1}(V_i)$. For every $i \in I$, except $i_1, \ldots, i_m$, $f_i^{-1}(V_i) = X$ (because $V_i = X_i$). Since f_i is continuous for each $i \in I$, $f_{i_j}^{-1}(V_{i_j})$ is open in X, $j = 1, \ldots, m$. Therefore

$$f^{-1}\left(\underset{I}{X} V_i\right) = f_{i_1}(V_{i_1}) \cap \cdots \cap f_{i_m}(V_{i_m}),$$

which is open in X since it is the intersection of finitely many open sets. Hence, by Proposition 7, Chapter 4, f is continuous.

Proposition 3. Suppose $\{s_i\}$, $i \in I$, is a net in the product space $X_J X_j$. Then $s_i \to y$ if and only if $s_i^j \to y_j$, where y_j is the jth coordinate of y and $\{s_i^j\}$, $i \in I$, is defined to be $\{p_j(s_i)\}$, $i \in I$.

The proof of Proposition 12 of Chapter 6 may be used *verbatim*.

Proposition 4 (*Tychonoff*). Let $X_I X_i$ be the product space of the nonempty family of nonempty spaces $\{X_i, \tau_i\}$, $i \in I$. Then $X_I X_i$ is compact if and only if each component space is compact.

The proof of Proposition 13 of Chapter 7 may be used *verbatim*; alternatively, one may use the proof of Proposition 16.

The propositions we would expect not to generalize are those which deal with the cardinality of a basis for the product topology. For example, Proposition 4(d), Chapter 7, does not generalize, as the following example shows.

Example 1. Let I be an uncountable set, and let $X_i = \{0, 1\}$ for each $i \in I$. Give each X_i the discrete topology. Then certainly each X_i is second countable. But the product space $X_I X_i$ is not second countable. This is true since the subbasis $\mathcal{S}$, as described in Definition 2 for the product topology on $X_I X_i$, contains uncountably many distinct members of the form $X_I V_i$, where $V_i = X_i$, except for precisely one $i \in I$. The family of such members of $\mathcal{S}$ can be shown to be a minimal subbasis for the product topology; thus the product topology could not be second countable. We could also prove that the product space $X_I X_i$ is not second countable as follows: If $X_I X_i$ is second countable, then $X_I X_i$ is Lindelöf, and hence every open cover of $X_I X_i$ has a countable subcover. But the collection of $X_I V_i$, where $V_i = X_i$, except for precisely one i, is an open cover of $X_I X_i$ which has no countable subcover.

The generalization of the notion of a product space also expands our horizons as to the possible uses of product spaces.

Example 2. Given spaces X, τ and Y, τ', then Y^X was used to denote the family of all continuous functions from X into Y (*cf.* Proposition 1, Chapter 11). Actually, Y^X more appropriately stands for the family of *all* functions from X into Y, that is, the product set Y^X. Since Y is a space, Y^X can be given a topology as a product space; the family of continuous functions from X into Y is a subspace of this space. This in turn leads to the possibility of putting topologies on the set of homotopy equivalence classes of functions from X into Y (using the identification topology) and of even giving a topology to fundamental groups. In point of fact, more important topologies than the product topology are used on Y^X, but having a generalized notion of product has awakened us to this possibility.

EXERCISES

1. If $\{X_i, \tau_i\}$, $i \in I$, is a countable family of spaces, show that there is a natural correspondence between the product space of this family as defined in this appendix and the product space as defined previously. In other words, prove that the product spaces formed in both ways are actually homeomorphic.
2. Formulate and prove a generalization of Proposition 20, Chapter 4.
3. Prove or disprove: The product of any family of nonempty first countable spaces is first countable.
4. Show that the product of a family of discrete spaces may not have the discrete topology. Does the product of any family of spaces with the trivial topology necessarily have the trival topology? Is it possible for an infinite product of infinite spaces to have the discrete topology?
5. Prove that the product topology on Y^X (Example 2) is equivalent to the topology generated by saying that any net $\{f_i\}$, $i \in I$, in Y^X converges to f if and only if $f_i(x) \to f(x)$ for all $x \in X$.

INDEX OF SYMBOLS

INDEX

A CATALOG OF SELECTED

DOVER BOOKS

IN SCIENCE AND MATHEMATICS

ROTARY-WING AERODYNAMICS, W.Z. Stepniewski. Clear, concise text covers aerodynamic phenomena of the rotor and offers guidelines for helicopter performance evaluation. Originally prepared for NASA. 537 figures. 640pp. 6⅛ x 9¼.
64647-5 Pa. $16.95

DIFFERENTIAL GEOMETRY, Heinrich W. Guggenheimer. Local differential geometry as an application of advanced calculus and linear algebra. Curvature, transformation groups, surfaces, more. Exercises. 62 figures. 378pp. 5⅜ x 8½.
63433-7 Pa. $9.95

INTRODUCTION TO SPACE DYNAMICS, William Tyrrell Thomson. Comprehensive, classic introduction to space-flight engineering for advanced undergraduate and graduate students. Includes vector algebra, kinematics, transformation of coordinates. Bibliography. Index. 352pp. 5⅜ x 8½. 65113-4 Pa. $9.95

A SURVEY OF MINIMAL SURFACES, Robert Osserman. Up-to-date, in-depth discussion of the field for advanced students. Corrected and enlarged edition covers new developments. Includes numerous problems. 192pp. 5⅜ x 8½. 64998-9 Pa. $8.95

ANALYTICAL MECHANICS OF GEARS, Earle Buckingham. Indispensable reference for modern gear manufacture covers conjugate gear-tooth action, gear-tooth profiles of various gears, many other topics. 263 figures. 102 tables. 546pp. 5⅜ x 8½.
65712-4 Pa. $14.95

SET THEORY AND LOGIC, Robert R. Stoll. Lucid introduction to unified theory of mathematical concepts. Set theory and logic seen as tools for conceptual understanding of real number system. 496pp. 5⅜ x 8¼. 63829-4 Pa. $12.95

A HISTORY OF MECHANICS, René Dugas. Monumental study of mechanical principles from antiquity to quantum mechanics. Contributions of ancient Greeks, Galileo, Leonardo, Kepler, Lagrange, many others. 671pp. 5⅜ x 8½.
65632-2 Pa. $14.95

FAMOUS PROBLEMS OF GEOMETRY AND HOW TO SOLVE THEM, Benjamin Bold. Squaring the circle, trisecting the angle, duplicating the cube: learn their history, why they are impossible to solve, then solve them yourself. 128pp. 5⅜ x 8½. 24297-8 Pa. $4.95

MECHANICAL VIBRATIONS, J.P. Den Hartog. Classic textbook offers lucid explanations and illustrative models, applying theories of vibrations to a variety of practical industrial engineering problems. Numerous figures. 233 problems, solutions. Appendix. Index. Preface. 436pp. 5⅜ x 8½. 64785-4 Pa. $11.95

CURVATURE AND HOMOLOGY, Samuel I. Goldberg. Thorough treatment of specialized branch of differential geometry. Covers Riemannian manifolds, topology of differentiable manifolds, compact Lie groups, other topics. Exercises. 315pp. 5⅜ x 8½. 64314-X Pa. $9.95

HISTORY OF STRENGTH OF MATERIALS, Stephen P. Timoshenko. Excellent historical survey of the strength of materials with many references to the theories of elasticity and structure. 245 figures. 452pp. 5⅜ x 8½. 61187-6 Pa. $12.95